THE STRATEGIC
MANAGEMENT OF
TECHNOLOGICAL
INNOVATION

THE STRATEGIC MANAGEMENT OF TECHNOLOGICAL INNOVATION

Edited by

Ray Loveridge

Aston University

and

Martyn Pitt

University of Bath

ECONOMIC AND SOCIAL
RESEARCH COUNCIL

E|S|R|C

Work Organisation Research Centre

JOHN WILEY & SONS

Chichester · New York · Brisbane · Toronto · Singapore

HD
45
S87
1990

Other Wiley Editorial Offices

John Wiley & Sons, Inc., 605 Third Avenue,
New York, NY 10158-0012, USA

Jacaranda Wiley Ltd, G.P.O. Box 859, Brisbane,
Queensland 4001, Australia

John Wiley & Sons (Canada) Ltd, 22 Worcester Road,
Rexdale, Ontario M9W 1L1, Canada

John Wiley & Sons (SEA) Pte Ltd, 37 Jalan Pemimpin #05-04,
Block B, Union Industrial Building, Singapore 2057

Library of Congress Cataloging-in-Publication Data:

The Strategic management of technological innovation / edited by Ray
   Loveridge and Martyn Pitt.
       p.   cm.
   Includes bibliographical references.
   ISBN 0 471 92499 7
   1. Technological innovations—Management.   2. Strategic planning.
   I. Loveridge, Ray.   II. Pitt, Martyn J. (Martyn John), 1948–
   HD45.S847   1990
   658.5′14—dc20   89-70548
                                                                 CIP

British Library Cataloguing in Publication Data:

The strategic management of technological innovation.
   1. Technological innovation. Management aspects
   I. Loveridge, Ray   II. Pitt, Martyn
   658.4′063

   ISBN 0 471 92499 7

Printed and bound in Great Britain by Biddles Ltd, Guildford, Surrey

Contents

Contributors

Jan Barbé,

Vlerick School Voor Management, University of Ghent, Sint Pietersnieuwstraat 49, 9000 Gent, Belgium

Stephen Barley,

Cornell University, Ithaca, New York, USA

Roland Calori,

Groupe ESC Lyon, 23 Avenue Guy de Collongue, B.P. 174, 69132 ECULLY cedex, France

Christopher Carr,

Manchester Business School, University of Manchester, Booth Street West, Manchester, M15 6PB, UK

John Child,

Work Organization Research Centre, Aston University Business School, Aston Triangle, Birmingham B4 7ET, UK

Peter Clark,

Work Organization Research Centre, Aston University Business School, Aston Triangle, Birmingham B4 7ET, UK

Ken Clarke,

School of Management, University of Bath, UK, Claverton Down, Bath BA2 7AY, UK

Christopher DeBresson,

University of Western Ontario, Canada

Arnoud De Meyer,

INSEAD, Boulevard De Constance, 77305 Fontainbleau, France

Dirk Deschoolmeester,

Vlerick School voor Management, University of Ghent, Sint Pietersnieuwstraat 49, 9000 Gent, Belgium

John Freeman,

Cornell University, Ithaca, New York, USA

Bryn Jones,

School of Social Sciences, University of Bath, Claverton Down, Bath BA2 7AY, UK

Christian Koenig,

Départment Stratégie et Management, ESSEC, Avenue Bernard Hirsch, BP 105/95021 Cergy-Pontoise Cedex, France

Raymond Loveridge,

Work Organization Research Centre, Aston University Business School, Aston Triangle, Birmingham B4 7ET, UK

Rudy Moenaert,

Vlerick School voor Management, University of Ghent, Sint Pietersnieuwstraat 49, 9000 Gent, Belgium

Martyn Pitt,

School of Management, University of Bath, UK, Claverton Down, Bath VA2 7AY, UK

Michael Saren,

School of Management, University of Bath, UK, Claverton Down, Bath BA2 7AY, UK

Jacqueline Senker,

Science Policy Research Unit, Mantell Building, Sussex University, Brighton, BN1 9RF, UK

Christopher Smith,

Work Organization Research Centre, Aston University Business School, Aston Triangle, Birmingham B4 7ET, UK

Raymond-Alain Thiétart,

ESSEC, France

Howard Thomas,

Dept. of Business Administration, University of Illinois at Urbana Champaign, 1206 South Sixth Street, Champaign Illinois 61820, USA

Richard Whittington,

School of Industrial and Business Studies, University of Warwick, Coventry CV4 7AL, UK

Preface

This book derives from the networking of European and American scholars over a number of years. It has its origins in a three-day workshop held by the ESRC Work Organisation Research Centre at Aston Business School in 1987. Since then the themes explored at the workshop have become translated into multi-disciplinary research projects several of which are described in this volume. In some ways, then, the collection may be seen as contributing to an emerging movement towards the strategic understanding of technological change at the global level.

As the various chapters came together it became apparent that they formed a commentary on a decade of change in academic as well as managerial disciplines and outlooks. Specifically, that a shared emphasis on understanding major aspects of the strategic environment of organizations brings an awareness of elements in the approaches of separate disciplines that provide points of conceptual integration as well as differentiation. Most studies reported here reveal the *emergent* nature of strategic direction and managerial modes under conditions of transformational uncertainty. Decision making is seen as a learning activity wherein potential risks and benefits have (sometimes painfully) to be *discovered* if organizations are to mitigate the former and appropriate the latter.

How far this accent on transience will remain a permanent feature of the analysis offered by social scientists must be part of the current debate. What is already clear is that the transience of micro phenomena has forced structural analysis into ever more holistic assumptions and generalizations. Whilst offering useful insights into the general direction in which whole economies or industrial sectors may be expected to move in the face of technological change, such explanations must be complemented by a processual understanding of the design of operational strategies at the level of the firm. In our choice of contributions we, as editors, have attempted to juxtapose a range of perspectives while emphasizing those that focus on what WORC researchers describe as 'the firm in sector' approach.

At this point it is traditional to offer editorial thanks to those who have carried out the routine and onerous work. In this instance we do so with great sincerity, knowing that neither the Workshop nor this book would have taken on the form that they have without the creative organization of Pam Lewis, Caroline Jury, Jane Winder, Pat Clark and particularly the late Beryl Marston. In the editing and processing of several drafts of each chapter and in the compilation and lay-out of the final text we owe an especial debt to Caroline Arnold. The Co-ordinator of Research at Aston Business School, Mike Tricker, provided moral support and resources, as did our various colleagues and comrades at home and on campus. Diane Taylor has guided our errant pursuits on behalf of John Wiley with great tact and patience. We are grateful to them and to all the managers, designers, and providers of goods and services whose activities are the subject of this book.

Ray Loveridge
Martyn Pitt

Introduction: Defining the Field of Technology and Strategy

Ray Loveridge

Aston University

and

Martyn Pitt

University of Bath

Few textbooks on strategic management do not include the management of change as one of their central foci. Up to a few years ago, such texts were written in a manner that suggested that the restoration of environmental order and of a stable managerial frame of analysis marked the end of a successful exercise in regulated adaptation. Most such prescriptions emphasized the importance of recognizing the existence of a limited range of archetypal patterns of strategic context and of choosing from a similarly finite number of organizational designs to produce a best fit. Named case histories provided testimonials to the success of the authors' prescriptions (bespoke adjustments were always necessary in what were after all no more than ideal types). By contrast, contemporary accounts of change emphasize the fluidity of their managerial context and the

The Strategic Management of Technological Innovation. Edited by R. Loveridge and M. Pitt
©1990 by John Wiley & Sons Ltd

need for continuous flexibility in adjustment to a dynamic and ever-changing strategic environment. Case histories can often provide false friends for would-be prescribers unless treated modestly and open-endedly. Particular examples of good house-keeping may well have fallen victim to a leveraged buy-out or to a joint-venture that deprived them of their source of competitive advantage before the prescriptions could reach their intended audience.

It is, then a great deal easier for the scholar living through such 'interesting times' to resort to broadly holistic explanations of global trends rather than to seek an understanding of the rapidly changing minutiae of managerial decision-making. Yet it has become clear that the mechanisms by which organizations adapt their operations to changes in strategic context and vice versa—that is, the means by which firms create or invent changes that impact on their external context—are central to an understanding of changes in the wider sphere of the economy and society as a whole. Much of the widespread sense of crisis that emerged in Western capitalism at the beginning of the 1980s, and that had spread to the planned economies of the Socialist world by the end of the decade, stemmed from a loss of belief in the permanency of institutionalized structures of consumer demand and in the efficacy of the operational technologies designed to serve these domains. The challenge that came from the erosion of these structures in the late 1960s was met most effectively by 'newcomers' from the East who shared none of the assumptions of Western strategic planners and were able better to perceive and to meet the needs of new consumers and to realize the potential of new technologies.

Perhaps most important to the success of the new entrants has been their recognition of the strategic need to define and appropriate the benefits of technology in terms of both the social and technical requirements of organization. More particularly, strategists have become aware of the centrality of the culture of the organization and of the sector from which it derives its operational norms. Modes of combining the right tools with the techniques and appropriate task situations define the firm's operational know-how in a manner that is distinctive and not to be found solely in the handbook of a production engineer, cost accountant, or even the merchandiser of new equipment. Yet, in the right mode, it appears able to set a direction for the progressive development of both product and process or, in the wrong one, for a negative spiral of 'too little and too late'.

It must be said that the strategic aspects of technological design and of its roots in the culture of the organization were recognized by pioneering managements in the West long before they were conjoined in the literature. Twenty years ago, Donald Schoen could say that the literature on the management of technological innovation was sparse, that relatively little had been written about the *total* (organizational) process by which firms translate a technical advance or invention into exploitable products, processes or services (Schoen, 1969). Schoen's diagnosis was that strategy and technology research, each a legitimate domain of inquiry, had tended to develop in parallel and all too often in isolation. Interested parties frequently did not explore the linkages between innovation and corporate strategy systematically or regarded them as not meriting special attention.

A decade later Kantrow (1980) endorsed Schoen's position, claiming none the less that there was now a fast-growing awareness by *managers* of the need to incorporate technological issues within strategic thinking and decision-making. We agree with Kantrow, believing that for many, perhaps a majority of senior level managers, the role of technology in strategic management is a *de facto* preoccupation and has probably always been so. By way of illustration we cite three recent sources. Donaldson and Lorsch (1983) studied twelve US corporate managements. In three 'technical products' corporations the centrality of technological concerns in top managers' shared beliefs is striking. For them technology appears to have two important roles, firstly in *defining and orientating* the firm, as exemplified in statements like 'pay-offs will always come from inventions' and 'technology is our driving force, in fact and perception'; secondly in creating and sustaining the basis of *strategic advantage*, as indicated in beliefs such as 'market share comes through technically superior products' and 'go anywhere, but only where you have a technological edge'.

In our second illustration, 150 strategic decisions by top managers in 30 UK organizations were studied (Hickson *et al.*, 1985, 1986). The single most important category of decision (23 cases) concerned process technology, typically whether to invest in new assets and facilities to boost competitive standing. There were also 28 decisions to add or modify products and services offered, many involving technological change in one form or another. Thus, technological innovation was a significant aspect of strategic decision-making in this sample.

Thirdly, in a fascinating text entitled *Strategies ... Successes ... Senior Executives Speak Out* (McLimore and Larwood, 1988) three dozen chief executives and senior officers of major US and European Corporations discussed their management preoccupations. Virtually *all* cited the generation or application of technological innovation as a major factor in the historical and expected future development of their firms.

Similarly, there is widespread interest in technology/strategy links in managerially-orientated journals such as *Fortune* and *Management Today*. So we feel confident that awareness of technological change in terms of its pervasiveness, dynamism, difficulty of managing and strategic significance is high among practitioners. In contrast, admittedly with some honourable exceptions, research-orientated management writers have only recently chosen to address squarely the significance of technology strategic decision-making. In consequence there is a relative paucity of systematic, research-supported findings in this area in journals such as *Academy of Management Journal*, *Administrative Science Quarterly* and *Strategic Management Journal*. Still along with Schoen and Kantrow, we express optimism that the position *is* changing: various streams of thought *are* converging and the linkages of technologies and strategies increasingly recognized. Clearly, however, more work—and more recognition of it—is needed to establish and validate useful theory and practical implications.

MANAGING TECHNOLOGICAL INNOVATION: THE MACRO PERSPECTIVE

At the macro level of analysis, economic change is often characterized (in neo-Schumpeterian terms) essentially as the result of entrepreneurial, new (small) firms exploiting novel technological competences to the detriment of the existing, mature, inertia-ridden large firms for whom specific innovations are exogenous, i.e. largely beyond their scope to contain or appropriate. Entrepreneurs who compete effectively on this basis create structural and knowledge-deficiency *crises* for the 'old guard' which frequently prove terminal. These technological discontinuities are said to give rise to long-wave or 'Kondratiev' cycles of economic activity (Freeman, Clark and Soete, 1982). Schumpeter acknowledged the possibility that by virtue of successful long-run strategic postures large firms may become endogenously innovative, reducing their uncertainty over—and vulnerability to—exogenous innovation (Phillips, 1971). More recently Rothwell (1983) argued that both mechanisms coexist: he described the US semi-conductor industry as having evolved to a state of 'managed' innovativeness, whilst still benefiting from the considerable innovative power of small firms typically 'spun-off' from the big ones. The co-existence of large and small firms is a productive state of 'dynamic complementarity'. He also provided UK data showing that innovativeness is particularly associated with very large (10,000+ employees) and very small firms (under 200 employees). Their 'innovation-push' as dynamism, often characterized as product-led, is also complemented by 'demand-pull' or market pressures in high technology sectors of the economy (Schmookler, 1966).

Meta-technologies

Some technologies such as those involving computing, information, communications, automation and micro-electronics-based competencies, can be termed 'meta-technologies'. These vital forms are by no means confined to enterprises in high-technology sectors. As the term implies, they have the power to transform any sector of the economy in which they are applied. 'Crisis-visionary' works by Abernathy, Clark and Kantrow (1983), Drucker (1969), Piore and Sabel (1984), Servan-Schreiber (1967), Tofler (1970) and others may have alerted top managers to the significance of meta-technological change but prescriptions for action are not always evident or operable, save in hindsight. To be fair, the processes by which innovations diffuse and become exploitable within and across industrial sectors are complex and unpredictable (Rosenberg, 1976). Furthermore, the exploitation of a particular idea or invention typically depends on the convergence of multiple technological capabilities in the firms where application might be feasible, so its impact depends on the idiosyncrasies of firms and markets exposed to it and, crucially the timing of the exposure. Boisot (1986) also pointed to the culturally-bounded aspects of the diffusion and codification of technological knowledge.

Many important difficulties are raised at the macro level of analysis, not least anticipating the sources of new ideas for exploitation and forecasting their potentialities; predicting their rate of diffusion across sectors and geographic boundaries; understanding the role of multinational firms on the diffusion process; the impact of globalized technologies on country-centred industries and firms; the potential of innovations for *strategic surprise* and for *redefining* sectoral boundaries, with all that that implies for de-stabilizing the balance of competitive forces; and the role of governmental policies allied to external agencies such as universities, science and technology parks and research institutes in facilitating technological innovations in a deliberate, prudent manner.

From a research standpoint, many of these problems have proved intractable or have yielded contradictory findings. We suspect also that for many interested parties macro-economic theorizing on innovation is perceived to be distinctly less relevant that a sector-specific perspective, to which we therefore turn.

THE FIRM-IN-SECTOR PERSPECTIVE

In practice, technological innovation (as indeed any form of environmental change) affects firms via the mediating action of the competitive or collaborative context of the industry sector in which it operates. We emphasize, therefore, the importance of the firm-in-sector unit of analysis.

None the less, we recognize at the outset important differences of treatment adopted by our contributors, broadly classifiable as the micro-economic 'theory of the firm' perspective and the managerial/behavioural perspective, which itself takes various forms including the relatively recent attempt to identify a 'strategic management paradigm' (Hofer and Schendel, 1978). Whilst the micro-economic perspective acknowledges the role of individual firms, it is more concerned to explore regularities in sectoral equilibria and patterns of change; behavioural perspectives desire to 'open the black box' of the firm and make sense of managerial processes inside.

Another important distinction between various forms of firm-in-sector treatment is that of cross-sectional, steady state analysis versus longitudinal/temporal (evolutionary) evaluation of firms and sectors. Technological innovation poses a wide range of analytical problems which can be roughly grouped into (i) boundaries and structures (ii) sector evolution and (iii) technology as means to competitive advantage.

Boundaries and structures

Chandler (1962) for example, claimed that many firms control the means of production and innovation through vertical integration of activities in the 'value adding chain' via a unitary (U-form) organization structure (e.g. Ford in the 1920s). Some subsequently restructure in a multi-divisional (M-form—e.g. as pioneered by General Motors). Basing their work on Chandler's theory, Rumelt (1974) and others tested the impact of structuring and operational diversity on firms' economic performance. The consensus was that divisionalization and related diversification across industry sectors generally made economic as well as organizational sense.

But Chandler's thesis that the structuring of firms—hence that of the sectors in which they operate—develops predictably over the long term according to a managerial logic of control, has been challenged. Williamson (1975) sought alternative *contingent* contractual explanations of *de facto* vertical and horizontal sectoral structures using notions of organizational inter-dependencies based on a need for specific assets provided by other parties, of a bounded rationality in choice of partners and of opportunistic behaviour towards customers and suppliers. Caves and Porter (1977) explained industrial structures in terms of barriers to entry and mobility arising from the scale of firms' technology-specific investment and cumulative experience. The 'uncertain imitability' of these technical skills and resources appears as the key for Lippman and Rumelt (1982). Differences in firm performances are also related to intra-sectoral structuring in strategic groups of organizational sets (Cool and Schendel, 1988; McGee and Thomas, 1986; Newman, 1973). The closure of such sets is marked by the contestability of their boundaries by would-be entrants (Baumol, Panzar and Willig, 1982; Hatten and Hatten, 1987) and by the internal strength of their networks (Hakansson, 1987; Jarillo, 1988; Thorelli, 1986). Johanson and Mattsson (1987) contrasted network and transaction cost theories, concluding that network constructs are particularly helpful in exploring dynamic aspects of industry sector structuration.

Teece (1985) has drawn much of these approaches together in an attempt to direct *strategic management* attention to the emergence of a 'robust and technologically sensitive theory of the firm with strong normative implications'. He argued that the production and utilizations of technological know-how is a central economic activity, generating managerial options for innovation and change. These choices intermittently alter the boundaries of the firm at technologically separable interfaces; they affect cost structures in the firm and create considerable inter-firm dependencies, creating complexity and uncertainty. Teece strongly refuted the exogeneity of technological innovation by affirming its importance as a managerially initiated source of comparative advantage which leads to the erection of barriers to isolate or inhibit competition.

In the literature on entrepreneurship we see similar arguments based on the phenomenon of growth and innovation poles (Perroux, 1950; DeBresson, 1989) as exemplified by the high-technology industries centred on Palo Alto in California (Rogers and Larsen, 1984; Miller and Cote, 1985) and industrial districts in Italy (Brusco, 1982, 1986). These geographically-localized concentrations of technology-specific resources and knowledge expose firms to cooperative and competitive relationships that enhance and coordinate their mutual development. Innovations may be exogenous to the individual firm, yet endogenous to the community and culture of which they are a part.

Research on sectorally specific aspects of technological innovation has adopted various methodologies. A well known example is the innovation-episodic approach associated particularly with the sector-comparative SAPPHO program at Sussex University (Rothwell *et al.*, 1974; Freeman, 1982). Subsequently there have been an expanding number of contributions (e.g. Cooper and Schendel, 1976; Langrish, 1969; Myers and Marquis, 1969; Saren, 1979, 1986). Most studies have attempted to identify structural and other factors which support successful innovations in particularized contexts; whilst foci and findings differ in detail, most results suggest that successful innovation requires, at minimum, the convergence of appropriate levels of effective demand, with a technological competence and creativity within individual firms.

Sector evolution

The sector-specific temporally evolving perspective is much associated with the work of Abernathy. This suggests that industry sector competence typically evolves in a relatively predictable fashion as measured by variations in product and process innovativeness (Abernathy, 1978; Utterback and Abernathy, 1975, 1978). This may be seen to assume an evolving collective competence or *trajectory* shared by industry members or productive units (Dosi, 1982; Freeman, 1982, p. 218; Nelson and Winter, 1977). More precise predictions attempt to chart the actual and expected parametric performance of a technology over time in the form of an S curve (Foster, 1986).

Abernathy's work focused on the automobile industry and stimulated other work in this sector (e.g. Altshuler *et al.*, 1984; Whipp and Clark, 1986) and elsewhere (e.g. Braun and MacDonald, 1978; Dosi, 1981, 1982) in semiconductors. Whipp and Clark (1986, p. 29) have argued that sector life-cycles are promoted by natural selection processes allocating new resources to those firms achieving high economic returns at others' expense by exploiting innovations. Positing natural selection in mature industries populated by powerful, oligopolistic firms is controversial, though there is evidence for the process in fragmented and emerging populations (Hannan and Freeman, 1977). Natural selection mechanisms within sectors is also supported by evidence from many observers of periods of incremental change punctuated by dramatic, 'frame-breaking' technological discontinuities (e.g. Tushman and Anderson, 1986).

McGee and Thomas (1985) documented the evolution of the reprographics industry over a fifty-year period. They found that by combining structural, i.e. intra-sectoral strategic firm groupings, deriving from broadly similar technological postures and temporal perspectives, they could explain survival patterns and performance differences amongst firms.

However, as Abernathy acknowledged, unidirectional models of evolving technological competence over-simplify complex realities; in automobiles sources of innovation are simultaneously exogenous and endogenous, so the possibility of *dematurity* in such established sectors must be taken seriously. Industries may be seen to develop 'innovation poles' which form and reform over time around different focal firms in a way which will affect the internally constructed 'design hierarchies' of other firms (Whipp and Clark, 1986).

Grinyer and Spender (1979) and Nelson and Winter (1977, 1982) respectively portrayed sectoral development as evolving changes in the 'portfolio' of generalized, shared beliefs about how to compete (recipes) and specific technological knowledge and competences (repertoires) enjoyed by constituent firms. Individual firms experience stringent practical limits on their ability to enhance their repertoires by innovating, which therefore constrains their freedom either to create and implement internal technological changes or to appropriate the benefits of breakthroughs made by others. Teece (1985) also emphasized the technological significance of repertoires, quoting Nelson and Winter (1982): 'a firm's capabilities are defined very much by where it has been in the past and what it has done ... a function of deeply engrained repertoires'. Hence technological emulation by a follower is necessarily subject to its ability to unravel the leader's cumulative tacit knowledge, skills and resources as encapsulated in the notion of a firm-specific 'technological trajectory' (Pavitt, 1986, p. 175).

Sustaining competitive advantage

The various distinct, but related conceptual approaches to analysing sector development converge around the role of technology in securing and sustaining comparative advantage. Many of the now well-known propositions about the relative competitive advantage of firms were first articulated—or comprehensively repackaged—by Porter (1980, 1985), though little in his earlier work treats technological innovation explicitly. His later work is more complete in this respect, assigning technological innovation a significant, albeit subordinate role, to be emphasized when it; (i) offers 'first mover advantage' to early innovators or (ii) buttresses the firm's overarching competitive strategy of low cost leadership or specialization or (iii) is unequivocally the key to survival. (How strategists decide this in practice is another matter!)

For Porter, technology strategy seems to be about exploiting *innovations*, rather than a stream of *innovatory behaviour* over time. Whilst a firm has, or should have, a technology strategy (i.e. a coherent posture toward technological innovations), the on-going act of innovating through technology is not portrayed as a core element of competitive strategy, by no means a universally accepted position.

Clarke (1987) for example, developed the concept of 'transilience', a nominal measure of the joint impact of an innovation on firms' market linkages and production systems. Since a given innovation will frequently affect competing firms differently in one or both respects, it follows that transilient technological changes may alter and disrupt the competitive status quo. Its significance lies in the scope it offers analysts of industry structures to evaluate the potentially dematuring affect of innovatory behaviour. Clarke underlined sustained technological prowess as central to a firm's long-run competitive strength:

> 'investment in new technology involves the development, nurturing and replenishing of the firm's productive and creative capabilities . . . its application requires a thorough understanding of the business and the way that the firm competes and the way that it intends to compete over the longer term.' (1980)

Ramanujam and Mensch (1985) suggested systematic ways of facilitating this understanding. Ford (1988) and Frohman (1982) however, sounded cautionary notes. Ford pointed out that firms implement strategies around clusters or bundles of technologies. If they are to sustain competitive advantage they need to distinguish technologies in which they have a *distinctive competence* from those that are *basic*, albeit necessary to competitive survival and those which confer no advantage, hence should be *bought-in*. Frohman pointed to the need to blend technological with broader managerial and organizational competences.

Other authors have attempted to explore the complementarity of developments taking place in the varying aspects of strategy development and the impacts of technological innovation on each one. For example Balakrishnan and Wernerfelt (1986) discuss the mutual impact of technical change, trends in vertical integration and states of market competition. Other workers have also been more circumspect in their evaluation of the contribution made by technological innovation to a firm's competitive advantage than perhaps were some writers earlier in the decade (Butler, 1988; Lieberman and Montgomery, 1988; McFarlan, 1984; Wightman, 1987; Williams, 1983). Indeed Dunford (1987) has suggested evidence that a common form of competitive response was systematic suppression of technological exploitation by 'blitzkrieg' patenting, take-overs and other defensive activities. Risk and the evaluation of risk has been a continued theme in a decade of change (Steele, 1983) over which there has been a growing acceptance that failure was a necessary complement to innovation and the ability to absorb or deflect the inevitable costs of failure was as necessary to the successful adaptation or transformation of firms as their ability to appropriate the benefits of success.

These themes are taken up throughout this volume under three headings, (i) Strategies in context, (ii) Structures, boundaries and alliances, (iii) Crisis, learning and adaptation.

Strategies in context

In this section we explore the notion of appropriate strategies in context. By context, we mean the circumstances locating the firm with respect to environmental opportunities and threats, organizational skills and resources as well as sector-specific issues. Notwithstanding some important criticisms of Abernathy's temporal framework for describing evolution of the productive unit (and frequently extended to apply to the sector context of such units), it is a well-known and useful way of categorizing the overall context of the firm.

Calori discusses effective strategies in emerging 'high technology' industries, where the firm's products are still at a relatively early stage of evolution. He emphasizes the importance of 'differentiation' strategies, but unlike Michael Porter, argues that differentiation is not incompatible with being a low-cost source, at least in the first wave of product development. Subsequent waves attract late entrants, often bigger firms who enter via acquisition. Still, other, less formal modes of collaboration exist and can work well.

Carr looks at the automotive components industry world-wide, with particular reference to the UK firms in this sector. Like Calori, he expresses reservations over 'standard prescriptions for competitiveness as proposed by Porter, B.C.G. and others. He argues that the decline of the UK sector is largely a failure to recognize that the industry is now a truly international industry. Hence, at least until recently, the strategies of UK firms have been inappropriate for competing in world terms, given the strength of competition presented by the Americans, the Germans and above all, the Japanese. In particular, the Japanese have shown all their competitors the benefits to sustaining competitive advantage through long-run momentum in incremental process and product innovation.

Moenaert *et al.*, too, look at strategies for competing in mature industry contexts. They build on the work of Abernathy and colleagues, arguing for serious consideration to be given to the dematurity concept. A key question is how firms with ageing technologies can appropriate new technologies from outside the sector in which they compete, effectively dematuring their activities. This process they term 'technological turnaround', implying choices not only about technology *per se*, but structural managerial process/style too.

Senker illustrates what appears to be a most effective dematurizing strategy on the part of the major UK supermarket chains. By initially developing a competence in quality control in processed foods, some chains have subsequently built on these competences in a highly adaptive fashion. Thus, the leading chains now innovate in product development and specifications and in packaging for safe transmission of processed foods along the 'cold distribution chain'. Those supermarket groups who have not actively developed these competences are now suffering a significant competitive disadvantage, as consumers become increasingly sophisticated and competition is no longer focused on standardized products and low prices.

Finally, in this section, Loveridge discusses the strategic successes and failures of Lucas Industries in a range of automotive and aerospace industry sectors. Building on conceptual work by a range of interpretive writers such as Spender and Johnson, he examines the role of executive assumptions and beliefs in the positioning of the firm and how it tries to compete. Loveridge draws the distinction between the shared managerial 'mind-set' or frame and the preferred style or mode of action (formula). In addition, he distinguishes between strategic and operational patterning of beliefs and actions. He argues that for an initiative to succeed there must be congruence between frame and formula at both strategic and operational levels. Unfortunately, achieving congruence is, arguably, one of the hardest managerial tasks in strategy implementation. One reason for this is given as the choice of reference groups made by strategic and operational managers. In this case the sectoral roles of Robert Bosch Gmb and the Bendix Corporation are seen as providing important comparators for Lucas strategists over a period of half a century.

Structures, boundaries and alliances

The structural implications of strategic decisions and actions are increasingly recognized in the literature. This subject draws together a wide range of interests in respect of how the firm should (and does) choose to organize, in particular where to draw its boundaries. In high technology industries more than most, managers take for granted that they can or should control all aspects of technologies on which their survival and profitability depends.

Thus, as Freeman and Barley demonstrate, a plethora of alliances and collaborative arrangements are manifest in a novel and highly uncertain area like biotechnology. The sum of these arrangements define the ecological niche occupied by the firm and its viability in the long run. They posit the convergence of two important fields of study, population ecology theory and social network theory and suggest an avenue of research which is both powerful and rigorous in delineating and ultimately, it is hoped, explaining successful strategies in context. Central to these ideas is the mutual inter-dependence of firms and a rejection of the notion of the firm as a quasi-closed and independent system.

Koenig and Thietart reach similar conclusions, starting from the institutional economic perspective much associated with Williamson. They study various forms of economic organization amongst large firms in the European Aerospace Industry. Notwithstanding the manifest benefits of collaboration, structural arrangements within and across firms have proved critical in determining whether particular collaborations have been effective. Managerial skills too, have been critical in getting results in a highly politicized environment of operation.

Whittington examines the manner of conducting industrial research and development in the UK in recent years. He suggests there is strong evidence for the fragmentation of R&D effort. By this he means; (i) a shift from large, centralized units to small units close to production facilities and users (ii) an increased willingness to expose in-house R&D staff to the rigours of the market and/or (iii) the subcontracting of specific projects to external agencies where this is strongly indicated and feasible. Whittington explains this externalization of R&D effort as part of a wider process of disintegration aiming to substitute market mechanisms for bureaucratic controls. Whilst, it may have short-run benefits, there are also long-run concerns for the strategic viability of firms, especially when technology is a major plank in their competitive position.

Saren, too, is concerned with the strategies and processes of research-proactive innovation, particularly with the role of inter-firm networks in the diffusion of new ideas and the appropriation of technological innovations in new contexts citing the example of laser technology for precision gauging applications in the engineering and process industries. He contends that existing accounts of innovating have ignored or undervalued the *interactive* aspects of the diffusion process.

Finally in this section, Clark and DeBresson explore the structural concepts and implications of innovation-design capabilities and innovation poles. They argue that to understand the present strategic competences of a firm, it is necessary to track its evolving innovation-design capability over time in its particular sector of activity. Further, innovation is in some respects antithetical to the quest for efficiency, at least in so far as innovation is construed as an activity directed at effectiveness. One way of resolving this dilemma is for the firm to participate in collaborative network relationships creating an innovation pole. Access to and exploitation of an innovation pole is a major factor in appropriating the benefits of technological innovation, whether by the firm or its fellow network members. Failure to sustain access is inimical to the firm's prospects for survival and prosperity.

Crisis, learning and adaptation

The thrust of this section is the argument that effective strategies of technological innovation benefit from, indeed may necessitate, an adaptive learning orientation strategists and managers would do well do come to terms with the reality that they are rarely in full control of the firm's destiny. Thus adaptive, opportunistic knowledge appropriation as a means of effective learning is the key to technological transformation.

Pitt draws an analogy between fundamental strategic change driven by technological innovation and crisis management. Many sources have noted the distinction between revolutionary and evolutionary change. Pitt suggests that fundamental changes arise in transitions which test managerial competences in the fashion of organizational crises. However, crises are of several forms, and managers are encouraged to adopt the transformational mode of behaviour variously referred to as *constructed* crisis. By being proactive, managers have the best prospect of retaining the initiative and the balance of power in and around the firm, thereby facilitating and securing periodic transformations between feasible and acceptable organization states.

By contrast Clarke and Thomas reiterate the significance of an apparent recursiveness in strategic decision-making and the relatively limited frame within which technological choices appear to be made. They suggest that an understanding of that frame demands an analysis of the organization's culture and of its accumulated competencies which together provide a varying capacity for change. They conclude with a plea for the unravelling of the complex inter-connections between context, culture and competencies in a manner that is pursued by the authors of our concluding chapters.

Jones points to a fundamental paradox in that new, more flexible production technologies offer enormous possibilities for strategic and operational benefits, yet to secure these, managers must cede much control to lower levels of management and staff than often they feel able to do. Further, instead of using inherent flexibility to secure strategic advantages commensurate with new organizational competence, most firms operate within the conventional Fordist paradigm, ignoring many possible benefits of flexible response. Thus, Jones concludes, with the exception of Japanese firms (though it is still too soon to be dogmatic on this point), flexible specialization has yet to realize the potentialities mapped out for it by Piore and Sabel.

Child and Smith discuss the emerging crisis of competitive disadvantage witnessed in Cadbury Ltd during the 1970s. They document and interpret the quite fundamental reorientation that has occurred subsequently. This wide-ranging and comprehensive survey builds on many of the propositions and agenda developed earlier in this book, in respect of both the substance of the socio-technical changes and the processes by which they were implemented. The authors emphasize the need to understand transformation holistically and in relation to what may be termed the objective, cognitive and systemic characteristics of the firm-in-sector.

Loveridge concludes this section with an examination of the way a new generic or 'meta' technology—in this case information technology—emerges and gains a foothold in an industrial sector or set of sectors. He distinguishes between strategic learning and operational learning. The early stages of innovation are dominated by a incrementalism in strategic decision-making caused by a lack of technical comprehension and an insufficiently developed strategic logic. The latter has generally emerged in the context of environmental threat and has been driven by the need to defend or extend the organization's domain rather than by the creative exploitation of the technical capability of the new technology. It may require future stages of development in social and technical comprehension for a viable basis for a logical incrementalism to emerge that will fully exploit the potentiality of IT.

Summary and conclusions

A broad-ranging, enquiring book like this inevitably raises more questions than it can hope to provide answers to. Pitt attempts to synthesize its contents in terms of a few overarching, fundamental questions to which future research and, of course, best practice should continue to address themselves. These are:

- How best can firms forecast the advent of novel technologies and interpret their significance?
- How can mature businesses appropriate the benefits of new technologies without being exposed to unacceptable levels of technical or commercial risk?
- What lessons for effective strategic technological transformation of the firm lie in the historical tracking and future prediction of sectoral structural developments?
- What kind of managerial styles and processes will achieve and sustain a creative, inventive intra-firm environment appropriate to a strategy of technological leadership?
- How should the firm respond to aspects of technological innovation that give rise to ethical dilemmas?

REFERENCES

Abernathy, W.J. (1978). *The Productive Dilemma: Roadblock To Innovation in the Automobile Industry*, Johns Hopkins Press, Baltimore.

Abernathy, W.J., Clarke, K.B., and Kantrow, A.M. (1983). *Industrial Renaissance: Producing a Competitive Future for America*, Basic Books, New York.

Altshuler, A., Anderson, M., Jones, D., Roos, D., and Womack, J. (1984). *The Future of the Automobile*, George Allen & Unwin, London.

Balakrishnan, S., and Wernerfelt, B. (1986). 'Technical change, competition and vertical integration', *Strategic Management Journal*, 7 (4), 347–60.

Baumol, W.J., Panzar, J.C., and Willig, R.D. (1982). *Contestable Markets and the Theory of Industry Structure*, Harcourt Brace Jovanovich, New York.

Boisot, M. (1986). 'Markets and hierarchies in a cultural perspective', *Organization Studies*, 7, 135–58.

Braun, E., and MacDonald, S. (1978). *Revolution in Minature: The History and Impact of Semi-conductor Electronics*, Cambridge University Press, Cambridge.

Brusco, S. (1982). 'The Emilian Model: productive decentralization and social integration', *Cambridge Journal of Economics*, 6(2), 167–84.

Brusco, S. (1986). 'Small firms and industrial districts: the experience of Italy', in D. Keeble and F. Wever (eds) *New Firms and Regional Developments in Europe*, Croom Helm, London.

Butler, J.E. (1988). 'Theories of technological innovation as useful tools for corporate strategy', *Strategic Management Journal*, 9(1), 15–30.

Caves, R.E., and Porter M.E. (1977). 'From entry barriers to mobility barriers', *Quarterly Journal of Economics*, 91, 241–62.

Chandler, A.D. (1962). *Strategy and Structure*, M.I.T. Press, Cambridge.

Clark, K. (1987). 'Investment in new technology and competitive advantage', in D. Teece (ed.), *The Competitive Challenge*, Ballinger, Cambridge Mass.

Cool, K., and Schendel, D. (1988). 'Performance differences among strategic group members', *Strategic Management Journal*, 9, 3, 207–224.

Cooper, A.C., and Schendel, D. (1976). 'Strategic responses to technolgical threats', *Business Horizons*, Feb., 61–79.

DeBresson, C. (1989). 'Breeding innovation clusters: a source of dynamic development', *World Development*, 17, 1.

Donaldson G., and Lorsch, J. (1983). *Decision Making at the Top: The Shaping of Strategic Decision*, Basic Books, New York.

Dosi, G. (1981). 'Institutions and markets in high technology industries: an assessment of government intervention in European microelectronics', in C.F. Carter (ed.) *Industrial Policies and Innovation*, Heinemann, London.

Dosi, G. (1982). 'Technological paradigms and technological trajectories: a suggested interpretation of the determinants and directions of technological change', *Research Policy*, **11**, 147–62.

Drucker, P.F. (1969). *The Age of Discontinuity*, Heinemann, London.

Dunford, R. (1987). 'The suppression of technology as a strategy for controlling resource dependence', *Administrative Science Quarterly*, **32**, 512–25.

Ford, D. (1988). 'Develop your technology strategy', *Long Range Planning*, **21**, 5, 85–95.

Foster, N. (1986). *Innovation: The Attackers Advantage*, Macmillan, London.

Freeman, C. (1982). *The Economics of Industrial Innovation*, Pinter, London.

Freeman, C., Clark, J., and Soete, L. (1982). *Unemployment and Technical Innovation*, Pinter, London.

Frohman A.L. (1982). 'Technology as a competitive weapon', *Harvard Business Review*, Jan/Feb, 97–104.

Grinyer, P.H., and Spender, J.C. (1979). 'Recipes, crisis and adaptation in mature businesses', *International Studies of Management and Organization*, **9**, 113–23.

Hakansson, H. (ed.) (1987). *Industrial Technological Development: A Network Approach*, Croom Helm, London.

Hannan, M.T., and Freeman, J. (1977). 'The population ecology of organizations', *American Journal of Sociology*, **82**, 929–64.

Hatten, K.J., and Hatten, M.L. (1987). 'Strategic groups, asymmetrical mobility barriers and contestability', *Strategic Management Journal*, 8(4), 329–42.

Hickson, D.J., Butler, R.J., Cray, D., Mallory, G.R., and Wilson, D.C.(1985). 'Comparing 150 decision processes', in J. M. Pennings, (ed.) *Organization Strategy and Change*, Jossey-Bass, 35–63, San Francisco.

Hickson, D.J., Butler, R.J., Cray, D., Mallory, G.R., and Wilson, D.C. (1986). *Top Decisions: Strategic Decision Making in Organizations*, Basil Blackwell, Oxford.

Hofer, W., and Schendel, D. (1978). *Strategy Formulation: Analytical Concepts*, West, St Paul.

Jarillo, J.C. (1988). 'On strategic networks', *Strategic Management Journal*, 9(1), 31–42.

Johanson, J., and Mattsson, L.G. (1987). 'Interorganizational relations in industrial systems: a network approach compared with the transaction cost approach', *International Studies of Management and Organization*, **17**, 1, 34–48.

Kantrow, A.M. (1980). 'The strategy-technology connection', *Harvard Business Review*, July/Aug., 6–21.

Langrish, J. (1969). *Innovation in Industry: Some Results of the Queen's Award Study*, Research Report No. 15, Department of Liberal Studies in Science, University of Manchester.

Lieberman, M.B., and Montgomery, D.B. (1988). 'First mover advantages', *Strategic Management Journal*, **9**, Special Issue (Summer), 41–58.

Lippman, S.A., and Rumelt, R.P. (1982). 'Uncertain imitability: an analysis of interfirm differences in efficiency under competition', *Bell Journal of Economics*, 13(2), 418–38.

McFarlan, F.W. (1984). 'Information technology Ccanges the way you compete', *Harvard Business Review*, May/June, 98–103.

McGee, J., and Thomas, H. (1985) 'Making Sense of Complex Industries', in D. Schendel, N. Hood, and J.E. Vahlne (eds), *Global Strategies*, Wiley, New York.

McGee, J., and Thomas, H. (eds) (1986). *Strategic Management Research: A European Perspective*, Wiley, Chichester.

McLimore, J.F., and Larwood, L. (1988). *Strategies . . . Successes . . . Senior Executives Speak Out*, Harper and Row, New York.

Miller, R., and Cote, M. (1985). 'Growing the next silicon valley', *Harvard Business Review*, July/Aug, 114–23.

Myers, S., and Marquis, D.G. (1969). *Successful Industrial Innovation*, National Science Foundation, 69/17.

Nelson, R., and Winter, S.G. (1977). 'In search of useful theory of innovation', *Research Policy*, **6**, 1, 36–77.

Nelson, R., and Winter, S.G. (1982). *An Evolutionary Theory of Economic Change*, Harvard University Press, Boston.

Newman, H.H. (1973). *Strategic Groups and the Structure–Performance Relationship: A Study With Respect to the Chemical Process Industries*, unpublished Doctoral Thesis, Harvard University.

Pavitt, K. (1986). 'Technology, innovation and strategic management', in J. McGee, and H. Thomas (eds), *Strategic Management Research: A European Perspective*, Wiley, Chichester.

Perroux, F. (1950). 'Economic space: theory and applications', *Quarterly Journal of Economics*, **LXIV**, 89–104.

Phillips, A. (1971). *Technology and Market Structure*, Lexington.

Piore, M.J., and Sabel, C.F. (1984). *The Second Industrial Divide: Possibilities for Prosperity*, Basic Books, New York.

Porter, M.E. (1980). *Competitive Strategy*, Free Press, New York.

Porter, M.E. (1985). *Competitive Advantage*, Free Press, New York.

Ramanujam, V., and Mensch, G.O. (1985). 'Improving the strategy–innovation link', *Journal of Product Innovation Management*, **4**, 213–23.

Rogers, E.M., and Larsen, J.K. (1984). *Silicon Valley Fever: Growth of High Technology Culture*, Basic Books, New York.

Rosenberg, N. (1976). *Perspectives on Technology*, Cambridge University Press, Cambridge.

Rothwell, R. (1983). 'Innovation and firm size: a case for dynamic complementarity; or, is small really so beautiful?', *Journal of General Management*, 8(3), 5–25.

Rothwell, R., Freeman, C., Horlsey, A., Jervis, V.T.P., Robertson, A.B., and Townsend J. (1974). 'SAPPHO updated—Project SAPPHO phase 2', *Research Policy*, **3**, 258–91.

Rumelt, R. (1974). *Strategy, Structure and Economic Performance*, Harvard University Press, Cambridge Mass.

Saren, M.A.J. (1979). *The Characteristics of the Innovating Firm*, unpublished Doctoral Thesis, University of Bath.

Saren, M. (1986). 'The role of strategy in technological innovation: a reassessment', in I.L. Mangham (ed.), *Organization Analysis and Development*, Wiley, Chichester, 125–65.

Schmookler, J. (1966). *Invention and Economic Growth*, Harvard University Press, Cambridge Mass.

Schoen, D.R. (1969). 'Managing technological innovation', *Harvard Business Review*, May/June, 156–68.

Servan-Schreiber, J.J. (1967). *Le Defi Americain* (The American Challenge), Editions Denoel, Paris.

Steele, L. (1983). 'Managers' misconceptions about technology', *Harvard Business Review*, Nov./Dec., 133–40.

Teece, D.J. (1985). 'Applying concepts of economic analysis to strategic management', in J.M. Pennings (ed.) *Organization Strategy and Change*, Jossey-Bass, San Francisco, 35–63.

Thorelli, H.B. (1986). 'Networks: between markets and hierarchies', *Strategic Management Journal*, 7(1), 37–52.

Tofler, A. (1970). *Future Shock*, Bodley Head, London.

Tushman, M.L., and Anderson, P. (1986). 'Technological discontinuities and organizational environments', *Administrative Science Quarterly*, 31, 439–65.

Utterback, J.M., and Abernathy, W.J. (1975). 'A dynamic model of product and process innovation', *Omega*, 3(6), 639–56.

Utterback, J.M., and Abernathy, W.J. (1978). 'Patterns of industrial innovation', *Technology Review*, 7, 41–7.

Whipp, R., and Clark, P. (1986). *Innovation and the Auto Industry*, Pinter, London.

Williams, J.R. (1983). 'Technological evolution and competitive response', *Strategic Management Journal*, 4(1), 55–65.

Williamson, O.E. (1975). *Markets and Hierarchies: Analysis and Antitrust Implications*, Free Press, New York.

Part 1
Strategies in Context

1
Effective Strategies in Emerging Industries

Roland Calori

Lyon Graduate School of Business, Lyon, France

This chapter defines and then discusses the concept and reality of emerging industries. It makes distinctions between the technological intensity of different emerging contexts, the role of technology in securing competitive advantage and the nature of effective strategies, taking account of these and other contingent factors. It also examines the large firm–small firm interaction and the structural and managerial imperatives to remain viable as the industry grows and matures over time.

TYPES OF EMERGING INDUSTRIES

Several structural characteristics of industries lead to possible typologies for descriptive or prescriptive purposes:

— the maturity of the industry (Hofer, 1975),
— the degree of concentration (versus fragmentation) according to industrial economists,

The Strategic Management of Technological Innovation. Edited by R. Loveridge and M. Pitt
©1990 by John Wiley & Sons Ltd

— the possible sources of differentiation through marketing and/or research and development (Boston Consulting Group 1981),
— the global (versus local) scope of the competition (Porter, 1985a).

The concept of an 'emerging industry' was propounded by Porter (1980). The main characteristics of emerging industries are their *technological newness and uncertainty*. Emerging industries may be in the introduction or growth stage, and they have several high differentiation opportunities based on research and development (R&D). By analogy with the Abernathy and Utterback (1975) product-process evolution model we would say that emerging industries are in the 'fluid phase' and that no 'dominant product or process design' has emerged in the competition. When some dominant product and process designs become evident, this technological maturity turns the competitive system into a mature one. In this sense, this concept is different from the well known product life-cycle theory which is based on the market growth rate.

Technological uncertainty has some other major structural consequences which are also characteristics of emerging industries: high strategic uncertainty (high rates of new entrants and exits), many embryonic companies and spin-offs, first-time innovative buyers, high initial costs but steep cost reduction. As emerging industries represent a high potential for growth, employment, productivity gains and positive trade balance (U.S. Department of Commerce, 1983), Government interventions may be significant enough to shape the competition and develop the market, especially in high technology strategic sectors such as telecommunications, robotics, energy, etc.

In fact, most emerging industries also are *high technology* industries, based on a technological innovation meeting a new or existing customer need. By high technology, here, we mean high technological entry barriers:

— the minimum critical mass of research and development is high,
— the R&D/sales revenue ratio is high,
— the technological uncertainty (products and processes) is high,
— the technological know-how is hard to find on the labour market.

According to the technological 'entry ticket', two generic kinds of emerging industries should be distinguished: high technology emerging industries and low technology emerging industries. For instance, solar heating (in the 1970s), sail surfing or video games (at the end of the 1970s and beginning of the 1980s) were low technology emerging industries. On the other hand, fibre optics (at the beginning of the 1980s), photovoltaics and most of the industries in the robotics and biotechnologies sectors are high technology emerging industries.

This dichotomy between 'High Tech' and 'Low Tech' certainly is an over simplification. In fact, technological barriers should be measured on a continuum, besides, in a 'Low Tech' emerging industry one may find some specific small segments which have a higher technological intensity. For instance, 'parabolic concentration solar panels' in the solar heating industry or 'international competition sail surf boards' in the sail surf business. The objective here is simply to make a rough distinction.

The study of several emerging industries shows a significant difference between the *'pioneers age'* of the industry and the age of *'second wave entrants'*. In the pioneers age the market is limited to one application and/or one customer group (for instance—providing power for satellites in the pioneers age of the photovoltaics industry—the military telecommunications applications in the fibre-optics industry). In the second wave new applications and customer groups add to the original one. Consequently, the business becomes more heterogeneous and the competition involves many more competitors (for instance providing power for water pumping, signalling, pocket calculators, remote houses or villages, telephone booths in the second wave of the photovoltaics industry).

The transition between the pioneers age and the second wave has to be managed carefully: the competitors should get involved in the most promising applications and marketing skills have to be developed to complement technological skills.

Combining these 2 dimensions (early/late stage and low/high technology) gives 4 types of emerging industries which are illustrated in Figure 1.

In the pioneers age of high technology emerging industries, competitors

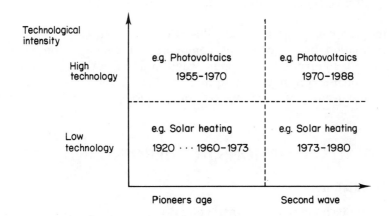

Figure 1.   Four generic types of emerging industries

may be small technological companies or big companies. In the second wave high technology emerging industries are capital intensive and often global industries. Financing the *technological ticket* and taking the risk of uncertain technology require high financial resources. In the second wave the products are marketed internationally to pay back these high investments. At the other end of the spectrum, low technology emerging industries in the pioneers age are more fragmented with several small independent competitors surviving.

These significant differences between emerging competitive systems have to be considered. Consequently, we shall stress effective strategies which are common to all emerging industries, but we shall also underline which strategies are specific to the generic situations shown in Figure 1.

In emerging industries, competitors may lose a lot of money for several years in spite of governments subsidies. For instance, in the photovoltaics industry the R&D budget often was *above* sales turnover during the 1970s. In fact managers do not expect short-term profitability in this kind of competitive setting. First of all, given the high rate of failures, success means surviving. In order to survive in the early stage of the industry the technological position is crucial—not to get involved in a dead-end technology—then, in the second wave, market share gain becomes crucial to improve competitiveness and achieve a strong position before the transition to maturity. So the question is: how to survive and gain market share in an emerging industry?

At the end of the second wave there is a transition phase between emergence and maturity. Dominant product designs appear, some new competitors, generally big companies, come into the market with high investments. This transition phase has to be managed very carefully, some incumbent firms may focus or niche, some may compete directly with the new entrants, some cooperate with new entrants, some are bought, some are shaken out.

This chapter proposes and illustrates some successful strategies—differentiation as a source of competitive advantage, combining big and small companies in various ways and networking—and effective management styles. But in these most unstable businesses surprises often occur and feeling of gambling (i.e. on technology), remains particularly strong.

STRATEGIC OPTIONS IN EMERGING INDUSTRIES

Differentiation

With high technological and marketing diversity, and steep cost reduction, emerging industries fall into the *specialization* competitive type according to the Boston Consulting Group (B.C.G., 1981). In emerging industries the successful survivors followed a differentiation strategy. In our work we have not found any successful cost leadership strategy as defined by Porter (1980). But differentiation is not necessarily linked with higher costs in emerging industries. In fact, some competitors succeed in *combining* differentiation and cost advantage, with cost advantage being a consequence of differentiation and technological innovation in a given market segment. This mechanism is described in Figure 2.

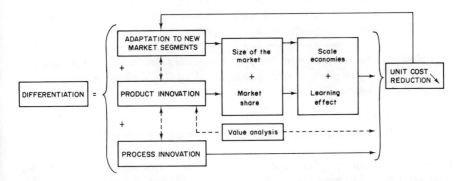

Figure 2.   Differentiation strategies in emerging industries and their impact on costs

For instance in 1983, Domilens, a small newly founded company, entered the ocular implants business with a new CAD/CAM process (Ramus, 1985). This process gave Domilens a competitive advantage in consistency of quality and delivery times (differentiation) which were key success factors. Besides, this new process gave Domilens a 30% cost advantage over the incumbent firms. Thanks to this *total advantage*, the newcomer took a significant share of the market within 3 years.

Some common strategic profiles of survivors appeared from a field study in 2 structurally different emerging industries (Calori, 1985); *photovoltaics*: high technology, concentrated and global; and *solar heating*: low technology, fragmented and local. All these strategic profiles derive from a differentiation-based competitive advantage. Three strategies are *focus-differentiation* (strategies 1, 2 and 4) as defined by Porter (1985b), one strategy (3) is differentiation across several industry segments (cf. Figure 3).

The pioneers (like Solarex in the photovoltaic industry, Corning Glass Works in the fibre-optics industry, Unimation then Trallfa in the robotics industry) have a differentiation strategy based on innovation. But the mass of *successful* followers in the second wave also entered the business with a marginal innovation and a differentiation strategy. For instance, in the fibre optics industry, after C.G.W. and Western Electric, the Japanese

Strategy 1	Highly focused firms, with average product innovation but with high quality for low price.
Strategy 2	Focused firms, exporting a high share of their production, co-operating with other firms of the industry to reach superior technology and distribution.
Strategy 3	'Couples' or 'constellations' of big and small firms serving a wide market with several technologies, backward vertically integrated, with high R&D expenses, high product innovation and high quality for average price.
Strategy 4	Firms focused on one production/distribution function (whether up or down the value chain), high quality for high prices and relatively high product innovation.

Figure 3.   Four strategic profiles of survivors in emerging industries

firms entered the industry some 3 years later with the V.A.D. process, 5 years after the first French company developed 2 original processes P.C.S and A.L.P.D. for applications outside telecommunications. Innovative licensees also provide examples of second wave innovations: for instance Lafarge-Coppee (Orsan), which bought a licence for amino acid production from Ajinomoto, improved the process by ultrafiltration through thin material membranes and by combining ammoniac and fumaric acids. Such differentiation strategies are linked with high market share survival (Calori, 1985; Cooper, 1979).

Focus on a few technologies and serve an international market

In the pioneers age of an emerging industry the industry scope is limited to a few technologies and applications, the few applications to which the new technology brings value. To innovate and pay back innovations companies need to reach a critical mass (R&D and sales), especially in high technology emerging industries. Focusing on a few technologies and applications and serving an international market are the appropriate conditions for reaching a critical mass (Strategy 2).

For instance, in the computerized tomography industry—scanners— (Abell, 1980), applications started in 1972 with brain scanning diagnosis, with a rectilinear pencil beam technology (first generation) for innovative teaching hospitals. Then, new applications (body scanning, new diseases, therapy planning and treatment), new generations of scanners (four generations coexisting in 1980) and new customer groups (all kinds of hospitals, clinics and radiologists) extended the initial market segment. Year after year, new technologies and new generations of products became competitive for new applications and the emerging industry spread (several technologies, several customer groups).

Even in the second wave era of high technology emerging industries successful companies focus on a few technologies and applications and/or sectors. For instance in the photovoltaics industry, the Japanese competitors have focused on amorphous silicon cells (or films) for low power applications, while European competitors focused on polycrystal silicon. In the robotics industry some companies have focused on robots for the *automobile* industry, others on spray painting (Trallfa) and some on assembly robots, like Olivetti or Sormel (Matra) which has specialized on cartesian robots for assembling small and medium volume products. In the fibre-optics industry, for instance, the focus of Quartz and Silice (now F.O.I.) on two new processes (P.C.S. and A.L.P.D.) for computer and control applications, created a competitive advantage. It gave the company the opportunity to license firms like E.B. Industries, Fibreguide Inc, Toshiba, Showa and Fibronics all around the world.

There are few competitors with a large *portfolio* of technologies and applications in high technology emerging industries. Such diversified competitors may result from mergers and joint ventures, generally at the end of the emerging phase of the industry. This focus strategy is a way to reach the critical mass for innovation and then to be competitive internationally.

Selling on a world-wide basis accelerates the pay back of the high initial investments. Successful companies in high technology emerging industries have an international market: a very high percentage of their turnover is done in foreign markets (from 40% to 90% according to Calori and Noel, 1986). For instance, in the photovoltaics industry, Solarex (the leader from 1975 to 1983 and a medium sized independent company at that time) had several licence agreements all around the world and exported the major part of its turnover.

However, in the pioneers age of high technology emerging industries, the opportunity of a *captive market* seems to be a condition for the success of the pioneering companies. By captive market we mean a segment of the market giving an *a priori* preference or exclusivity to the given supplier. It may be a national public demand or a demand coming from other companies belonging to the same industrial group. National public demand and preference were crucial in the fibre optics, the photovoltaics, the laser and more generally in the telecommunication sectors for instance. Car manufacturers and their sub-contractors were captive markets for the robots companies belonging to the same group. The impact of such captive markets is similar to the impact of the size of domestic markets on the industrial competitiveness of nations (Porter, 1985c). But after the pioneers age the successful companies are the ones which demonstrate their capacity to compete internationally.

International development and reaching a critical mass of R&D cost a lot. Big, powerful industrial groups have a competitive advantage in financing such investments; none the less, a few small innovative companies survive. Should *elephants* and *mice* join their efforts?

Combining the big and small: 'elephants and mice'

Innovating pioneers may be big—like the Bell Laboratories in photovoltaics and fibre optics—or small—like R2E and Apple in the pioneering age of the personal computer industry. The pioneering period is the most favourable to the emergence of new embryonic companies, created by technical entrepreneurs who have the technological know-how (cf. Schumpeter, 1954; Cooper, 1973). Spin-offs from established companies are often the source of such a proliferation. When the technological entry barriers are low (know-how and investment) the rate of birth may reach 30% a year as in the glorious times of the solar heating industry. Hundreds of artisans founded their own companies, manufactured and installed a few hundred solar water heaters in their various sunny regions. Because of the relative prices of the substitute technologies, the solar heating market started an early decline in the 1980s. Most of these small companies quickly disappeared, some bigger diversified ones survived. The same scenario happened in the sail surf industry which reached a peak in 1985.

Another shake-out scenario starts when powerful potential entrants notice the high growth prospects of the industry. Then, big companies come in—the *elephants* shake the *mice* (Strategy 3)—buy the best performing small competitors and often add their own R&D and marketing resources, raising entry barriers. The other small competitors are shaken out or have to focus on a niche either upstream in the value chain-for instance become a specialized research or control laboratory as in some bio-technology industries—or downstream in the value chain—for instance become a system engineering and maintenance company as in the photovoltaic industry (strategy 4). Success stories of small pioneers who became independent major competitors at the end of the emerging phase are extremely rare.

In the photovoltaics industry, small independent companies dominated the market for power generation in satellites until 1972. When terrestrial applications emerged in the middle of the 1970s the major oil companies (Atlantic Richfield, Sohio, Elf, Shell, etc.) and some electronics manufacturers (CGE, AEG, etc.) entered the business by acquisitions—first 'opening a strategic window' and then taking a position.

'LOGICS' OF ENTRY INTO HIGH TECHNOLOGY INDUSTRIES

Big companies entering an emerging industry may follow one of four logics: *technological progress, technological diversification, budding* or *growth,* according to their basic skills and goals. Figure 4 gives an overview of these four logics and the positions of Sormel-Matra and Lafarge, two contrasting cases described below.

Skills / Goals	Exploit existing technological skills	New technological skills
Extend and develop technological competences	TECHNOLOGICAL PROGRESS e.g. Sormel	TECHNOLOGICAL DIVERSIFICATION
Exploit emerging/ growth market attractiveness	BUDDING	GROWTH e.g. Lafarge

Figure 4. Four logics for big companies when entering into new high technology industries

Sormel was founded to solve the automation problems of Yema, a Matra subsidiary in watch-making (goal: develop technological competences). The firm's know-how in micro-mechanics, linear transfer chains, handling and rotative decks, opened up the possibility of new applications in the projects managed by Matra for automation of assembly (skills: exploiting some existing technological competences). Then this *technological progress* logic turned a *budding* logic (cf. Morin, 1985; Gest Euroconsult, 1985) and now 95% of the turnover of Sormel is done outside the Group. The *budding* logic consists of exploiting existing technological skills on promising new applications and markets. For instance, as in the case of some high quality glass manufacturers and electronics manufacturers who came into the fibre optics industry.

On the other hand, the entry of Lafarge into biotechnologies is typical of the pure *growth* logic. At the end of the 1970s all the businesses of Lafarge (cements, plasters, oven proof materials, etc.) were selling to mature or declining sectors of industry. The company was looking for new avenues for growth (goal: growth in new and growing attractive markets). Fermentation for food industries was selected in the biotechnology field. Lafarge did not have any technological skill in this area (amino acids). So it merged with the Belgium Group Coppee, a licensee of the Japanese leader of the industry, Ajinomoto. The same logic guided its entrance into the biovegetal business (genetics, seeds) in the mid 1980s through the acquisition of small companies: Wilson Hybrids, Harris Moran, Celpril and Sogetal.

The *technological diversification* logic corresponds to companies opening a window on a new technology (new to them) which *may* be threatening their core technology or which may be promising in the very far future. It was the logic of some oil companies when they came into the photovoltaic business.

This typology does not mean that companies having a technological progress logic do not care at all about *long-term* growth, and that companies which have a growth logic do not care at all about *technological progress*. The typology stresses what logic is dominant at a given point in time. Logics evolve as shown in the Matra case.

The contrasted cases of Matra and Lafarge illustrate the two extreme entry strategies:

— companies exploiting technological skills tend to enter early by internal development (this is also the strategy of most small companies),
— companies attracted by the potential market growth and without significant technological skills come late into the industry by acquisitions or merger.

Some empirical studies show that the combined action of a big company and a small company is generally successful (Calori, 1985). By a combined action we mean three possible scenarios (i) a big company buying a small one in the emerging business sector, (ii) a big company creating an independent subsidiary or (iii) a joint venture between a small and a big company. Strategy 3 described in Figure 3, couples or constellations of big and small firms serving a wide market, is related to higher market share (Calori, 1985) and most of the European leaders in emerging industries combine the big and small (Calori and Noel, 1986). The small firm brings its flexibility and technological know-how, the big one brings its marketing and financial power, all ingredients much needed in these uncertain competitive systems (cf. Hlavacek, Dovey, and Biondo, 1979; Quinn, 1979).

After some years, the result of this strategy may look like a constellation, with the parent company and several small ones serving complementary

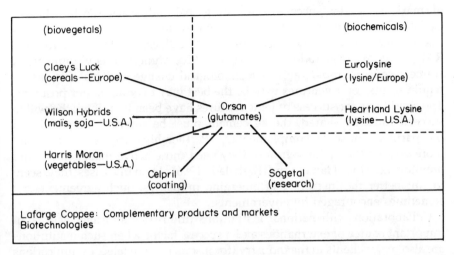

Figure 5. 'Constellations' in emerging industries, complementary products and markets. The Lafarge Coppee case in 1986

functions, products or markets. Figure 5 gives a simple description of Lafarge Coppee's constellation in the biotechnology sector in 1986.

This constellation form is a way to preserve the flexibility of each Strategic Business Unit and the flexibility of the whole portfolio in the new industry. Such a network between the parent company and subsidiaries or between partners in a formal joint venture is often extended to other partners.

NETWORKS AND ALLIANCES

Hamilton (1985) who studied several industries in the biotechnology field, stressed the importance of networks and alliances for middle size competitors. In high technology emerging industries, networks and alliances are built with three groups of partners (Calori and Noel, 1986):

— other competitors,
— universities, public research institutions, national ministries and agencies,
— clients/customers.

Licensing, collaboration agreements, partnerships, and joint ventures between competitors help to share technological risks and to reach a critical mass in R&D (common or complementary R&D programmes). They help to enhance the portfolio of products (in the case of a technological focus strategy employed by individual collaborators) and to enhance the portfolio of markets—a key success factor in global industries (Porter, 1985a). Collaboration between competitors is a way to reduce the technological uncertainty (vulnerability to technological deadlocks) and/or to reduce the R&D/Sales turnover ratio. As Harrigan (1985) argued: in environments of scarce resources, rapid rates of technological change and massive capital requirements, joint ventures may be the best way to attain better positions. The conditions for success of joint ventures have been identified: the will to succeed must be shared, the agreement must be carefully prepared and the partners must be complementary. Partnerships are less formal and more volatile than joint ventures, they need more active management and monitoring (Doz, Hamel, and Prahalad, 1986). In many cases they seem to suit better the instability of emerging industries. Such agreements are sometimes encouraged by governments.

Collaboration with national institutions (ministries and agencies) is an important source of externalities and a success factor when such institutions are also major clients in the industry (for instance in the telecommunications or energy markets). Collaboration with universities and public research institutions is a way to keep close to the scientific forefront and to secure access to highly qualified personnel.

For instance, in several sectors of the telecommunications industry, alliances were built between major companies: Matra (Fr) and Ericsson (Swe), Alcatel (Fr) and I.T.T. (USA) and new alliances are building in anticipation of the Single European Market (1992): Siemens (West Germany) and G.T.E. Sprint (USA); Philips (Holland) and A.T.T. (USA); Stet (Italy) and A.T.T. (USA), etc. The European public research programme 'RACE' will contribute up to 600 millions Ecus to pre-competitive research on telecommunications and national public telecommunications research centres are major actors. At the industry level, the French leader in fibre optics, F.O.I., has contracts with C.N.E.T., the French public research centre on telecommunications and collaborates with several universities. The network of other competitors includes the collaboration with Corning Glass Works on telecommunications applications and with the companies which bought the PCS and ALPD process licences.

Collaboration with clients is the third branch of the network; it is the major source of product innovations according to Von Hippel (1977), especially for industrial product idea-generation. It also helps to improve the competitiveness of new products by incremental innovations. Besides, these vertical alliances between suppliers and clients regulate markets along the vertical chain. Such agreements are much needed when the firms are focused at a single level of the vertical chain as is the case in many emerging industries.

Collaboration with clients and process engineering companies is crucial in the robotics industry; many companies rely on the experience of the 'parent' organization, which is often their major client (General Motors, F.I.A.T., Toyota, Renault). Companies which produce some elements of more complex systems also depend on this vertical collaboration, for instance photovoltaic panels have to be adapted to the application and the complete system (water pump, telephone booths, satellites). In the genetics industry partnerships with clients are launched to exploit genetic engineering innovations (cf. Genentech's strategy). Collaboration with clients is also important in low technology emerging industries, for instance in the sail surf industry the collaboration with wholesalers and retailers, with champions and schools and the resulting network appeared to be a key success factor.

We want to stress in conclusion that the three branches of the network are complementary conditions for success in emerging industries (Figure 6). A given company is the centre of gravity in its network. In the Flexible Manufacturing Systems industry the Japanese company Dainichi Kiko had 28 agreements with diverse partners when it moved from robots to complete factory automation systems!

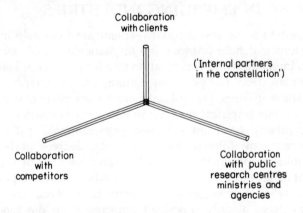

Figure 6.   A three branches network

In the photovoltaics industry the three branches of the network of company 'X' illustrate the functions and the complexity of the network. The Agence Francaise pour la Maitrise de l'Energie provided an important share of the R&D budget of X and access to demonstration equipment programmes managed by the E.E.C. and the French government. About ten collaboration or sub-contracting programmes were going on with universities and the Centre National de la Recherche Scientifique. Company X had chosen the polycrystal silicon solution since 1980, developed in a joint venture with a French chemicals industrial group. When the amorphous silicon solution (at first developed by some American and Japanese competitors) became a threat, X started a joint venture with a Japanese company to share R&D budgets and risks with a reciprocity agreement on findings. Company X had also close cooperation with clients in the vertical chain: a water-pump manufacturer, a television company and several telecommunications equipment manufacturers, to improve the systems and secure a share of the market.

In emerging industries such networks develop in line with changes in competitive positions. So flexible alliances (agreements, partnerships) seem to develop quicker than formal joint ventures. In some *strategic* sectors, governments are even encouraging such collaboration and the emergence of transnational *mutual organizations* (for instance in the aerospace industry, see Koenig and Thietart, this volume) for macro-projects. But the logic of networks and alliances is also valid for small companies which depend on leverage effects; for this reason they have to run the risks of such inter-organizational links.

MANAGEMENT PRACTICES AND STYLES IN EMERGING INDUSTRIES

The final question is: do specific management practices help in taking the 'right' decisions and help improve the implementation of strategies? The literature on innovation and organization (see for instance Aiken and Hage, 1971; Burgelman, 1985; Cooper, 1981; Quinn, 1979; Lawrence and Lorsch, 1967) gives some answers. We will only stress some particular points which appeared from our empirical research in emerging industries.

At the company level no common organizational pattern emerges probably because of different histories, sizes, degrees of heterogeneity of the portfolio of products and markets and because of differences in management styles of the parent companies when there is a parent. At the Group level, when a large industrial group is involved, the *constellation form* that we have already described appears to be the most common organizational solution. When the number of companies in the constellation grows, a division is created with very few staff.

Surviving companies in emerging industries also have diverse planning practices. Among embryonic independent successful companies, some have a business plan, others have a rather incremental approach. The only common point is a notion of where the company wants to go in terms of markets to address and technologies to try to develop, along with some qualitative concerns over key success factors. Such strategic thinking is sometimes implicit and sometimes explicit. In the case of companies belonging to a parent organization, the strategy is explicit and annual budget procedures are systematic. It seems that parent organizations keep strategic control of their subsidiaries involved in emerging industries.

The subsidiaries enjoy a relatively high autonomy on R&D programmes and product/market portfolios. But alliances are decided by the parent company. The investment level and annual budgets are negotiated, but the parent controls the *decision* on levels of investment and on budgets. Reporting procedures on budgets are very similar to those in other kinds of businesses. But, once again, successful exceptions to this rule can be found. In any event, when the technological uncertainty and the strategic uncertainty decrease, with the maturation of the industry and the maturation of newly founded firms, these management practices evolve. Independent, technology-oriented embryonic companies, for instance, quickly turn to more formal management practices (sales administration, cost control, etc.) when they face their first growth crisis.

Organizational forms may diverge but companies in emerging industries appear to have a typical organizational climate. When compared with other companies in other competitive systems, four characteristics (or shared values) seem to be particularly strong: personal commitment, team spirit, mutual trust and rewards for success (Calori and Noel, 1986). These values strengthen and reinforce each other and are probably shaped by the high challenges posed by emerging industries, a combination of uncertainty and high potential gains.

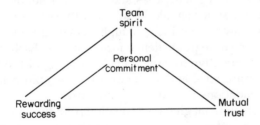

Figure 7.   Organizational climate in emerging industries

The contrast is even more striking when comparing the organizational climate of an emerging business with a mature business in the same diversified company, even when strategic planning practices are the same. The results are common to small independent companies and subsidiaries in large groups. Personal commitment, which is the key value, is emphasized by phrases like: 'high motivation at work', 'satisfaction in work more than in money', 'accepting long working hours', 'enthusiasm' ... Some other factors strengthen these values and norms of behaviour: strong internal training, the multidisciplinary skills of the personnel, the architecture (proximity), 'open door' practices, flat structures and project teams ... with 'leaders' at the top.

However, these cultural profiles diverge on one point which raises questions for further research: the acceptance of mistakes which is high in some companies and low in some others. These characteristics sound like common sense, but we believe that many companies involved in emerging industries can improve their management practices in relation to these priorities.

CONCLUSIONS

Porter (1980) came to the conclusion that it is hard to find 'rules of the game' in emerging industries. We agree on that; we did not find that companies which failed in such competitive systems did not follow the strategic paths that we have described nor did we find that following them is a sufficient condition for success. In emerging industries failure may come from a wrong technological bet leading to a technological deadlock. Anyway, we believe that the strategic paths that are proposed here could guide managers in the evaluation of the position of their emerging business compared with other competitors: differentiation position, focus, international presence, power/flexibility, network branching and positions of the partners in the network, management practices which could strengthen personal commitment, team spirit, mutual trust and success rewarding.

Four types of emerging industries should be considered (Figure 1) and strategic recommendations should be adapted to each type as appropriate. We have tried to show the specificity of high technology emerging industries which are the main strategic challenge for the future. Each industry has its own system of competitive forces; strategic thinking at the industry level, at the strategic group level and at the firm level should complement such general recommendations for emerging industries.

Further research on managing the transition from the end of the emerging phase to a mature stage where dominant product and process designs have appeared, should also prove useful. Defining types of industries and effective strategies in each type is one important step towards a contingency theory of the strategic management of technological innovation.

REFERENCES

Abell, D.F. (1980). *Defining the Business*, Prentice Hall, Englewood Cliffs.

Abernathy, W.J., and Utterback, J.M. (1975). 'A dynamic model of product and process innovation', *Omega*, **6**, 639–656.

Aiken, M., and Hage, J. (1971). 'The organic organization and innovation', *Sociology*, **5**, 63–82.

Boston Consulting Group (1981). 'Concepts avances de strategie pour les années 1980', *Conferences pour Dirigeants, Paris*.

Burgelman, R.A. (1984). 'Designs for corporate entrepreneurship in established firms', *California Management Review*, **3**, Spring, 154–66.

Burgelman, R.A. (1985). 'Managing the New Venture Division: research, findings and implications for strategic management', *Strategic Management Journal*, January–March, **6**, 39–54.

Calori, R. (1985). 'Effective strategies in emerging industries', *Long Range Planning*, **18** (3), 55–61.

Calori, R., and Noel, R. (1986). 'Successful strategies in French high technology companies', *Long Range Planning*, **19** (6), 54–65.

Cooper, A.C. (1973). 'Technical entrepreneurship: what do we know?', *R and D Management*, **3** (2), 59–64.

Cooper, R.G. (1979). 'The dimensions of industrial new product success and failure', *Journal of Marketing*, **43**, Summer, 93–103.

Cooper, R.G. (1981). 'Staffing the innovative technology based organization', *Sloan Management Review*, **22**, 19–34.

Doz, Y., Hamel, G., and Prahalad, C.K. (1986). 'Strategic partnerships: success or surrender?', *Strategic Management Society Conference, Singapore* (14–16 Oct.).

Gest Euroconsult (1985). 'Grappes technologiques et strategies industrielles', *C.P.E.*, **57**.

Hamilton, W.F. (1985). 'Corporate strategies for managing emerging technologies', *Strategic Management Society Conference, Barcelona* (10–14 Oct.).

Harrigan, K.R. (1985). *Strategies for Joint Ventures*, D.C. Heath, Lexington Books, Mass.

Hlavacek, J.D., Dovey B.M., and Biondo, J.J. (1979). 'Tie small business technology to marketing power', *Harvard Business Review*, **55**, January, February, 106–16.

Hofer, C. (1975). 'Toward a contingency theory of business strategy', *Academy of Management Journal*, **18** (4), 784–810.

Lawrence, P.R., and Lorsch, J.W. (1967). *Organization and Environment*, Irwin, Homewood, Il.

Morin, J.M. (1985). *L'excellence technologique*, Publi Union, Paris.

Porter, M.E. (1980). *Competitive Strategy*, Free Press, MacMillan, New York.

Porter, M.E. (1985a). *Competition in Global Industries*, Harvard Business School Press, Boston Mass.

Porter, M.E. (1985b). *Competitive Advantage, Creating and Sustaining Superior Performance*, Free Press, Macmillan, New York.

Porter, M.E. (1985c). 'The industrial competitiveness of nations', *Strategic Management Society Conference, Barcelona* (10–14 Oct.).

Quinn, J.B. (1979). 'Technological innovation, entrepreneurship and strategy', *Sloan Management Review*, Spring 1979, 19–30.

Ramus, V. (1985). *The Domilens Case*, Centrale des Media Pedagogiques, Paris.

Schumpeter, J.A. (1954). *History of Economic Analysis*, Oxford University Press, New York.

U.S. Department of Commerce (1983). *An Assessment of U.S. Competitiveness in High Technology Industries*, International Trade Administration, Washington D.C.

Von Hippel, E.A. (1977). 'A customer active paradign for industrial product idea generation', *Sloan School of Management, Working Paper* 935/77.

2

Turnaround Strategies for Strategic Business Units with an Ageing Technology

Rudy Moenaert, Jan Barbé, Dirk Deschoolmeester

Vlerick School of Management

and

Arnoud De Meyer

Insead

Technology is an important asset in every company. It plays a manifest role in determining the competitive position and the economic outlook of the company (Porter, 1980, 1982; Abernathy, Clark and Kantrow, 1983a; Tushman and Anderson, 1986). The presently observed technological turbulence most definitely points in the direction of technological change as a major factor affecting a firm's competitive posture. Strategic planning must more than ever take into consideration the technological environment.

Assuming that technology is indeed important, there are some issues that must be addressed by every organization thinking about strategic planning. How does the present set of technologies used by the organization influence

The Strategic Management of Technological Innovation. Edited by R. Loveridge and M. Pitt
© 1990 by John Wiley & Sons Ltd

its competitive advantage within the industry? Is this likely to change as a result of the emergence of new technologies? And if so, how does the organization best acquire the new technologies (De Meyer and Van Dierdonck, 1984)? Otherwise stated: what technologies do we need today, what technologies will we need tomorrow, and how can we internalize alien technologies? This chapter attempts to provide some new insights on these issues.

Firstly, it explores and clarifies some important concepts, namely the product life-cycle, the technological life-cycle and the technological S-curve. These concepts are not only useful tools in the analysis of the competitive position of the organization, but help as well in the planning process. Secondly, this chapter proposes a contingency framework for the internalization of an invading new technology. This framework is the outcome of a three year case study research project done by the authors (Barbé et al., 1988). The objective of the second part of the chapter is to take a new look at the factors which need to be considered when organizing for technological turnaround.

Throughout the chapter, technology is used in the sense of a practical application of scientific or engineering knowledge (Ketteringham and White, 1984). A technological innovation is then any product launched by the organization, or any process introduced in production, for which the innovating unit had to familiarize itself with one or more new technologies, or with a new combination of existing technologies (De Meyer and De Clercq, 1983).

MODELS OF PRODUCT AND PROCESS IN THE FIRM

Three models

Various models with a direct or indirect reference to technology have been developed over time. The present discussion will be limited to three models that are generally considered to be helpful tools in the analysis of technology and strategy within the firm, namely the product life-cycle, the technological life-cycle, and the technological S-curve. Another reason for this selection is that the above concepts have already been empirically validated.

The product life-cycle pictures the evolution of the sales and profits in an industry over an extended period of time. Several distinct periods have traditionally been identified. The most recurrent segmentation comprises four stages. The product life-cycle starts with the launch of a new product. Sales usually increase only slowly. Once the potential customers have been convinced, demand increases sharply and the market grows rapidly. This ultimately results in market saturation. Sales will reach their maximum during the maturity stage. When the product loses customer appeal, this

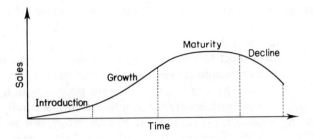

Figure 1. The product life-cycle (adapted from Levitt, 1965)

marks the beginning of the market decline stage, during which sales will decrease and finally fall to zero (Levitt, 1965). More detailed segmentations have been proposed in marketing literature. They do not alter the basic model in a significant manner, but display nearly always fine tuning modifications (e.g. Wasson, 1978).

The technological life-cycle pictures the absolute frequency of product and process innovations in a productive unit (Utterback and Abernathy, 1975; Abernathy and Utterback, 1982). The unit of analysis, the productive unit, is a product line and its associated production process. The 'Abernathy and Utterback model' highlights the interaction between product, process and technology during the life-time of a productive unit. During the initial 'fluid' stage, product innovations will tend to be predominant as market needs are ill-defined and only vaguely stated. Flexibility, entrepreneurship and informal relations are characteristic features of such organizations. During the transition stage, the frequency of process innovations sharply increases. This occurs as the result of the emergence of a dominant design in the market place. A dominant design is the generally accepted embodiment of all the relevant characteristics required by the users, at an acceptable price. Performance criteria have from that moment on been standardized in the industry and price becomes the new critical factor for success. Emphasis will logically shift from product to process innovations. Once process innovations are distinctly predominant over product innovations, the productive unit has reached the specific stage. This stage usually implies a highly structured organization with definite goals and rules. The productive unit will subsequently move to the mature stage, during which, innovative activity declines to a very low level. It must be noted that later discussions have questioned the intrinsic rigidity of the original model (Porter, 1982; De Bresson and Townsend, 1981; Utterback and Kim, 1985).

The technological S-curve pictures the performance capability of a technology with respect to the invested R&D-resources (Twiss, 1979; Sahal, 1981; Foster, 1986). Three stages can be distinguished. During the emerging stage of a technology, initial efforts bring about a merely marginal performance improvement. Once a critical mass of knowledge has been accumulated a rapid and almost exponential growth takes place. This development stage will be followed by the maturity stage, as further improvement with respect to the performance criterion slows down, due to technological and economic causes. Two remarks need to be made here. Firstly, a single technology can occupy different positions along different S-curves as different parameters are taken into account. Secondly, note that we consider an increase in the *performance* of technology, not an increase in the *penetration* of a technology (e.g. Ford and Ryan, 1981; Ketteringham and White, 1984), which closely resembles the product life-cycle of an individual technology.

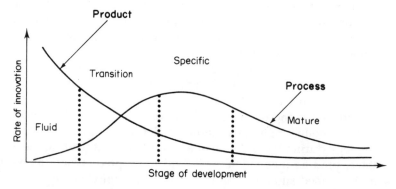

Figure 2. The technological life-cycle (adapted from Utterback and Abernathy, 1975; Abernathy and Utterback, 1982)

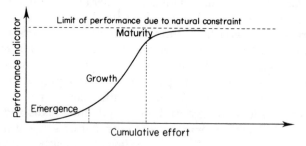

Figure 3. The technological S-curve (adapted from Twiss, 1979)

An analysis of the differences among life-cycle models

Confusion exists about the very content of each of these life-cycles. This is nicely illustrated by the concept of maturity. In their analysis of business performance, Anderson and Zeithaml (1984) described the maturity stage as 'products or services familiar to a vast majority of prospective users; technology and competitive structure reasonably stable'. In the same respect, the following observation made by Dowdy and Nikolchev (1986) seems highly normative: 'For technology-based industries, the time frame to reach stability and maturity corresponds to the rate of change both in the level of technological uncertainty and the buyer's needs'. We argue that industry maturity and technological maturity are two separate concepts and do not necessarily coincide in time. Technological maturity may occur in the early stages of the product life-cycle and technological turbulence can emerge during the more mature stages of the product life-cycle (see Cady (1985) for examples from the information industry).

A first plausible explanation for this confusion may be that the areas of innovation (technological life-cycle, technological S-curve) and marketing (product life-cycle) have touched each other only on rare occasions. A second plausible explanation concerns research methodology and scale construction: surrogate measures are used to measure technology (Pfeffer, 1978). These surrogate measures often resemble the measures applied to assessing industry maturity (see for instance the measures of technological maturity presented in Roussel (1984) and Cady (1985) as compared to those of industry maturity in Bleicher, Bleicher, and Paul (1984)). It is only recently that attempts have been made to measure technology by means of intrinsic characteristics rather than by applying surrogate measures (Souder and Srivastava, 1985).

However, in our opinion three elements stand out which allow us to make a clear distinction between the above life-cycle models, namely the measured dimension, the unit of analysis, and the starting point of every life-cycle. Firstly, each concept measures a different variable. The product life-cycle measures sales and profits generated by a specific product or group of products. The technological life-cycle plots the rate of innovation in a productive unit. The technological S-curve portrays the performance of a single technology with respect to a particular performance parameter.

Secondly, there is a difference in the unit of analysis of the three models. A product life-cycle is usually constructed at the level of the industry. The industry can be considered as the aggregation of all strategic business units within a specified field of activity. A technological life-cycle is typically defined at the level of one single strategic business unit, belonging to a specific industry. And the unit of analysis of the technological S-curve is not an economic entity at all, but a specific technology. Throughout the analysis we treat the concepts of 'strategic business unit' and 'productive unit' as synonyms. The analysis by Abell (1980) of strategic business units, and the analysis by Abernathy and Utterback (1982) of productive units, showed that these authors determined the boundaries of the respective concepts on both demand-side and supply-side conditions. A workable definition of a strategic business unit must take into account the supply-side conditions. Similarly, a workable definition of a productive unit needs to include demand-side conditions. Hence we suggest that these concepts can be used interchangeably.

The third important difference deals with the starting point of each life-cycle. The emergence of a new product life-cycle is marked by the start of a new technological life-cycle. The reverse however is not necessarily true. Situations can occur where a new technological life-cycle is started in more evolved stages of industry evolution. Technological de-maturity, which will be the subject of the second part of this chapter, underscores this. The take-off of a new technology does not necessarily coincide with any of the previously mentioned starting points. A technological S-curve is, above all, technology driven, implying that R&D activities can and do happen outside the strategic business unit and even outside the industry. Strictly speaking, the relative absence of industrially competitive behaviour as a driving force in technology development makes it possible for a technology to emerge at any point in time, depending on the initiators' objectives and consequent efforts.

A life-cycle approach to planning technological strategy

The stage of development of an industry is an important consideration in business strategy (Anderson and Zeithaml, 1984). From a proactive point of view, strategic recommendations can be formulated on the basis of the product life-cycle. As technological strategy needs to be integrated in the total business strategy, the product life-cycle is a useful tool in determining the business unit's technological strategy. An analysis of the industrial history, industry sales trends and comparison with the performance of the considered business unit serves as a valuable guide in determining the research and development programme (Levitt, 1965). Not only does insight into the actual product life-cycle permit management to plan for the future, it also helps in defining specific technological objectives when going for a particular strategy. Innovative activity can be related to the type of competitive advantage the management hopes to establish, be it cost leadership or differentiation (Porter, 1980, 1982).

Moore and Tushman (1982) outlined a trajectory of innovative activity throughout the product life-cycle. During the introductory stage of industry evolution, innovative activity will be focused on product innovation. When the market starts to expand rapidly, a technological core can be determined. This enables the business unit to tighten the planning for each discrete technology. Finally, as low cost and high quality become indispensable elements to success in a maturing market, technological investment pc...y will gradually be focused on process innovations.

The generality of the product life-cycle is its major shortcoming as a planning tool in determining technological strategy. It pictures sales evolution for the whole industry. This is too all-embracing and can obscure important technological trends. Solely relying on industry sales and profit trends may even prove hazardous. For example, Moore and Tushman (1982) stated that product innovation will be minimal during the maturity stage. This need not necessarily be true. Important counter-examples can be given, such as new bank services (out-of-branch cash dispensers, home banking), new medical diagnostic instruments (ultrasound, CT-scanning, nuclear magnetic resonance), and new telecommunication devices (telefax, cellular radio). As Abell (1980) said:

> 'In reality, the product should be considered simply as a physical manifestation of the application of a particular technology to the satisfaction of a particular function of a particular customer group. The choice is one of technologies, functions, and customers to serve, not of products to offer. The product is the result of such choices, not an independent decision that results in such choices.'

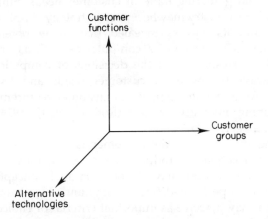

Figure 4. The three dimensions defining a business (from Abell, 1980)

This explains why management should not restrict its analysis to the product life-cycle in order to determine its technological strategy. Product life-cycle evolution is the *result* of the positioning or repositioning of the industry population along the above three dimensions. It must be mentioned here that the definition by Abell of technology, namely, the way customer needs are satisfied, extends beyond the boundaries of technology as defined in this chapter. Technology as defined by us is only one element in the total transformation process described by Abell, which also includes such elements as human resource management, logistics, and marketing. Nevertheless, technology is a major driving force behind the product life-cycle.

The technological life-cycle complements the analysis of industry sales evolution in a substantial way. Despite the fact that the time patterns of these two life-cycles do not necessarily coincide, the technological life-cycle is a powerful tool when evaluating the appropriateness of a particular strategy. From a normative view, one can argue that a *fit* must be established between the pattern of technological innovation, i.e. the specific character of the product/process innovation mix in the business unit, and the time-dependent focus of technological competition in the industry. The downfall of EMI in the CT-scanning market can at least partly be elucidated by its ignorance of the technological strategy deployed by competitors. The focus of research and development of EMI was still on product innovation, whereas most competitors had already achieved manufacturing efficiency (Berggren, 1985).

The analysis of the technological life-cycle should not restrain management from gathering data on customer needs. Simply complying with the rest of the industry may be a wrong strategy. Apple was considered a maverick, until its success revealed that many potential customers valued the user convenience of a computer as high as processing speed or compatibility. Irrespective of the decisions of competitors, a business unit must always be aware of customer needs and try to spot new opportunities. An adequate response to customer requirements is a major factor determining innovative success (Rothwell *et al.*, 1974; Cooper, 1979; Maidique and Zirger, 1985).

Emphasizing the needs of the customer is a necessary ingredient for success in the marketplace. 'Products or services perform certain functions for the customer. Functions have to be separated conceptually from the way the function is performed (technology) and the attributes or benefits that a customer may perceive as important criteria for choice' (Abell, 1980). Having analysed, compared and evaluated the situation in both the industry and the strategic business unit, and having forecast tomorrow's customer needs, it now becomes possible to assess the potential of the different technologies. The benefits that are sought by the various customer segments equate to required dimensions on the ordinates of the different technological S-curves that might be drawn. Plotting the technological S-curves of different technologies against these dimensions provides an indication of the relevant key factors for success, helps identify the technologies with the highest potential (De Meyer, 1984; Roussel, 1984; Foster, 1986; Becker and Speltz, 1986; Turner, 1983) and indicates how R&D resources can be best allocated (Arthur D. Little, 1981; Dowdy and Nikolchev, 1986).

TECHNOLOGICAL DE-MATURITY

A technologically mature strategic business unit is one in which uncertainty about product performance and about appropriate process design has been reduced to such an extent that the details can be codified. Technological maturity can be defined at the industry level (Clark, 1983) as well as at the business unit level (Abernathy, 1978). Our research has focused on the latter, taking our cue from Abernathy:

> 'All the changes discussed might seem to imply that the automobile has matured as a product, but this is not the case. In the first place, the automobile is not the relevant unit of analysis: the productive unit is a better focus. As long as new productive units are being added, or as long as existing ones are resisting extreme states of development, the car has not matured' (Abernathy, 1978).

The above perspective is based on the technological life-cycle. Analogously, one can define *industry* maturity or maturity of a *technology* depending upon whether one considers respectively the product life-cycle or the technological S-curve. A mature industry is one in which all potential customer groups have been approached and where sales have reached a plateau; a mature technology is one that has become R&D-inelastic with respect to the relevant performance criteria. Further investments of R&D resources in the technology will generate only marginal improvement. A definition with respect to R&D input has been chosen, as technological development plotted over time does not necessarily follow the nicely S-shaped curve. Intermediate plateaus could create the illusion that the ultimate plateau has been reached (a nice example illustrating this point can be found in Wiseman (1983)).

Technological maturity does not imply that innovation has come to an end within the business unit. In their seminal work, Abernathy, Clark and Kantrow (1983b) introduced the concept of technological de-maturity:

> 'In terms of productive unit evolution, it [technological maturity] is the process by which the drive toward standardization inoculates the unit against competitively significant innovation. That which had been "open" gradually becomes "closed" [the product attributes]. But this movement toward closure, toward inoculation, and toward immunity does not necessarily run in a single direction. It can be reversed. [...] At the heart of any reversal is a major change in the established relationship between technology and market preferences [...] No two industries will respond to changes of this sort in identical ways, but it is true nonetheless, that a common thread does link all such shifts in preference and technology. The need for re-establishing the "fit" among them upsets the usual patterns of purchase and supply, and prompts the re-emergence of repeated searches and learning on both sides of the market. This in turn constitutes a reversal of industrial development towards maturity; that is, a return to an earlier stage of productive unit evolution. It is exactly this kind of reversal we have in mind when we speak of industrial de-maturity'.

We define technological de-maturity as the technological innovation by which the product or process performance in a technologically mature strategic business unit is significantly improved by the introduction of new technologies, an improvement or turnaround unattainable by further investments in the old technology. We define this concept at the level of the strategic business unit, not at the level of the industry. De-maturity as conceived by Abernathy and Clark (1985) implies a rather general trend. Although technological de-maturity may be a general phenomenon within a particular industry, every business unit makes an independent decision with respect to the set of technologies it will implement. We hence do not rule out the possibility that only a limited number of business units opt for technological turnaround. The focus is on the approach taken by a particular business unit.

Clark (1983) identifies three possible causes of reversal in the trend to maturity: changes in demand preferences, supply-side changes in technology, and changes in the prices of substitute or complementary products. A major change in any of these conditions may render the existing set of technologies obsolete. When facing such changes, the business unit may have to make a technology 'jump' (De Meyer and Van Dierdonck, 1984) and acquire a new set of promising technologies. De-maturity is then a redefinition strategy (Abell, 1980) in terms of technologies. It moves the business unit away from the maturity stage into earlier stages of the technological life-cycle. Product innovation has become pervasive again within the graphics industry, as a result of the introduction of electronics. Similarly, the ready-to-wear clothing industry is marked by a major increase in process innovation, because of information technology.

Renewed competition

A study by Cooper and Schendel (1976), in which 22 companies facing technological threats were analysed, revealed that major new technologies are often introduced by companies outside the mature industry. The new technologies were relatively expensive and initially showed important deficiencies. For instance, despite their high cost, the first transistors had *limited* power capability, frequency response and temperature tolerance. Generally, as a result of such teething troubles, the substitution of an old by a new technology did not occur in a straightforward manner. The observed time-span between the commercial launch of the new technology and the moment when its sales exceeded those of the old technology ranged from five to fourteen years. Technological substitution is seldom an invasion. Rather, the new technology creates its own market and substitutes the older technology incrementally in existing markets.

Notwithstanding the rather slow diffusion rate of major new

technologies, its impact cannot be denied. 'By its very nature, epochal or disruptive innovation—whatever its degree of technical novelty—makes obsolete existing capital equipment, labour skills, materials, components, management expertise, and organizational capabilities' (Abernathy, Clark and Kantrow, 1983b). In maturity the competitive value of technology has been quasi-nullified. In other words, existing technologies become base technologies, degraded from a competitive weapon to a required basic input mastered by almost every competitor, with competitive advantage being achieved via other elements such as a better service or a unique image.

Technological de-maturity clearly marks the start of a new competitive arena. The fact that an organization masters a new technology will determine its profitability and its growth in the short run, and its chances for survival in the long run. This is pre-eminently illustrated by the response of the vacuum-tube producing firms to the transistor technology (Soukup and Cooper, 1983; Foster, 1986). Empirical research shows however that threatened companies do not respond in a logical way. On the contrary, the emergence of a new technology is likely to be countered by renewed attention to the existing technology (Cooper and Schendel, 1976). As a result, the old technology will often be optimized with respect to its key performance parameters only after the appearance of a new technology (Utterback and Kim, 1985). In addition to this success, the technologically mature organization can, by banking on its experience, market the old technology more efficiently than the new competitors. As the risk in innovation can be very large (Williams, 1983), why then should an organization develop a new and unknown technology, if it thinks it can sustain its competitive advantage, apparently almost without any risk? Technological turnaround is hence considered to be a very expensive strategy as compared to the alternative of continued investment in the existing technology.

Becker and Speltz (1983) observed that precisely at the midpoint of the technological S-curve, when R&D-productivity starts to decline, economic returns will be optimal. Sales and profits will be highest during this so-called management comfort zone. In contrast, innovation absorbs many resources and necessitates the acquisition of new skills and infrastructure. This may result in the gradual liquidation of existing infrastructure and resources (Harrigan and Porter, 1983; Littler and Sweeting, 1984). This is known as the productivity dilemma (Abernathy, 1978): the trade-off management has to make is between the proven technology of today and the uncertain technology of tomorrow. Top managers receptive to innovation are a very strong impetus for a sound technological strategy (Frohman, 1982), whereas the technological inertia of others is an often-encountered inhibitor of innovation (Kotter and Schlesinger, 1979; Strebel, 1987). Not only do financial and strategic issues come into play, but also personal and emotional considerations may hinder the decision to innovate. Since their competence with the old technology has often helped top management to climb the managerial ladder, unlearning can be a troublesome step for these people (Galbraith and Kazanjian, 1983).

The history of the Ford Model T illustrates the danger of a strategy aimed at preserving the position of the old technology. Ford heavily invested in its low-cost position without recognizing that customer requirements had changed over the years, creating a de-maturing affect. The innovative activity being directed towards process, not product development, endangered the competitive position of Ford (Abernathy and Wayne, 1974). The substitution of latex foam by polyurethane foam technology provides us with another case. The latex foam industry still experienced a strong annual growth, but its mature technology offered little prospect for improvement:

> 'Had the latex foam industry recognized the vulnerability implicit in a aging technology, it might have applied R&D strategies whose results would have materially prolonged its life and profitability, though in the long run the greatly superior economics of polyurethane resilient foam would have prevailed. Distracted by the strong growth characteristics of the industry, latex foam producers failed to recognize its technological age and were unprepared to deal with the onslaught of new technology' (Roussel, 1984).

The allocation of all resources to the improvement of mature technologies overlooks the potential in emerging technologies (Tushman and Anderson, 1986; De Meyer, 1984; Foster, 1986). The inherent weakness of such a strategy is its short-term orientation. When the old technology has approached its natural limit, the new technology will probably still offer many opportunities for substantial improvement.

STRATEGIES FOR ORGANIZING TECHNOLOGICAL TURNAROUND

Several important issues need to be addressed by an organization when one of its business units goes through an era of technological turnaround: technological forecasting, technology assessment, internalization, implementation and integration, market entry and so on. Our research has focused on the issue of technological internalization, i.e. the strategies and organizational structures designed by the decision making authority in order to acquire and develop the alien technology.

Some existing frameworks describe the strategy a company should choose when it internalizes a new technology. Killing (1980) determined the trade-off between a joint-venture and a licensing strategy on the basis of two factors, the relatedness of the technological diversification to the existing business and the scale of the project. De Meyer and Van Dierdonck (1984) complemented the technology construct of Perrow (1974) (i.e. the analysability of a technology versus its variability), with the concept of exposure, in order to determine the internalization design. Roberts and Berry (1985) took into account two factors for organizing the entry of a new market, the degree of familiarity with the market and the degree of familiarity with the technology. Boisot (1986) linked two characteristics of knowledge, the degree of knowledge diffusion and the degree of knowledge codification, to the ratio of returns on external to internal transaction styles.

What all these models have in common is a 'matrix' solution. In such an approach internalization is seen as pro-active and intended (Mintzberg and Waters, 1985). However, as all of the cited models use different variables to arrive at the optimal solution, we can assume that each model gives us only a fraction of the solution. Confronted with an ageing technology our basic proposition is that an appropriate organizational strategy is needed for the successful internalization of a new technology. Our aim has been to develop a contingency model (Woodward, 1965; Galbraith, 1973) to enable managers to determine the appropriate organizational strategy for de-maturing business units.

Owing to the lack of existing comprehensive theory, we have used the case study research method, as this is a very useful method for exploratory purposes (Kerlinger, 1973; Yin, 1984). Recent case study examples have contributed to the understanding of the management of innovation (Berggren, 1985, Gobeli and Rudelius, 1985, Clarke, 1983; Maidique and Zirger, 1985). Eighteen case studies have been analysed, all within Belgian organizations (Table 1). With the exception of case 15, all organizations had already started and in some cases had completed the internalization.

We distinguished six basic organizational strategic designs for technology internalization (De Meyer and Van Dierdonck, 1984; Moenaert *et al.*, 1986); (i) internal development, (ii) contract research, (iii) intercompany co-operation, (iv) joint-venture, (v) acquisition, and (vi) purchase of technology. This was the basic criterion for the selection of case studies. It helped to introduce heterogeneity within the sample (Cook and Campbell, 1979), but of course, it represents a rather coarse classification. Within each group there exists a range of possibilities. For instance, in the case of joint ventures, a distinction can be made between venture capital, venture nurturing and new-style joint ventures (Roberts, 1980). Furthermore, the six categories are not mutually exclusive and most actual internalization designs are an amalgam of these six pure types. In this respect, we have classified the case studies in Table 1 into dominant strategies based on the relative size of investment mode. Finally, the internalization design will tend to evolve over time. An organization can for example internalize a new technology by purchasing it, and then gradually start its own development.

Table 1.   Survey of the case studies

Industry	New technology	Internalization design
1. Clothing	Computer-aided design	Purchase
2. Clothing	Computer-aided design	Co-operation
3. Clothing	Computer-aided design	Contract research
4. Coating	High solids	Internal development
5. Coating	Enzymatic technology	Co-operation
6. Coating	Special purpose coating	Internal development
7. Photochemics	Micro-electronics	Internal development
8. Photochemics	Micro-electronics	Acquisition
9. Photochemics	Micro-electronics	Acquisition
10. Reprographics	Thermal coating	Acquisition
11. Reprographics	Dielectric technology	Co-operation
12. Ceramics	Neo-ceramics	Contract research
13. Contractor	New material	Internal development
14. Additives	Biotechnology	Joint venture
15. Telecom.	Gallium arsenide	Contract research
16. Rubber	New material	Acquisition
17. Rubber	New material	Contract research
18. Dairy	Biotechnology	Contract research

HOW ORGANIZATIONS INTERNALIZE NEW TECHNOLOGIES

We identified three different stages in the process of technological turnaround (Figure 5). The first stage occurs when top management perceives the existence of a new technology. The process of uncertainty reduction is subsequently started, during which an evaluation will be made of the relevance of the new technology. By gathering and analysing information, the organization wants to find out what the distinctive characteristics of the technology are and to estimate its potential impact on the industry. If, on the basis of this information, management decides to internalize the new technology, the process of internalization then begins.

The case studies show that the earlier a company perceives a new technology *vis-à-vis* its competitors, the more strategies will be competitively possible. Likewise, the sooner a company attempts uncertainty reduction or the sooner it decides to internalize the new technology as compared to its competitors, the more organizational choice a company has. In the coating industry an important technology in the field of enzymes was first perceived by researchers working at a research institution. This was communicated to the companies which participated

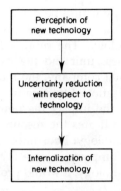

Figure 5. The process of technological turnaround (Barbé *et al.*, 1988)

in the research institution. Competitors who had not participated in this R&D programme perceived the technology later and could not at that stage participate in the research, nor take any individual advantage of it (case 5). A company operating in the medical diagnostic industry had to co-operate with a foreign partner because the uncertainty reduction was started very late. By the time it was convinced of the relevance of the new technology, other companies had already launched products on the market and acquired image and customer loyalty. Internal development had consequently become infeasible (case 9). A company operating in the dairy industry decided to internalize a new technology while its major competitor had already co-operated with a university and another company. This alliance excluded the first company from co-operation with the same university on that subject (case 18).

The lesson is clear: information must be gathered and processed as early as possible, before entry barriers have been erected by competitors. Organization is not only crucial in the final stage of internalization, a company must organize itself to perceive a new technology early *vis-à-vis* its competitors, and start uncertainty reduction as soon as possible. Planning for networking and gatekeeping activities is important. This can be achieved by participating in other companies or by managing formal and informal contacts with universities, research institutions, customers, suppliers, competitors and other companies.

This does not necessarily imply that a business unit must opt for an early entry. Rather, it means that an organization must keep up to date with technological trends. Following an early perception, a telecommunications company decided to postpone the internalization of gallium arsenide technology. Uncertainty-reducing research at a Belgian university had clearly indicated that the competitive advantage to be gained from implementing this technology would be questionable (case 12).

The uncertainty reduction process informs management about the relevance of a new technology. The more relevant a technology is for a particular strategic business unit and the more important the strategic business unit within the total company, the more resources generally will be allocated to its internalization. Effective uncertainty reduction requires an adequate calculation of the impact the new technology may have on the existing business and indicates the resources which must and/or can be invested in the new technology. Poor information often leads to wrong decisions. If the ceramics company in case (12) had thoroughly inquired about the new ceramics technology, it would probably not even have started the internalization.

A new technology which has a great impact on the internal linkages in an organization (Abernathy and Clark, 1985) probably makes it necessary to set up a separate unit to develop the technology. A technology that deviates too much from the internal know-how base requires new resources and skills. It is difficult to start development of such a technology in the existing structure. A separate unit needs to be established (e.g. an internal corporate venture). The transfer of technology from such an independent unit can be bothersome (Hopkins, 1975; Sykes, 1986). An important cause is the 'culture clash' between the more organic venture unit and the more mechanistic parent company. One important mechanism to avoid this consists of the early planning and communication of objectives and guidelines to both the unit and the parent company. Technology transfer and eventual re-integration of the separate unit will then be enhanced.

When a decision has been made to internalize a specific technology, the organization will often take into account a multitude of variables. Our research shows that these variables can be classified in four groups: internal resources, external resources, technology variables, and strategic values. *Internal resources* are the internally available means a business unit has at its disposal to develop a particular technology. We can distinguish five types of resources, i.e. human resources (people), technical resources (R&D equipment and infrastructure), financial resources (R&D funds), exposure (experience with a specific technology) and organizational slack (spare capacity of resources). *External resources* are the outside human, technical or financial resources a company can appeal to. *Technology variables* are those that characterize a new technology, namely its stability (rate of evolution), its availability (the degree up to which applications and know-how can be bought in the market place), its threshold level (the minimum investment necessary for development), its analysability (the ease with which scientific research can get insight into the underlying principles), its variability (the number of exceptions encountered when working with the technology),

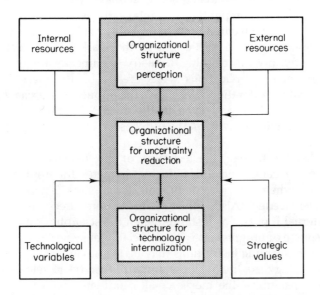

Figure 6. A framework for organizing technological turnaround (Barbé *et al*., 1988)

its complexity (the number of scientific disciplines the technology is based on), and its scientification (the degree to which scientific research is the source of technological progress). *Strategic values* include variables such as the cost/benefit considerations made by management, the relative technological position the organization occupies *vis-à-vis* its competitors, the organizational culture of the organization and its technological leadership or followership style. This last category is probably the least exhaustive one.

The nature of the technology can necessitate an organic structure (Burns and Stalker, 1961). Generally speaking, the less stable or the more variable a new technology is, the more organic the organizational design will be. A high variability and a low stability are often characteristic features of emerging technologies. In order to keep up with technological evolution, the company must in such a case set up an organic structure. When contracting external parties, diversity is sought and long commitments are avoided. This enables the organization to tap relevant information from a broad network and increases flexibility. 'In essence, variety in innovative activity is best stabilized through variety in R&D activity' (Sahal, 1983).

The more the technology climbs the technological S-curve and the more stable or the less variable the technology is, the more mechanistic the approach can be. As it gradually becomes possible to make more accurate forecasts, the organization can focus its R&D programme and limit its external contacts. At the extreme of this continuum, a fully developed technology will be widely available, so it may make sense to purchase the technology. The competitive advantage to be gained from internal development of such a technology will probably not be great, as both know-how and applications can be bought in the market place. For instance, when the ready-to-wear clothing industry became aware of computer-aided design, many applications were already on the market. It was a largely financial consideration which gave the impulse to purchase a specific system (cases 1 and 2), and even in the event of contract research, the involved parties relied heavily on the purchase of already existing hardware technology (case 3).

When either the human, the technical, or the financial resources of an organization are below the threshold requirements for implementing the technology or when exposure or organizational slack is too low, outside input will be needed. Also, other elements can direct an organization towards external inputs. For instance, when the scientific community is an important source of progress, as with artificial intelligence, some form of co-operation with scientific institutions may be the obvious way to internalize the new technology. As the business unit will then try to select the partner with the best resources, the merits of all candidates need to be weighed. Such an evaluation is not as easy as it might seem. In order to evaluate a future partner adequately, a company must be knowledgeable about a specific technology. But it is precisely the shortage of relevant resources, skills and experience which induced it to look for outside support. To make a good decision, the company needs to have some exposure to the new technology. There must at least be one person in the company who is knowledgeable. If not, the organization should acquire information from an independent technology consultant.

Our research discovered that management often minimizes partnership risks not by gathering better information, but by limiting the number of potential partners. Firstly, organizations are often inclined to involve the outsider operating in the context where they have first perceived the new technology. Secondly, when a company seeks external support, it will prefer partners it has co-operated with before. This bounded rationality (March and Simon, 1958) derives from relative ignorance of the organization. As management tries to minimize risk, it often chooses those companies or institutions which have made a positive impression on earlier occasions. And what is more convincing than that an external organization either drew attention to the new technology, or that it has provided excellent support during previous development efforts? There are however some caveats to proceeding thus. Relying on initial appearance or past experience often turns out to be hazardous. Our data have taught us that just because another company or institution helped the organization to spot a new technology, it does not necessarily imply the external party is a centre of excellence with respect to that particular technology. Similarly, the fact that a particular party has proved to be a valuable partner in prior technological ventures is poor indication of the quality of their input in the present situation.

The role of the individual in technological innovation should not be underestimated (Maidique, 1980). Some individuals, having particular competences, can assume important roles in technological turnaround. Our research confirmed the proposition of De Meyer and Van Dierdonck (1984) that the less codified a new technology is, the more the technology will be tied to certain individuals. Before internalization is started, an organization can examine to what extent it is possible to analyse and codify the technology it wishes to internalize. If the technology cannot easily be defined in terms of laws, rules and formulas, the technology will usually be strongly tied to individuals' know-how. The success of internalization will heavily be determined by the selection of the right people. For example, industry experts indicated that the codification of software development for graphics purposes is low. The fact that an acquired company did not have one of these so-called 'gurus' hindered some projects (case 8). Some forms of externalization (e.g. contract research) may be risky in the case of a poorly codified technology as the experts may not be strongly linked to the contracting organization. In the event that these people leave the external organization, the entire project would be endangered.

A *complex* technology requires also that the company pays attention to the role of the individual. By complexity, we mean the number of scientific disciplines a technology is based on. Projects aimed at the development of complex technologies will often be structured in different sub-projects, each dealing with a specific set of disciplines. The role of the technological integrator then becomes vital. This integrator must be familiar with all those disciplines if he is to make the right decisions. The absence of such a talented individual was another basic reason for some of the setbacks encountered while developing software for image processing (case 8).

We have already shown that management does not always act rationally. Deeply rooted company policies, an established company culture or historically grown beliefs heavily influence the decision outcome. Top-management is often biased towards a certain approach to internalization because the particular strategy being considered might have the best cost-benefit outlook, or bears the least risk, or helps to build an innovative image: 'The fact that the two variables, technology and structure, occasionally vary together, depends more on the strategy of the political contest than on determinate relationships derived of organizational efficiency' (Pfeffer, 1978). In this respect, perhaps the most important contribution of our research is that choices always exist and that the way an organization makes its choice will determine internalization success. Management may indeed have to un-learn some of its traditional planning.

CONCLUSIONS

Innovative success is determined by how well the organi   onitors
the technological environment and how well it responds   logical
opportunities and threats. The product life-cycle, the techno   e-cycle
and the technological S-curve are valuable tools in planni: logical
strategy, as these models give comprehensive insight into tl   ; forces
behind product and process evolution.

An important development is the turnaround fror   logical
maturity towards technological de-maturity. The design   opriate
organizational strategies and structures is essential for   ing in
turnaround. A new technology can be successfully interna   various
ways. Analysis of case-studies has helped to uncove   fferent
variables which should be considered when planning fu. ....logical
turnaround. Organizations considering technological turnaround can apply
the framework as a diagnostic instrument for analysing the merits of various
organizational strategies.

However, much remains unanswered. Firstly, the model needs more
validation. Further research could aim to prove or disprove some of
the propositions, but also reveal which variables are more likely to be
included in the managerial decision-making process. Secondly, considering
the pay-off from early perception and early and thorough uncertainty
reduction, more needs to be known about how to organize the processes
of perception and uncertainty reduction. Thirdly, emerging concepts in
information gathering are networking activities and strategic alliances.
Research must be directed to provide answers on how to set up and
manage such strategies. Finally, taking into account the different modes
of technological turnaround, how does one manage the transition from
internalization towards exploitation? This is likely to be different depending
upon the internalization approach the organization has chosen.

An understanding of the important elements and their impact on
technological innovation contributes to the successful management of
technological turnaround. In view of continuing technological turbulence,
this is an area that deserves further attention. The *study* of *innovation*
requires innovation as well!

ACKNOWLEDGEMENTS

The authors gratefully acknowledge support from the Intercollegiate Centre
for Management Science (Brussels). We also wish to thank Filip Caeldries
for his contribution to an earlier version of this text.

REFERENCES

Abell, D.F. (1980). *Defining the Business: the Starting Point of Strategic Planning*, Prentice-Hall, Englewood Cliffs.

Abernathy, W.J. (1978). *The productivity Dilemma: Roadblock to Innovation in the Automobile Industry*, John Hopkins University Press, Baltimore.

Abernathy, W.J., and Utterback, J.M. (1982). 'Patterns of industrial innovation', in *Readings in the management of innovation* (eds M.L. Tushman, and W.L. Moore), pp. 97–108, Pitman, Boston.

Abernathy, W.J., and Clark, K.B. (1985). 'Innovation: mapping the winds of creative destruction', *Research Policy*, **14**, 3–22.

Abernathy, W.J., Clark, K.B., and Kantrow, M.A. (1983). *Industrial renaissance. Producing a competitive future for America*, Basic Books, New York.

Abernathy, W.J., Clark, K.B., and Kantrow, M.A. (1983b). 'Mature industries can be revitalized', *Research Management*, **26**, July–Aug., 6–7.

Abernathy, W.J., and Wayne, K. (1974). 'Limits of the learning curve', *Harvard Business Review*, **52**, Sept.–Oct., 109–19.

Anderson, C.R., and Zeithaml, C.P. (1984). 'Stage of the product life cycle, business strategy, and business performance', *Academy of Management Journal*, **17**, 5–24.

Arthur D. Little. (1981). *The strategic management of technology*. Arthur D. Little, Cambridge.

Barbe, J., Moenaert, R., Deschoolmeester, D. and De Meyer, A. (1988). *The Turnaround of companies with an aging technology* (report to the ICM). Vlerick School voor Management, Ghent, Belgium.

Becker, R.H. and Speltz, L.M. (1983). 'Putting the S-curve concept to work', *Research Management*, **26**, Sept-Oct, 31–33.

Becker, R.H. and Speltz, L.M. (1986). 'Making more explicit forecasts', *Research Management*, **29**, July-August, 21–23.

Berggren, U. (1985). 'CT-scanning and Ultrasonography: a comparison of two lines of development and dissemination', *Research Policy*, **14**, 5–24.

Bleicher, K., Bleicher, F., and Paul, H. (1984). 'Innovation in mature industries: can smoke stack industries rise again?', in *Managing Amidst Tensions and Conflicts in a Global Economy* (eds I. Waters, and T. Hustad), pp. 9–24, Indiana University, Indiana.

Boisot, M.H. (1986). 'Markets and hierarchies in a cultural perspective', *Organization Studies*, **7**, 137–40.

Burns, T., and Stalker, G. (1961). *The Management of Innovation*, Tavistock, London.

Cady, J.F. (1985). 'Marketing strategies in the information industry' in *Marketing in an Electronic Age* (ed. R.D. Buzell), 249–78, Harvard Business School Press, Boston.

Clark, K.B. (1983). 'Competition, technical diversity, and radical innovation in the U.S. auto industry', in *Research on Technological Innovation, Management and Policy*, (ed. R.S. Rosenbloom), 103–49, Jai Press, Greenwich.

Cook, T.D., and Campbell, D.T. (1979), *Quasi-experimentation: Design and Analysis Issues for Field Settings*, Houghton Mifflin, Boston.

Cooper, A.C., and Schendel, D. (1976). 'Strategic responses to technological threats', *Business Horizons*, **19**, 1, 61–9.

Cooper, R.G. (1979). 'The dimensions of industrial new product success and failure', *Journal of Marketing*, **43**, Summer, 93–103.

De Bresson, C., and Townsend, J. (1981). 'Multivariate models for innovation— looking at the Abernathy–Utterback model with other data', *Omega*, **9**(4), 429–36.

De Meyer, A. (1984). *Technologiemanagement op het strategische niveau*. Paper presented at the Vlaams Technisch-Wetenschappelijk Kongres 'Technologie en Strategie', Antwerp.

De Meyer, A., and Van Dierdonck, R. (1984). *Organizing a Technology Jump or Overcoming the Technology Hurdle*. Paper presented at the Fourth Annual Strategic Management Management Society Conference, Philadelphia.

De Meyer, A., and De Clercq, J. (1983). 'Technische innovatie en organisatie', *Tijdschrift voor Economie en Management*, **28**, 165–85.

Dowdy, W.L., and Nikolchev, J. (1986). 'Can industries de-mature? Applying new technologies to mature industries', *Long Range Planning*, **19**(2), 38–49.

Ford, D., and Ryan, C. (1981). 'Taking technology to market', *Harvard Business Review*, **59**, March–April, 117–26.

Foster, R.N. (1986). *Innovation; the Attacker's Advantage*, Summit Books, New York.

Frohman, A.L. (1982). 'Technology as a competitive weapon', *Harvard Business Review*, **60**, Jan.–Feb., 97–104.

Galbraith, J. (1973). *Designing complex organisations*, Addison-Wesley, Reading, Mass.

Galbraith, J.R., and Kazanjian, R.K. (1983). 'Developing technologies: R&D strategies of office product firms', *Columbia Journal of World Business*, **18**, Spring, 37–44.

Gobeli, D.H., and Rudelius, W. (1985). 'Managing innovation: lessons from the cardiac-pacing industry', *Sloan Management Review*, **26**, Summer, 29–43.

Harrigan, K.R., and Porter, M.E. (1983). 'End-game strategies for declining industries', *Harvard Business Review*, **61**, July–Aug., 11–120.

Hopkins, D.S. (1975). 'The role project teams and venture groups play in new product development', *Research Management*, **18**, January, 7–12.

Kerlinger, F.N. (1973). *Foundations of Behavioral Research*, Rinehart and Winston, London.

Ketteringham, J.M., and White, J.R. (1984). 'Making technology work for business' in *Competitive Strategic Management* (ed. R. B. Lamb), 498–519, Prentice-Hall, Englewood Cliffs.

Killing, P. (1980). 'Technology acquisition: license agreement or joint venture', *Columbia Journal of World Business*, **15**, Fall, 38–46.

Kotter, J.P., and Schlesinger, L.A. (1979). 'Choosing strategies for change', *Harvard Business Review*, **57**, March–April, 106-14.

Levitt, T. (1965). 'Exploit the product life cycle', *Harvard Business Review*, **43**, Nov.–Dec., 81–94.

Littler, D.A., and Sweeting, R.C. (1984). 'Business innovation in the U.K.', *R&D Management*, **14**(1), 1–9.

Maidique, M.A. (1980). 'Entrepreneurs, champions and technological innovation', *Sloan Management Review*, **21**(2), 59–76.

Maidique, M.A., and Zirger, B.J. (1985). 'The new product learning cycle', *Research Policy*, **14**, 299–313.

March, J.G., and Simon, H.A. (1958). *Organizations*, Wiley, New York.

Mintzberg, H., and Waters, J.A. (1985). 'Of strategies, deliberate and emergent', *Strategic Management Journal*, **6**, 257–72.

Moenaert, R., De Meyer, A., Barbe, J., and Deschoolmeester, D. (1986). *Analysing the Issues Concerning Technological De-maturity*, Insead Working Paper, 86/10, Insead, Fontainebleau.

Moore, W.L., and Tushman, M.L. (1982). 'Managing innovation over the product life-cycle', in *Readings in the Management of Innovation*, Pitman, Boston.

Perrow, C. (1974). *Organizational Analysis; a Sociological View*, Tavistock, London.

Pfeffer, J. (1978). *Organizational design*, A.H.M. Publishing Corporation, Illinois.

Porter, M.E. (1980). *Competitive Strategy. Techniques for Analyzing an Industry*, Free Press, New York.
Porter, M.E. (1982). *The Technological Dimension of Competitive Strategy*, Harvard Business School Working Paper 82-19, Harvard Business School, Boston.
Roberts, E.B. 91980). 'New ventures for corporate growth', *Harvard Business Review*, **58**, July–Aug., 134–42.
Roberts, E.A., and Berry, C.A. (1985). 'Entering new businesses: selecting strategies for success', *Sloan Management Review*, **26**, Spring, 2–17.
Rothwell, R., Freeman, C., Horlsey, A., Jervis, V.T.P., Robertson, A.B., and Townsend, J. (1974). 'SAPPHO updated—project SAPPHO phase II', *Research Policy*, **3**, 258–91.
Roussel, P.A. (1984). 'Technological maturity proves a valid and important concept', *Research Management*, **27**, Jan.–Feb., 29–34.
Sahal, D. (1981). *Patterns of technological innovation*, Addison-Wesley, London.
Sahal, D. (1983). 'Invention, innovation and economic evolution', *Technological Forecasting and Social Change*, **23**, 213–35.
Souder, W.E., and Srivastava, P. (1985). 'Towards a scale for measuring technology in new product innovations', *Research Policy*, **14**, 1515–160.
Soukup, W.R., and Cooper, A.C. (1983). 'Strategic response to technological change in the electronic component industry', *R&D Management*, **13**, 219–30.
Strebel, P. (1987). 'Organizing for innovation over an industry cycle', *Strategic Management Journal*, **8**, 117–24.
Sykes, H.B. (1986). 'Lessons from a new ventures program', *Harvard Business Review*, **64**, May–June, 69–74.
Turner, W.D. (1983). 'The coming challenges in European retail banking', *The McKinsoy Quarterly*, Autumn, 65–76.
Tushamn, M.L., and Anderson, P. (1968). 'Technological discontinuities and organizational environments', *Administrative Science Quarterly*, **31**, 439–65.
Twiss, B. (1979). *Management of Technological Innovation*, Longman, London.
Utterback, J.M., and Abernathy, W.J. (1975). 'A dynamic model of process and product innovation', *Omega*, **3**, 639–56.
Utterback, J.M., and Kim, L. (1985). *Invasion of a Stable Business by Radical Innovation*. Paper presented at the Flemish Scientific Economic Congress 'Innoveren en Ondernemen', Antwerp.
Wasson, C.R. (1978). *Dynamic competitive strategy and product life cycles*, Austin Press, Austin, Texas.
Williams, J.R. (1983). 'Technological evolution and competitive response', *Strategic Managment Journal*, **4**, 1, 55–65.
Wiseman, P. (1983). 'Patenting and inventive activity on synthetic fibre intermediates', *Research Policy*, **12**, 329–39.
Woodward, J. (1965). *Industrial Organization: Theory and Practice*, Oxford University Press, London.
Yin, R.K. (1984). *Case Study Research: Design and Methods*, Sage, Beverly Hills.

3

Strategic Prescriptions which Undervalue Innovation: Lessons from the Automotive Components Industry

Christopher Carr

Manchester Business School

Portfolio management techniques may yield prescriptions in direct conflict with the requirements of an effective strategy for innovating. Businesses finding themselves classified in strategic portfolio terms as 'cash cows' or 'dogs', for example, may find themselves denied necessary investment and suffer morale problems when employees suspect that their own innovative efforts are not fully valued.

Yet we cannot ignore the possible implications of empirically derived relationships such as the effect of market growth, or product life-cycle position and market share on profitability and cash flow (Schoeffler *et al.*, 1974; Hedley, 1977). Given financial pressures and the requirement to boost

The Strategic Management of Technological Innovation. Edited by R. Loveridge and M. Pitt
© 1990 by John Wiley & Sons Ltd

earnings growth on pain of takeover, particularly in Britain and America, companies may find it necessary to review their activities via portfolio techniques (Haspeslaugh, 1982). We begin, therefore, by assessing the degree to which such approaches are appropriate; in some circumstances, it is suggested, an enhanced emphasis on *strategies for innovation* will be justified.

Porter (1987, p. 52) warned against naive acceptance of strategic portfolio prescriptions and his own appraisal of competitive forces (Porter, 1980) potentially enabled such issues to be addressed more discriminatingly. For example, where an industry's structural characteristics encourage fragmentation, large market share may offer little strategic advantage. Companies might then be wiser to invest in innovations aimed at cost reduction rather than in building market share.

Yet Porter also emphasized issues of product-market positioning and market power. Negotiating power was highlighted (indeed the case of vehicle component suppliers was used as 'a classic example of the challenges of selling to a powerful buyer' (Porter, 1983, p. 275)), whilst product and process innovations were both regarded as relatively poor sources of sustained competitive advantage in being readily imitable by competitors (Ghemawat, 1986). Such positions imply a somewhat secondary role for innovation; moreover they do not easily reconcile with arguments that a failure to appreciate the full strategic significance of manufacturing and technology has often caused declining Western competitiveness in the face of rivals from Japan (Skinner, 1978; Wheelwright, 1981; Abernathy, Clarke, and Kantrow, 1981; Hayes and Abernathy, 1980; Garvin, 1983; Patel and Pavitt, 1987).

The vehicle components sector is rich in complexity and affords opportunities to compare rival companies in different countries. In an earlier study (Carr, 1985) thirty manufacturers in Britain were 'matched' with thirty in Germany, the U.S.A. and Japan in six components sectors. Field research in all four countries was carried out between 1981 and 1983, being complemented by desk research. This evidence is combined with more recent data to enhance longitudinal testing of earlier findings. For a fuller account, see Carr (1990).

THE COMPETITIVE CONTEXT OF BRITISH VEHICLE COMPONENT MANUFACTURERS

The general decline in the competitive fortunes of U.K. vehicle component manufacturers is fairly clear up to about 1983. The question of subsequent recovery in the context of a more favourable sterling exchange rate and expanding U.K. vehicle production levels is more controversial. Longer term trends in the sector's profitability and international trade performance

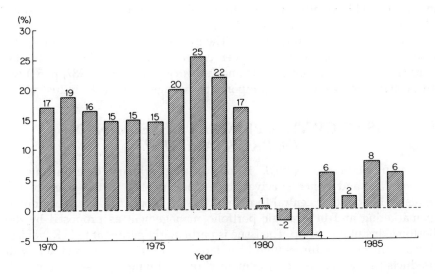

Figure 1. Profitability of U.K. vehicle component manufacturers. *Source*: Carr (1990)

(shown in Figure 1 and Table 1) give an indication of the severity of decline and the fragility of recovery.

Ford, like most other vehicle assemblers, has reduced the number of its European suppliers, from 2500 to about 900 in the last five years and has further similar plans. International rationalization is proving a severe test of competitive strategies. It is important to examine not only financial and market performance, but also the extent to which suppliers are meeting customers' key purchasing criteria in respect of quality, cost, flexibility and technical excellence. An earlier study's gloomy prognosis (Carr, 1985), based on these factors and in spite of expected improvements in market and exchange rate conditions, was vindicated by subsequent withdrawals of some of Britain's major companies from four product sectors examined

Table 1. Trade in motor vehicle parts and accessories (£m at 1980 prices using the producer price index for vehicles and parts)

Year	Exports	Imports	Balance	Imports/Exports
1970	1900	460	1440	0.24
1975	2360	810	1550	0.34
1980	2040	990	1050	0.49
1985	1910	1850	60	0.97
1986	1890	2130	−240	1.13

Source: Memorandum submitted by the Society of Motor Manufacturers and Traders Ltd (M76) to Britain's House of Commons Trade and Industry Committee (1987, pp. 330).

in detail. R.H.P. totally divested from rolling bearings, as did T.I. from silencers. G.K.N. virtually divested out of forgings by means of a new joint venture company with British Steel. The modern Lucas/Smiths automotive instrumentation factory has also been closed. More generally, as stated by Ford (House of Commons, Trade and Industry Committee, 1987, p. 51) we are seeing 'large gaps in the component coverage of the U.K. industry'.

IS GROWTH A PRIMARY DETERMINANT OF PROFITABILITY?

The assumption that *growth* is one of the primary determinants of profitability, and hence survivability—when related to the stage of the industry life-cycle, is central to many ideas on product-market positioning in marketing and in strategic portfolio management as proposed by the Boston Consulting Group. The issue is particularly pertinent to U.K. vehicle component manufacturers who have faced a shrinking home market, with products that generally occupy 'mature' or 'declining' market sectors.

Empirical evidence in this industry, however, suggests that the growth assumption is unwarranted. Between 1973 and 1979 the U.K. market for vehicle components virtually *halved* as domestic car production fell from just under two million cars to just under one million, yet Figure 1 shows that suppliers' average return on capital employed (ROCE) remained remarkably *stable*, averaging around 18%. Even in the worst year during this period following the first oil crisis ROCE averaged 14.6%. Over the next few years U.K. car production remained stable at about one million cars, yet suppliers' profits *decline* was both dramatic and sustained. ROCE is the product of two variables—sales/capital employed (a measure of utilization highly sensitive to market growth) and profit/sales (margin), but it is the latter which explains the decline in profitability. This decline may be attributed to customers, such as Rover, successfully subjecting most components to a four year price freeze after 1979 (Bessant *et al.*, 1984, p. 61). The U.K. wholesale price index rose 46% during this period but suppliers were unable to pass on their own cost increases.

The critical factor was an unprecedented rise of over 40% in Britain's manufacturing labour costs relative to those of rival countries, shown in Figure 2, caused by the combination of an appreciating currency and differential inflation rates. Despite poor productivity, U.K. vehicle component prices were fully internationally competitive in 1978 (Price Commission, 1979) so their *cost* levels were broadly competitive at this time in view of the profit levels indicated in Figure 1. Yet immediately after 1979, following the rise in the inflation-adjusted exchange rate, U.K. component suppliers were clearly uncompetitive on costs. Comparing prices internationally, Rover justifiably argued that despite its savage

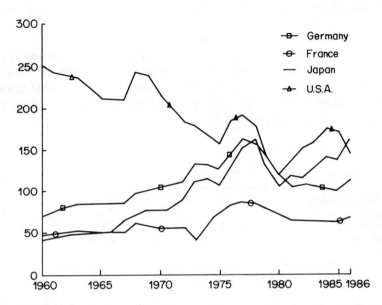

Figure 2. Trends in manufacturing earnings in the U.K. and rival countries (U.K. = index 100)

component price squeeze it was nevertheless subsidizing UK suppliers to the tune of $30 million p.a. (Bessant *et al.*, 1984, p. 64). In this sector, therefore, remaining competitive on cost is far more critical to profitability than market growth or decline; in an international market one must also keep a careful watch on real exchange rate movements owing to their effect on relative cost positions. From a longer-term viewpoint one must also examine other fundamental factors affecting cost competitiveness, such as productivity and important non-price factors such as quality.

In recent years companies have begun to adopt turnaround measures to restore profitability, so even without other environmental changes substantial improvement in average ROCE might reasonably have been expected. The real exchange rate position has rectified and U.K. component prices are once again broadly internationally competitive, albeit with slight deterioration in recent years (House of Commons, Trade and Industry Committee, 1987). But why has the recovery been so meagre and hesitant? Market growth offers little explanatory power, for U.K. car production is once again booming and output levels have increased from 1.0 to 1.4 million and are forecast to climb even higher. Thus for a satisfactory explanation of sector profitability we are forced to look elsewhere.

If growth affords little explanation for trends in sector profitability, does it account for *variations* in profitability among different vehicle component companies displaying different rates of sales growth? Regressing average ROCE against sales growth produced the results shown in Table 2. With the proviso that since many firms are involved in non-automotive activities, the use of company-specific data cannot be entirely rigorous, there appears a slight tendency for profitability to correlate more positively with sales growth the longer the time horizon. But none of these results are statistically significant, offering little support for the proposition that firms' sales growth is a key determinant of profitability.

Table 2. Results of regressing ROCE against sales growth

Time period	1975–86	1981–86	1984–86
Sample population	44	64	106
Constant	11.2	7.8	14.6
X-coefficient	+28.3	+0.2	−12.7
R-squared	0.028	0.020	0.003
T-value	1.08	1.12	−0.59

Source: ICC. Growth figures after inflation, simple % pa.

The automotive instrumentation sector was examined in more detail to establish whether U.K. company performance was better in a market segment displaying much higher than average growth. The conclusion was that underlying performance was nevertheless weak and vulnerable (Carr, 1985, Section 5.3).

IS SCALE A PRIMARY DETERMINANT OF PROFITABILITY?

The notion that *size* may be a critical determinant of performance (via scale economies and experience effects) is still based on the presumed importance of market share and various other size-related strategies, building on economies of scope as opposed to more specific benefits of genuine specialization. Luffman and Reed (1985) suggested that size may have become increasingly important. Many companies operating internationally face this strategic dilemma. For U.K. vehicle component manufacturers the question is particularly pertinent: Sleigh (1988) argues, though without empirical evidence, that scale is of such critical importance in this industry that companies with automotive sales under £10m are so disadvantaged as to be unlikely to survive.

Carr (1985) examined profit and sales performances of 66 companies offering continuous data between 1975 and 1982 in six size classifications (based on sales in 1975). Continuous data back to 1970 were also available

Table 3. Profitability and sales growth performances by size classification

Size groups (descending)	ROCE 1970–74 (%)	Growth 1970–75 (% p.a.)	ROCE 1975–82 (%)	Growth 1975–82 (% p.a.)	ROCE 1983–86 (%)	Growth 1983–86 (% p.a.)
1–11	12.7	3.4	8.3	(2.9)	4.3	7.4
12–22	26.5	0.1	11.7	(2.9)	1.3	(1.1)
23–33	19.6	2.6	19.8	(3.7)	5.1	9.5
34–44	21.4	(0.7)	8.0	(1.2)	21.0	16.6
45–55	5.1	(3.2)	13.2	(2.6)	12.9	9.9
56–66	—	—	10.2	(0.7)	4.6	4.7

Source: ICC.

for 24 companies (mostly fairly large) and to 1986 for 45 of these companies; they have been analysed all on the same basis in Table 3.

The largest size group (firms 1–11) includes household name companies holding dominant U.K. market share positions. As a group, they were substantially out-performed by the next size group in every year between 1970 and 1978. Since then the position appears to have changed somewhat. Size may now have a slightly greater impact on performance, but these figures hardly support the proposition that larger companies inevitably win out in the longer term, either in profitability or sales growth.

Size rankings used in Table 3 are based on sales in 1975 and so become less satisfactory as time goes on. Some simple regressions of ROCE against LOG (SALES) were done year-by-year as well as looking at more extensive time periods. The reasons for taking LOG (SALES) are outlined in Whittington (1980). Table 4 shows the results of this analysis. Over this twelve-year period, any relationship would appear to be negative (though the correlation is not statistically significant), consistent with Whittington's (1980) earlier findings when investigating the population of U.K. companies as a whole. There may be some indication of a change to a positive relationship in recent years, but the degree of correlation is almost negligible.

Table 5 examines the characteristics of the most profitable twenty companies from the sample of 106 over the most recent three year period.

Table 4. Results of regressing ROCE against LOG (SALES)

Years	1986	1985	1984	1984–86	1981–86	1975–86
Sample population	106	106	106	106	64	44
Constant	−13.6	+6.2	−49.2	−15.4	+22.9	+16.0
X-coefficient	+6.8	+3.4	+13.2	+7.0	−3.01	−33.78
R-squared	0.010	0.002	0.018	0.012	0.004	0.038
T-value	1.02	0.41	1.40	113	−0.50	−1.28

Source: ICC.

Table 5. Twenty most profitable companies 1984–6

Company	Av. ROCE 1984–6 %	Growth ranking 1984–6	Size ranking* 1986	Products
Delanair	307	63	24	Car heaters
BMAC	114	55	106	Elec lighting accessories
Stadco	94	32	39	Sheet metal work
Britax PMG	71	54	82†	Lighting and electrical
Hardy Spicer	46	76	12†	CV joints
Billover	44	51	91	Accessories
TRW United Cars	42	31	49†	Sub-assemblies
Alfred Teves	41	18	33	Brakes
Crestoll Hldgs	39	15	23	Safety glass
Albion Pressed Products	37	25	65	Metal pressings
T.J. Filters	34	28	71	Filters
Tecamec	31	99	90	Hoses
Motaproducts	30	24	53	Accessories
Eaton	30	87	10	Axles, transmission
GKN Kent Alloys	30	64	37†	Wheels (AL)
Clearplas & Vacumet	28	72	81	Plastic parts
IHW	27	45	74†	Door hinges
Berg Manufacturing	27	80	79	Brakes
Morley Foam	26	26	28	Foam products
Widney	25	29	67	Proprietary product
Average rankings above		49	56	

* In descending order, 106 being the lowest figure. Size ranks 1–55 all have sales above £10m total sales, others below.
† Denotes further support from parent group with UK auto sales of over £10m.
Source: ICC.

If Sleigh's (1988) view were correct, one would expect a preponderance of companies with high size-rank in the list, which is not the case. Neither growth nor size rankings appear to be a consistent feature of the better performing companies. The typical company in Table 5 has performed slightly above average in sales growth, but is slightly below average in sales turnover. Apart from two G.K.N. subsidiaries, Hardy Spicer and Kent Alloy, others such as Lucas, Associated Engineering, Automotive Products and Chloride, which Sleigh considers are of the size needed to compete in international markets, do not rank in the top fifty most profitable companies! Moreover, these companies all command dominant domestic market shares in respect of their main products. Dunlop, another dominant market leader in the past, has necessarily been excluded from this sample, having shed its major tyre operations and subsequently been taken over.

Detailed studies of competition in automotive bearings, forgings, instrumentation and silencers (Carr, 1985) again do not support the notion that companies with dominant domestic market positions win out. R.H.P. divested its entire bearings activities when it was the U.K. market leader. G.K.N. virtually divested its entire forgings activity when it held almost half of an otherwise highly fragmented UK market. Lucas/Smiths major joint venture facility in instrumentation was closed, despite a dominant domestic market position, though a much smaller rival, A.B. Electronics, succeeded in growing thirteen-fold over the last five years. T.I. Cheswick, though it shared domestic market leadership in exhaust systems with its overseas rival Tenneco Walker and enjoyed good performance in the past, produced losses a year or so ago and these activities have now been divested. A further point is that in areas such as forgings, where very substantial international productivity differentials exist, some British plants studied were actually *larger* than those of their more highly productive Japanese counterparts.

It may be argued that the correct basis for assessing the effects of size and market share is no longer Britain, but Europe, if not the world. Today an international perspective is of much greater relevance. Yet this raises problems for those who argue that factors such as market share represent critical strategic perspectives. Almost all U.K. vehicle component companies command only very small shares of the European market. Are they *all* to be offered the same strategic prescriptions? In view of their widely differing performance levels, what can one say when those with even smaller market shares do so much better?

Many companies which have performed well appear to be classic 'dogs' in strategic portfolio terms, characterized by relatively low market shares and unexciting growth levels. Undue faith in the merit of a commanding domestic market position (or sometimes, one suspects, in size for its own sake), has led to an over-eagerness to harvest or divest those business units with unexciting product-market positions.

On the evidence of this industry, one can have little confidence in strategic portfolio prescriptions which emphasize size and growth rate rather than positive technological strategies—even where the latter entails investments in 'dogs'. At the very least, it seems unreasonable to divest without first examining the competitive characteristics in specific sectors via a more discriminating analytical approach such as Porter's.

The poor performance of large household name companies, many fairly loose conglomerations built up over many years, and the extent of their subsequent divestment programmes, seems to indicate that economies of scope are less important than those derived from genuine specialization. One might conclude that in this industry companies would have done better to invest in existing specialisms or at least stick to strategies of tightly related diversification—for example by exploiting technology synergies as G.K.N. did successfully with Hardy Spicer and Kent Alloys.

Christopher Carr

STRATEGIC MANAGEMENT: WHERE PORTER'S INSIGHTS ARE HELPFUL

Porter's (1980) structural model acknowledges that scale benefits are more important for some products than others, whilst still yielding other strategic insights with implications for firms' technological strategies.

Automotive bearings

Scale advantages are relatively more important in a sector such as automotive bearings—where Britain's major (and virtually only) contender has suffered severe competitive reversals—than elsewhere (Carr, 1985).

R.H.P. was formed in 1969 as a consolidation of Britain's three major bearings manufacturers. Its intended strategy was, in terms of Porter's generic strategies, one of industry-wide overall cost leadership, based on competing across the full product range, whilst reducing product variety to exploit cost reduction opportunities from modern high-volume production lines. Cost reductions for these types of products were real and substantial. Given its dominant domestic market share, this strategy would probably have been effective had competition remained more or less national. The impact of internationalized competition, however, was so rapid and ferocious as to render R.H.P.'s strategic position incoherent: only S.K.F. was in a position to employ a cost leadership strategy in an international context.

Other successful performers pursued international 'focus strategies'; Japanese companies like N.S.K. focused on specialist electrical bearings, gradually broadening from a very narrow product range; Timken (U.S.A.) focused on tapered roller bearings; both Torrington (U.S.A.) and I.N.A. (Germany) focused on needle bearings and on customer orientated 'engineering service'.

For over a decade R.H.P. floundered, 'stuck in the middle', hanging on to the illusion of an overall cost leadership strategy as progressively it was squeezed out of volume segments on price; finally it switched to an internationally oriented 'focus strategy' based on wheel pumps and automotive spindle bearings, ultimately to surrender on this front with the closure of its modern, specialized factory. Today its activities have finally been adjusted to a more coherent, albeit much reduced role as a specialist manufacturer, and have been sold off. Huge costs ensued from inappropriate competitive positioning and in hindsight R.H.P. would have been wiser to focus on technological innovations appropriate to a 'specials' producer offering a customized service.

Automotive forgings

Scale advantages are much lower in automotive forgings. Other factors identified in Porter (1980, ch. 9) have combined to produce a fragmented industry: high transport costs, relatively small order sizes, diverse market needs and perhaps most important, high exit barriers. G.K.N. Forgings' strategy of 'seeking dominance', gaining about 45% of the U.K. market, bringing in modern volume-oriented processes and the latest technological developments, ultimately ran into precisely the 'strategic trap' identified by Porter (1980, pp. 210–11). Even by the depressed standards of the sector, its profitability has been consistently poor over the last fifteen years and G.K.N. Forgings was in effect divested via a joint venture with British Steel. Given this sector's structural characteristics, Porter's approach (albeit with the benefit of hindsight) draws attention to the danger of overinvesting in volume-oriented research and development and production processes.

TECHNOLOGICAL INNOVATION IN PORTER'S APPROACH

Porter's (1980) warning that companies cannot afford to ignore broader forces of competition such as changing bargaining power among suppliers and customers is appropriate in this industry. U.K. suppliers assumed that they could achieve high profitability on the back of domestic customers such as Rover, despite the latter's mounting losses and declining international competitiveness; their own performances were devastated after 1979, as radically altered exchange rates and Rover's new international procurement policies gave it the upper hand.

Recognizing negotiating power as a strategic issue, Porter encouraged companies to find independent, defensible positions offering some degree of market power, rather than rely on innovatory sources of competitive advantage. This emphasis is consistent with his theoretical base, arising from the problems of imperfect competition. Yet the approach has proved counter-productive for many British and American suppliers and is in sharp contrast to the position of successful Japanese—and to a lesser extent German—suppliers.

Figure 3 illustrates the *trade-off* between an emphasis on market power and on incremental innovation. The larger U.S. and British competitors also maintain relatively independent stances *vis-à-vis* their customers, as advocated by Porter, attempting to secure some degree of market power. This position is located in the bottom left-hand corner of Figure 3. In contrast, both Japanese suppliers and assemblers recognize their essential interdependence (Anderson, 1981; Carr, 1989; Shimokawa, 1982a,b; Altshuler *et al.*, 1984) and this has been a significant feature of their developing competitiveness. Most Japanese suppliers have avoided options which might have increased their independence and bargaining power; in contrast to the U.S.A., assembler/supplier contracts are considered virtually irrelevant. From a short-term equilibrium perspective these subservient strategic postures seem disadvantageous, rendering suppliers vulnerable and depressing their profit margins; from a longer-term dynamic perspective, suppliers have gained competitive advantages in association with their flourishing assemblers. The relationship is so close and covers such a range of policy issues that it is beneficial to think in terms of a network of collaboration. Vehicle assemblers are themselves a 'vehicle' through which components have been sold throughout the world.

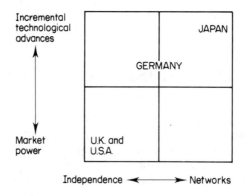

Figure 3. National differences in strategic themes

The Japanese industrial group system and the fact that acquisitions are virtually ruled out have constrained opportunities for changes in product-market positioning. So supplier companies have preferred to look almost exclusively to manufacturing and technological improvements as sources of competitive advantage—the top right-hand corner of Figure 3. By responding closely to customer industry demands, Japanese suppliers over the last fifteen years have relentlessly pursued incremental technological advances, securing improvements in productivity, quality and flexibility—advances which U.S. and British suppliers, lacking the same strategic imperative, have seriously neglected.

Large component companies in the U.S.A. for many years generally maintained strong enough negotiating positions to insist on large order sizes. If customers wanted smaller order quantities, price premiums were such that they were encouraged to go elsewhere, usually to smaller suppliers. The result was that large U.S. companies, cushioned by large production runs right up until 1980, felt little need to implement flexible manufacturing systems. Then the vehicle assemblers, under pressure from Japanese competition, quite suddenly demanded much smaller runs.

The sector context in the West is not entirely comparable with Japan; yet experience suggests that Western companies should be wary of placing any long-term reliance on those opportunities afforded through secure, independent niches. Large U.K. suppliers, historically able to exploit high bargaining power in the domestic market have, it seems, been lured into a dangerous complacency as longer-term performance gains evaporated after 1979. Even today profits have scarcely recovered to an acceptable level. Furthermore, some suppliers are now paying dearly for this past complacency in respect of quality: being struck off Ford's preferred supplier listing as a consequence is a very real threat to their continued viability.

INNOVATION AS A SOURCE OF SUSTAINABLE COMPETITIVE ADVANTAGE

Here it has been suggested that an emphasis on Porter's (1980) strategic prescriptions may be misleading or at least only a *partial* answer. The issue, then, is what other factors constitute exploitable sources of sustainable advantage in the longer term? If some of the alleged advantages of market power are marginal and short-lived in this industry, does innovating provide a more substantial and lasting source of advantage, or is Ghemawat (1986) correct to point out how readily innovations can be imitated by competition?

In fact the importance of innovation for the vehicle assemblers appears to be clear. All vehicle producers are actively reducing the number of suppliers, so that suppliers most likely to remain competitive will be those able to meet key purchasing criteria. Twelve European vehicle assemblers were recently questioned about their purchasing criteria (House of Commons, Trade and Industry Committee, 1987) and the following factors emerge consistently: (i) cost/productivity, (ii) quality, (iii) delivery (covering flexibility as well as reliability) and (iv) technical excellence (including development capability). Innovation is important in each area, though productivity-related innovation is of most relevance in reducing cost. The competitive impact of innovation in these four areas and its sustainability will now be addressed.

Productivity

International productivity differentials in automotive components have been substantial, e.g. over 40% in the case of automotive bearings dual-sourced in Britain and Germany. International productivity comparisons made by the author around 1983 in automotive forgings showed that British firms were less than half as productive as U.S. counterparts and less than one third that of the Japanese, whether measured in tonnes per man or sales per annum.

U.S. manufacturers benefited particularly from longer production runs, whereas in Japan there was more emphasis on flexibility with just-in-time systems of delivery. Scale did not seem to be the main issue: some U.K. factories were larger than those in Japan, though the latter were generally more modern and enjoyed some advantages of specialization. Overall, however, Japanese/U.K. productivity differentials largely reflected a high managerial commitment in Japan to programmes of incremental operational improvements sustained over the last twenty years.

Despite extravagant recent claims about 'productivity transformations' in U.K. companies, there is little evidence to suggest that such productivity gaps, emerging over many years, can be closed easily or quickly, given the momentum of Japanese productivity advances. Innovative Japanese gains in operational efficiency, it is suggested, cannot be easily or quickly imitated by competitors; they present a potent and sustainable source of competitive advantage.

Quality

Twenty-five years ago the reputation of British components for quality was extremely good whilst Japan's was mediocre; by 1983 the position had reversed. Unfavourable comparisons have also been made with Germany, for example, by one multinational who reported reject rates twice as high over the previous three years from a U.K. supplier as from its dual-source partner in Germany on the same product (Carr, 1985). Since 1983 U.K. suppliers have faced specific, highly demanding supplier quality rating targets and the risk of losing preferred suppliers status has forced them to treat quality as a strategic priority. In the eyes of their customers, they have generally made considerable progress, though in a number of sectors including forgings the situation is still considered patchy.

Two problems remain, however. Firstly, it is intensely difficult to keep pace with the sheer momentum and commitment to quality in rival countries such as Japan. Secondly, U.K. companies are still heavily over-reliant on ex-post inspection procedures, in contrast to Japan and Germany where programmes directed towards the more difficult target of 'built-in quality' were well established long ago. One U.K. company recently took drastic quality turnaround measures to achieve the required target score on Ford's Supplier Quality Rating System, but the high cost of its essentially ex-post measures is estimated at about 3.75% of turnover.

In summary, the quality advantages achieved by successful competitors like the Japanese, through their total and obsessional attention to operational improvements, provide a potent and *de facto*, sustainable source of competitive advantage.

Flexible manufacturing systems

Unlike their more independent counterparts in the U.S.A., Japanese supplier companies have rarely been able to insist on long production runs. A close relationship with assemblers has meant submitting to customer requirements for radically more flexible, just-in-time delivery service, despite the potentially high cost penalty in foregoing economies of scale. The practical consequence of this customized orientation is that for two decades or more Japanese suppliers have pursued innovative operational improvement programmes to minimize the cost penalties involved in such an orientation. In contrast, the strategic approach of U.S. companies led to similar developments being virtually ignored; they were instead content to deliver a fairly efficient and cost competitive service on the back of long production runs.

The first of two main flexibility themes pursued by the Japanese involved radical improvements in machine changeover times and their progress in comparison with US companies is notable. For example, U.S. changeover times on forge hammers three or four years ago were typically in the range of two to three hours compared with under half an hour in Japan.

The second major theme pursued by the Japanese to improve flexibility required improvements in factory through-time. One Japanese exhaust system manufacturer improved changeover times, then re-engineered its process routes to bring factory through-time down to about fifteen days, compared with typically about two months for U.K. manufacturers. Having done this, it still could not satisfy Toyota's requirements, because of thirty days response time from its own suppliers. It therefore worked closely with them to provide a coordinated just-in-time delivery service. With help from Toyota's own engineers and from a machine tool company in the same industrial group, the company claimed that the supply chain could respond to new instructions from Toyota within about ten days in total. Such close working relationships also encouraged the company to match its manufacturing capability closely to the needs of its particular customers. Some products did not require the same degree of flexibility, permitting segregated manufacturing operations which could utilize more volume-oriented manufacturing systems.

By 1983 'After Japan' programmes in the U.S.A. had led to immediate plans to effect reductions in changeover times. Sadly, the author found little evidence of forgers or other vehicle component suppliers in Britain (with the exception of T.I. Silencers in exhaust systems) systematically pursuing such programmes at this time.

Many British companies have paid a high price for choosing to operate relatively large factories and volume-oriented production processes and equipment. Many were very hard hit as batch sizes declined quite precipitously after 1980/81. Some firms made desperate attempts to improve the flexibility of their modern assembly lines by literally disassembling them. As a consequence, set-up time/operating time ratios above 100% were not uncommon and contributed substantially to declining financial performance.

Most of these companies now recognize that such issues were critical to their dramatic performance decline in 1980/81 (for a detailed account of Lucas's recent changes see Heller, 1987, pp. 183–90). They also recognize that restoring the position cannot be achieved quickly or piecemeal, but requires board-level strategic direction. Bold trade-offs can be involved. One brake manufacturer recently invested in six new machines to accomplish the same output that previously was produced on a single modern, volume-oriented machine. Productivity is lower, but smaller batch sizes can be made with reasonable efficiency. There is some cost disadvantage but they can now deliver variety and genuine customization.

As with quality and productivity, the problem is compounded by competitive progress overseas. Saab, for example, agrees that British suppliers' delivery flexibility is improving, but warns that Japanese standards are rising even faster: specifically they are seeking reductions in manufacturing lead times of 25% over the next four years (House of Commons, Trade and Industry Committee 1987). Whilst companies vary considerably, one visited recently by the author stressed the importance of flexibility, but had yet to begin improvements in changeover times or production flow routings to reduce factory through-times. Thus, competitive disadvantages on basic operational issues appear remarkably sustainable!

Radical versus incremental technological developments

From an engineering perspective, issues such as quality and technical excellence cannot easily be disentangled: quality is not merely a matter of surface finish and is affected by both product and process specifications. Issues of productivity, quality, delivery and technical excellence lie at the heart of the failure of British suppliers to close the gap even against direct European rivals in Germany after many years and despite relatively more favourable wage costs. Comparisons between the two countries (Carr, 1982) highlighted the competitive significance of Germany's close attention to technology enhancement and of their superior engineering skills. Even today if one asks why an independent English-speaking customer such as Volvo purchases 35% of its components from Germany and only 6% from

Britain, the unequivocal answer is that Germany is preferred because of the excellence of its products and reliability of delivery (House of Commons, Trade and Industry Committee, 1987, p. 307). Owing to changes in exchange rates, Volvo is actively searching for British components, but is critical of the low level of U.K. resources devoted to R&D, which in its own case are 10% of turnover. Ford and General Motors echo the same complaint, but more fundamentally they place the blame on business philosophies too financially orientated and lacking in commitment to detailed engineering excellence.

Technological developments in this industry have generally been subtle and incremental—an accumulation of numerous minor changes, each in themselves often scarcely visible to the outside observer. In terms of engineering support this usually calls for *strength in depth*, rather than undue reliance on a few high calibre R&D engineers and it is here that Britain is so far behind Germany and Japan–and why it will take so many years to catch up. More obvious advances, unless they are adequately patented, are relatively more imitable as Ghemawat (1986) suggested, since vehicle assemblers do apply considerable pressure for diffusion of know-how among suppliers; but these cases are not as common as is often assumed.

Occasional, radical product innovations can prove a potent source of sustained competitive advantage, so long as they are properly supported by a wider range of business strengths and a coherent overall business strategy. But it is simplistic to write off all such technological advances as easily imitable: Michelin's leadership in steel-belted radial tyres finally capped Dunlop's chances of survival in this industry. In contrast, one of Britain's few breakthroughs leading to a dominant international position was G.K.N./Hardy Spicer's constant velocity joints. Until its recent expiry, patent protection allowed for the development of a viable, international strategy. In stark contrast to the failure of many of G.K.N.'s unrelated acquisitions, its success with constant velocity joints attests to the positive role of the corporate centre when real commitment and vision are directed towards enhancing an S.B.U.s competitiveness. The patents G.K.N. acquired when it bought Hardy Spicer in the 1960s were no guarantee of success, but there was a clear relatedness with other activities and technological skills. Adequate funding enabled the development of German manufacturing operations and two modern, purpose-built plants in the U.S.A. Having backed this innovation, G.K.N. now operates in one of the very few vehicle components sectors where Britain stands a good chance of remaining internationally dominant. The proportion of group profits accounted for by this single product has been consistently high. On a smaller scale, had G.K.N. not backed Kent Alloys' new developments in low pressure aluminium die-casting in the early 1970s it is highly unlikely it would have sustained a successful niche in the specialist aluminium alloy wheels market. Apart from G.K.N.'s Hardy Spicer and Kent Alloys, virtually no other large British-owned vehicle component manufacturers appear in Table 5. In some situations therefore, providing enthusiastic support for technological developments can be absolutely critical to competitive success.

CONCLUSION

For U.K. vehicle component companies neither market growth nor market share correlate with performance. In general they should be extremely cautious about prescriptions yielded by portfolio approaches to strategic management which could implicitly undermine their commitment to technological innovation. Porter's approach is flexible enough to identify situations where issues such as market share are important, or situations where there is a danger of overinvesting in volume-oriented R&D or production technology.

Porter's reminder that other factors affect the intensity of competition, such as the power relationship between suppliers and customers, is highly pertinent to this industry. Yet there is an inherent bias in his theoretical framework towards issues of product-market positioning and market power, with the risk of underplaying the strategic significance of technological innovations. In effect, suppliers are encouraged to seek ways of achieving more defensible independent niches offering significant degree of negotiating power relative to customers. American and British suppliers who followed this approach seem to have been lulled into complacency, whereas their Japanese counterparts have gained important sources of *sustainable* competitive advantage via product and especially process innovations. On the other hand, benefits from increased negotiating power (eschewed by Japanese suppliers in their search for closer customer relationships) have proved short-lived and illusory for many Western suppliers.

The Japanese obsession with incremental technological innovation appears highly appropriate in this industry: their growing strength arises, it is suggested, from just such a policy emphasis, not from cultural factors *per se*. Sumitomo's turnaround of Dunlop's former European tyre operations within about three years is a salutary reminder of just what can be achieved. From losses estimated at $20m a year on $100m turnover, it is already in profit and production is up by 50%. Yet not one of the themes outlined in Clutterbuck's (1988) account relates to any major change in market positioning; instead Dunlop focused on building a real commitment throughout the organization to basic manufacturing improvements.

Finally, although major technological advances capable of really altering patterns of competition are less common, when they do arise and when they are adequately supported in terms of a coherent overall business strategy, they constitute a crucial 'window of opportunity' for future competitiveness. In short, sustained commitment to technological innovation in the context of a *coherent* corporate strategy pays off.

REFERENCES

Abernathy, W.J., Clarke, K.B., and Kantrow, K.M. (1981). 'The new industrial competition', *Harvard Business Review*, **59** (5), 68–82.
Altshuler, A., Anderson, M., Jones, D., Roos, D., and Womack, J. (1984). *The Future of the World Automobile: The Report of MIT's International Automobile Program*, George Allen and Unwin.
Anderson, M.L. (1981). *The Strategic Organization of the Japanese Automobile Groups*, mimeo, Cambridge, Mass, USA.
Bessant, J., Jones, D.T., Lamming, R.L., and Pollard, A. (1984). The West Midlands Automobile Components Industry: Recent Changes and Future Prospects, *West Midlands County Council Economic and Development Unit Sector Report No 4*, West Midlands County Council.
Bailey, P.J.H., and Farmer, D. (1981). *Purchasing Principles and Management*, Pitman, London.
Carr, C.H. (1982). A Comparison of British and German Companies Producing Components for the Automotive Industry, unpublished paper, Lord Rootes Fund research, University of Warwick.
Carr, C.H. (1985). The Competitiveness of UK Vehicle Component Manufacturers, unpublished doctoral dissertation, University of Warwick.
Carr, C.H. (1990). *Britain's Competitiveness: The Management of the Vehicle Component Industry*, Routledge.
Clutterbuck, D. (1988). 'How the Japanese rebuilt Fort Dunlop', *Director*, September, 94–8.
Garvin, D.A. (1983). 'Quality on the line', *Harvard Business Review*, **61** (5), 64–76.
Ghemawat, P. (1986). 'Sustainable advantage', *Harvard Business Review*, Sep.–Oct., 53–8.
Haspeslaugh, P. (1982). 'Portfolio planning: uses and limits', *Harvard Business Review*, **60**(1), 58–74.
Hayes, R.H., and Abernathy, W.J. (1980). 'Managing our way to economic decline', *Harvard Business Review*, July, 67–77.
Hedley, B. (1977). 'Strategy and the "Business Portfolio"', *Long Range Planning*, **10**(1), 9–16.
Heller, R. (1987). *The State of Industry. Can Britain Make It?*, BBC Books, London.
House of Commons (1987). *The U.K. Motor Components Industry. Third Report from the Trade & Industry Committee Session 1986–87. Report, Proceedings of the Committee, Minutes of Evidence and Appendicies*, HMSO, London.
Inter Company Comparisons (annual). *Business Ratio Report: Motor Components*, ICC, London.
Luffman, G., and Reed, R. (1985). 'The giant company comeback', *Management Today*, June.
Patel, P., and Pavitt, K. (1987). 'The elements of Britain's technological competitiveness', *National Economic Review*, November, 72–82.

Porter, M.E. (1980). *Competitive Strategy: Techniques for Analysing Industries and Competitors*, Free Press.

Porter, M.E. (1983). *Cases in Competitive Strategy*, Free Press.

Porter, M.E. (1987). 'From competitive advantage to corporate strategy', *Harvard Business Review*, May/June.

Price Commission (1979). *Prices, Costs and Margins in the Manufacture and Distribution of Car Parts*, HMSO.

Schoeffler, S., Buzzell, R.D., and Heany, D.F. (1974). 'Impact of strategic planning on profit performance', *Harvard Business Review*, March–April, 137–145.

Shimokawa, K. (1982a). 'Development of the supplier relationship in Japan—its innovation and production flexibility', paper presented at International Policy Forum of the MIT Future of the World Automobile Program, 16–20 May, Hakone, Japan.

Shimokawa, K. (1982b). 'The structure of the Japanese auto parts industry and its contribution to automotive process innovation', paper presented at International Policy Forum of the MIT Future of the World Automobile Program, 16–20 May, Hakone, Japan.

Skinner, W. (1978). *Manufacturing in the Corporate Strategy*, Wiley.

Sleigh, P. (1988). *The U.K. Vehicle Components Industry*, Economist Intelligence Unit, London.

Wheelwright, S.C. (1981). 'Japan—where operations really are strategic', *Harvard Business Review*, **59**(4), 67–75.

Whittington, G. (1980). 'The profitability and size of United Kingdom companies 1960–1974', *Journal of Industrial Economics*, vol. xxviii, no 4, June.

4

Technology and Competitive Strategy in Food Retailing

Jacqueline Senker

Science Policy Research Unit University of Sussex

Competitive success in food retailing in the U.K. has been achieved, since the late 1970s, by several multiples which have adopted a technological strategy as part of their overall business strategy. They have not only innovated in systems for the distribution and handling of food but have also been responsible, through close collaboration with their suppliers, for some important food manufacturing innovations.

Porter's (1980) theory of competitive strategy has provided a useful framework for interpreting this relationship between food retailers and their suppliers; it is a development of Galbraith's theory of countervailing power which recognized that competition can appear 'not on the same side of the market but on the opposite side, not with competitors but with customers or suppliers' (Galbraith, 1963, p. 125). However, the importance of incorporating technology into competitive strategy was slow to emerge into Porter's analyses, and there are still deficiencies in his treatment. This chapter reviews the development of Porter's ideas on the technological

The Strategic Management of Technological Innovation. Edited by R. Loveridge and M. Pitt
©1990 by John Wiley & Sons Ltd

dimension of competitive strategy, then presents evidence from recent research to highlight two weaknesses in Porter's approach: the failure to recognize the need for *cumulative* in-house technological competence in order to follow such a strategy and the virtual omission of a discussion of the *risks* and *uncertainties* of undertaking innovation.

THE TECHNOLOGICAL DIMENSION OF COMPETITIVE STRATEGY

Porter's early work recognized that competition not only arises within an industrial sector, but may involve buyers, suppliers of similar or substitute products or new entrants to the industry (Porter, 1980). Subsequently he suggested that technology as a strategic variable can change the competitive 'rules of the game' by impacting on all the forces driving competition. Technology can be employed to implement a firm's chosen strategy—whether it be cost leadership, differentiation or focus, but his discussion focused mainly on the relative merits of technological leadership or followership (Porter, 1983). This idea was later expanded to take account of the fact that technology is embodied in every 'value activity' within a firm. Value activities include not only core technologies encompassed within a firm's products and production processes, but also those in support activities, for instance procurement, office technology, transportation or design. Technological change can affect competition through its impact on any of these activities, the collection of which Porter calls the 'value chain'. Porter recommended that firms should identify all the technologies in their value chain, and then determine which technologies and potential technological changes have most significance for reinforcing the competitive advantage which they are seeking to achieve and sustain. He noted that 'in choosing among technologies to invest in, a firm must base its decision on a thorough understanding of each important technology in its value chain . . .' (Porter, 1985, pp. 67–68).

The main strength of Porter's contribution derives from its multifaceted approach to technological strategy which encompasses competitive threats and opportunities arising within and without an industrial sector. However, his approach lacks an appreciation of the necessity for cumulative in-house technological expertise in a *wide range* of technologies in order to achieve 'a thorough understanding of each important technology in its value chain'. The complexity of technology means that firms wanting to adopt technologies developed elsewhere, or to imitate existing products, requires them also to have relevant in-house expertise. Furthermore, technical change is a cumulative activity as Dosi *et al* (1988, p. 223) and Pavitt (1987) acknowledged:

'Technological activity in firms tends to build out incrementally from what they know already .... Technological search and choice in firms are therefore constrained by what they have already learned, and their technological activities tend to follow a "trajectory".' (Pavitt, 1987, p. 186).

Porter's approach also ignored the substantial risks and uncertainties involved in innovation.

Research recently carried out within eight leading British food multiples (leading, in terms of shares of the national food market) to discover *why* and *how* some of them were influencing their suppliers to innovate, confirmed Schumpeter's view that firms are affected both by their pasts and by their potential futures (Schumpeter, 1970, p. 84). The food retailers able to adopt a technological dimension to their competitive strategy when the opportunity arose were those with a long tradition of investment in technical expertise. They had always derived benefit from such investment but now it became a major weapon in their competitive armoury. Furthermore, they were able to deploy this expertise in a way which minimized the risks associated with innovation because they have been able to test-market new products at selected stores without bearing the heavy development and promotional costs faced by manufacturers. Additionally, their products are subject to imitation only when they appear on the shelves. Manufacturers' new products are in danger of imitation from the moment the 'sales pitch' is made to the retailer (Senker, 1988).

After outlining changes in food retailing competitive strategies since the late 1970s, evidence of food multiples' differing capability to adopt these new strategies will be presented, together with some examples of successful firms' achievements. This chapter concludes by discussing how food retailers have minimized the risks associated with innovation and thereby increased their potential for appropriating its benefits. It highlights the fact that the accumulation of technological expertise was often 'serendipitous', not consciously related to gaining competitive advantage. Evidence also seems to indicate that the decision to gain technological competence in food technology was not taken after a careful assessment of all the important technologies in the value chain. Rather it was a combination of factors— food legislation, company culture and by the 1970s, the problems and opportunities posed by operating larger sized stores. Whatever the *reasons* for acquiring competence in food technology, its internalization could be turned to competitive advantage as the basis of competition changed over time.

COMPETITIVE STRATEGY IN FOOD RETAILING

Retail food chains which introduced self-service supermarkets from the late 1940s made considerable savings on labour costs. The wish to pass on these savings to the customer in the form of lower prices created pressure which led to the eventual abolition of resale price maintenance (RPM) in the U.K. in 1964. Multiple food retailers could then compete for customers on the basis of price and their success with this strategy allowed them to gain market share.

Concentration in food retailing gathered pace in the 1960s, fuelled both by the spread of self-service supermarkets and the abolition of resale price maintenance. By the 1970s a difficult trading environment had developed for the multiples. The volume of food purchased was almost static and profit margins for packaged groceries were eroded by strong price competition. These conditions were exacerbated by inflation, increasing commodity prices and Government prices and incomes controls. The major grocery retailers responded by adopting strategies to increase sales volumes to compensate for low margins, often by merger or acquisition of less successful chains. Substantial economies of scale were achieved from increased sales volume, particularly through the exercise of strong purchasing power to negotiate considerable discounts from suppliers, but also in store operations. Another strategy for coping with slim margins was diversification into fresh and frozen foods and non-foods, and by opening larger stores to house the wider range of products.

The multiples found that larger stores had the advantage of lowering operating costs per square foot. As a result, many small units were replaced by outlets with increased selling area. Rapidly increasing floor capacity in the grocery trade created an over-supply of retail grocery services, which led to a major price war between the multiples. By 1977 selective price cuts and trading stamps were dropped in favour of across-the-board price cutting backed by extensive advertising (Akehurst, 1984). The cumulative effect of these developments further increased concentration in food retailing. U.K. multiples' share of grocery shop sales was estimated at 20% in 1950 (Beaumont, 1982, p. 9). By 1987, they accounted for over 78% of packaged grocery sales (AGB, personal communication).

Price competition is a valid strategy when the rate of price inflation of the goods being sold outstrips the inflation of selling costs. This was the case up till 1980, but subsequently when selling costs inflation became greater than inflation in the prices of the goods being sold, price competition lost its appeal as a competitive strategy. The failure of this strategy was confirmed in a speech given by the Chairman of Tesco's in 1985:

> 'It is very clear that no company will get a price advantage for long. A very low price advertised today is bettered within 24 hours by a competitor.
>   This is all very well short term, but in the long term it cannot be in anyone's interest' (Reported in the *Grocer*, 23rd November 1985).

With the disappearance of price competition, new competitive strategies were implemented by the largest multiples. They adopted a variety of strategies, primarily to differentiate themselves from one another. One strategy to increase market share, based on the belief that consumers were now more interested in novelty and quality, was to enhance store image by its identification with exclusive own label products and high quality fresh foods. This strategy demands in-house technological capability and it is not surprising that the companies first able to deploy it were those with a long tradition of company investment in technical expertise.

FOOD TECHNOLOGY IN RETAILING

The decision by food retailers to set up in-house food technology departments was largely influenced by food legislation. The Food & Drug Act, 1955, was similar to earlier food legislation in placing an obligation on retailers to ensure that the food they sold was not injurious to health or adulterated in any way. The new legislation also included regulations concerning minimum standards of composition for certain foods. No offence was committed until a retail sale had been completed. The only statutory defence for the retailer was that the food had been supplied by a branded manufacturer. The brand name on the invoice amounted to a warranty that the product could lawfully be sold. There was no such defence for own label products (i.e. products sold under the name of the store). Growth in the volume and range of own label goods since the 1960s places retailers under a very strong obligation to ensure that these goods meet the requirements of the law. Accordingly, multiples selling own label products made by external suppliers have employed food technologists to check on the hygiene of suppliers' premises and sample own label products to ensure they meet health and compositional standards.

Thus by 1970, food technology departments were found in most of the leading food multiples which sold own label products. One notable exception was Tesco Stores. It did not establish its own department until 1973, but this quickly grew to become one of the largest retail food technology departments in the country. The size of departments ranged from 3 or 4 technologists at the lower end of the scale up to as many as 100 technical staff. Although legislation can account for the existence of food technology departments in food retailers, it cannot explain the age or size of those which pre-date the Food & Drug Act. Large established food technology departments, like those in Sainsbury and Marks & Spencer, reflect company cultures which believe that benefits are gained from investment in technical expertise.

The size of multiples' food technology departments has continued to grow in recent years. What has caused this growth? Two factors, both linked to increased concentration in food retailing, appear to be influential: first, growth in the floor area of supermarkets and second, the exhaustion of price competition as a profitable strategy.

When food multiples' responded to the slim margins resulting from the major price war of the 1970s by diversifying into fresh and frozen foods and non-foods, and opening larger stores to house the wider range of products offered, this caused problems and created opportunities; specifically, how to maintain 'fresh foods' in peak condition and how to minimize wastage. Such problems were resolved by the application of technology. Similarly, with the demise of price competition in the 1980s, competitive strategy demanded an input from technologically qualified staff. In order to sell the exclusive own label products and high quality fresh foods which enhance store image and increase market share, retailers have been at the forefront of devising appropriate manufacturing, packaging and palletizing methods for the short-life, own-label, fresh and chilled products which are taking increasing market share from traditional packaged groceries.

Interviews carried out in 1984/85 with eight leading food multiples on the size and activities of their food technology departments, revealed the steady increase in size of most departments over time, as shown in Table 1.

Table 1.   Technical staff employed by food retailers

Company	Date laboratory founded	1970	1975	1980	1985
A	(late 1920s)	100	100	100	100
B	(1948)	30	43	56	69
C	(1973)	—	9	40	112
D	(unknown)	3	3	3	3
E	(1930s)	11	9	8	11
F	(1964)	8	14	20	27
G	(1970)	8	15	21	27
H	(1965)	4	5	8	12
Total		164	198	256	361

Details of the work carried out by these technological departments indicated that they could be divided into two groups—'accepters' and 'interveners'; *accepters* have small technical departments, whilst *interveners* have substantial in-house technological capability. The size of the technological department is one major factor affecting its capability to undertake various tasks, but there is much variation amongst companies. For instance, some companies shared facilities with the parent company's food manufacturing laboratory; another took advantage of a local Food Research Institute, to which it often turned for advice.

The accepters carry out the minimum of tasks necessary to cover themselves under food and drug law: they carry out quality control and sensory evaluation of own label products through spot checks on samples selected at random from the shelves of their stores, and factory inspection of prospective own label suppliers to ensure they comply with good hygienic practice. Some try to conduct regular follow-up inspections on current suppliers, but lack of personnel creates problems. In the procurement of own label products, accepters rely on manufacturers who are able to produce imitations of branded products. Manufacturers are provided with a layman's specification, typically including quality and price, with reference to an existing branded product, and might include taste criteria or mouth feel. Suppliers are expected to carry out any necessary development, which generally does not present a problem, given the information provided by the statutory list of ingredients on the reference product. Suppliers are chosen on the basis of samples produced, after which specification details are firmed up with the chosen supplier. The accepters also have limited recourse to external sources of technical expertise. They seek advice from the research associations to which they belong, to cope with legal, labelling or analytical problems, i.e. problems they could cope with themselves with a larger technical staff.

The interveners, in contrast, cover a much wider range of activities. In addition to carrying out the necessary tasks to safeguard themselves under food and drug law, the food technology departments provide in-house services to their stores and warehouses, adopt methods which impose requirements on their suppliers of own label products, and have diversified in new directions. Interveners' food technologists supply technical information for their buying and marketing departments, and in some companies they are even fully integrated into the buying department. Hygiene in stores and warehouses is checked, covering such matters as pest control and sanitation, and staff in stores are trained in correct food handling procedures. With the growing importance of the cold distribution chain (in which foods are handled at between 0 °C and 10 °C to avoid spoilage), tests are made throughout the chain to check that the correct temperature has been maintained.

In addition to these in-house services, interveners work closely with their suppliers to ensure the quality of own label products. Suppliers of own label products are required to carry out regular microbiological tests on production runs and supply results to the retailer. Retailers compare these results with tests on random samples taken from their stores. As well as quality control, these checks can improve standardization and reproducibility and give information on shelf life.

Interveners' food technologists liaise very closely with their suppliers. They spend a great deal of time in their suppliers' factories, providing analytical and technical help when necessary. Their technologists make visits to factories world-wide, which often makes them more knowledgeable about developments in food manufacturing than individual manufacturers. Similarly, factory visits provide the basis on which to assess the performance of available plant, which means they can advise their suppliers on the selection and development of process plant. Interveners use customer complaints as a source of information about failures in quality control and seek to develop procedures to overcome such problems. They carry out competitive testing with branded goods and competitors' own labels and are able to identify recipes from competitive products. They are active in developing new products in their test kitchens. In procuring own label products, they draw up rigorous technical specifications for both manufactured and fresh foods. Specifications vary in detail between companies and for different products, but an indicator of the type of matters that may be included in a specification is as follows:

> '[The specification] will have a description of the product to be provided, e.g. its appearance and taste. It will also have, for manufactured foods, a complete breakdown of ingredients by percentage composition and . . . the source of the ingredients; manufacturing details are likely to be included in the specification, e.g. times and temperatures of heat processes; nutritional composition . . . physical, chemical and/or microbiological standards are likely to be an important part of the specification and, most important, it is likely to include details of the quality control procedures to be carried out by the supplier . . .' (Griffiths, 1984, p. 14).

Interveners have also been responsible for passing on the information they acquire from their membership of research institutes and other sources to their suppliers. In-house technological capability not only affects the methods for procuring own label products, it also affects the choice of supplier. The interveners demand that their suppliers have high hygienic standards, modern production plant, in-house technical expertise and quality control capability. They prefer small firms, so long as they are able to supply the necessary volume, because they are more flexible and more ready to work closely with the retailer. Large companies are used for products which involve trade secrets, or where large volumes are required.

Most significant, however, is the fact that the retailers with large, established food technology departments—the interveners—are the companies which have been most able to implement the new competitive strategy of offering high quality fresh foods and exclusive own label products to their customers. They have been responsible, in collaboration with their suppliers, for developing new methods of distribution and storage and for introducing innovative products to market: frozen and chilled chicken; low fat milk; controlled atmosphere packaged (CAP) meat and gas-flushed fish; myco-protein pies; chilled pizza, chilled recipe meals (Senker, 1988). Retailers who have influenced these innovations all have strong food technology departments, but the benefits of such departments do not appear to accrue until several years after their establishment. Sainsbury, with long-standing food technology expertise related primarily to its manufacturing interests, was able to exploit this expertise for the benefit of its retailing activities. Marks & Spencer set up its food technology department in 1948, but it did not fully reap the benefits of this strategy until the 1960s. The example of Tesco, in particular, shows that even companies who aspire to be 'followers' need to have cumulative expertise. It set up a technology department in 1973 and recruited staff from Sainsbury. But it took several years to build up its technological capabilities and to alter consumer perceptions about the changed quality of its own label brands. However, when the nature of retailing competition changed in the 1980s, Tesco took the opportunity to follow Marks & Spencer's and Sainsbury's example of developing and selling innovative, high quality own label products. By contrast, the 'accepters' could only offer alternatives to branded products.

APPROPRIATING THE BENEFITS OF A TECHNOLOGICAL STRATEGY

Retailers who develop innovative own label products have various methods for appropriating the benefits of this activity. They do not appropriate the benefit of specific innovations directly, because the ready availability of process equipment, the lack of patenting and the ease of copying new food products means that there is only a limited lead-time before their competitors stock the same item. This is exacerbated by retailer dependence on their suppliers for the introduction of the innovation, and the limited power of exclusivity agreements (agreements entered into with suppliers which lay down the length of time which should elapse before an own label specification can be used in competitors' own label products. They are not legally binding because of the ease of copying most food products). However, in an environment where food multiples all sell more or less the same manufacturer brands of food at more or less the same prices, a store can differentiate itself from its competitors and build store reputation and market share by offering exclusive products—either high quality fresh foods, pioneering new products, or own label products of a higher quality than branded goods. The reputation gained from being involved in innovation confers on retailers the opportunity to appropriate benefits by increasing market share of total food sales. Because of the short lead-time enjoyed, the introduction of new own label products must be a *continuing* process.

Retailers have greater opportunities than manufacturers to appropriate the benefits of innovation. Exclusivity agreements and secrecy about new own label products until they appear on the shelves delays imitation time. Manufacturers' response time, unless protected by trade secrets, is eroded from the moment they make their sales pitch to the buying departments of multiples, because samples can be used as a basis for producing own label versions. Unlike manufacturers, many of whom produce commodity-like and relatively unchanging products for the mass market, retailers are able to offer a constantly changing range of foods, obtained from numerous suppliers, and are able to tailor the products to meet the emerging needs of the market segments they seek to attract. This activity is aided by methods of research which give them better information on changing consumer demand than food manufacturers can secure. Both undertake market research, but retailers can monitor what is actually happening in their stores. They can test-market new own label products in selected stores without having to bear the heavy launching and promotional costs incurred by food manufacturers. For all these reasons the leading retailers can not only appropriate the benefits of technological innovation, but can do so in ways that significantly reduce the commercial risks attached to such strategies.

FOOD TECHNOLOGY AND COMPETITIVE ADVANTAGE

Has the application of technological expertise by the 'interveners' (Sainsbury, Marks & Spencer and Tesco) given them competitive advantage? Multiples' shares of the food market can provide some evidence for this, but most assessments are inadequate because they are presented as shares of the packaged grocery market, which only accounts for 35% of grocery outlet turnover. They ignore provisions, fresh fruit and vegetables and omit the food sales made through large mixed businesses, such as Marks & Spencer, which explains the discrepancy between multiples' share of the total food market shown in Table 2, and their share of packaged groceries, mentioned earlier.

Table 2 assesses multiples' market shares by relating their food turnover to the national retail food trade as a whole. This approach has limitations in that the turnover figures for most food multiples includes non-food items, such as alcoholic drinks, sugar and chocolate confectionery or light bulbs. And for some multiples such as Asda and Tesco it includes turnover from clothing and household goods sales. This means that the share of the food market for multiples with large non-food sales tends to be overstated compared wwith multiples who concentrate on food sales or who break down their turnover to show the contribution from food, for instance Marks & Spencer and Sainsbury. These problems demand that the market shares presented in Table 2 be interpreted with caution. None the less, they show that Sainsbury and Tesco are clearly market leaders; the position of Marks & Spencer is more difficult to pin down, until it is recognized that (i) by comparison with the other companies its major activity is not food retailing, and (ii) it does not stock a full range of products. Given this interpretation it does appear that the application of food technology has given the interveners considerable competitive advantage.

Table 2. Retail multiples shares of the national food market, 1982

Company	%
Tesco Stores	8.5
J. Sainsbury	8.3
Asda	4.7
International Stores*	4.3
Fine Fare	4.0
Marks & Spencer	3.4
Argyll	3.1
Safeway	2.1
Kwik-Save	2.1
Waitrose	1.5

* International Stores market share is grossly overstates because it includes turnover from Argos Distributors.
Source: Business Monitor, 1982 Retailing and annual reports.

CONCLUSIONS

Technological strategy can be an important source of competitive advantage. This chapter suggests such strategy must be long term because it depends on *cumulative* technological expertise, and this point should be borne in mind by companies who wish to gain competitive advantage through the application of technology to any activity in the value chain—for instance by employing point of sale or inventory management techniques.

Firms with cumulative in-house computer software and hardware expertise will be in a better position to select, purchase, install and utilize such equipment—and thus gain competitive advantage—than firms which buy turnkey products off the shelf.

Firms must also recognize that corporate strategy should be directed not just at building technological strength, but also at developing methods to minimize the risks and uncertainties connected with adopting a technological dimension to competitive strategy.

REFERENCES

Akehurst, G. (1984). ' "Checkout": the analysis of oligopolistic behaviour in the UK grocery retail market', *The Service Industries Journal*, **4**(2), 189–241.

Beaumont, J.A. (1982). *The Multiple Grocery Trade. A Brief History*, Institute of Grocery Distribution, Watford.

Cranston, M. (1971). *Consumers and the Law*, 2nd edn., Weidenfeld & Nicolson, London.

Dosi, G., Freeman, C., Nelson, R., Silverberg, G., and Soete, L. (1988). *Technical Change and Economic Theory*, Pinter, London.

Galbraith, J. (1963). *American Capitalism*, Penguin, Harmondsworth.

Griffiths, J. (1984). *Future Trends in Retailing*, Paper presented to BAAS Conference, Norwich.

Pavitt, K. (1987). 'The objectives of technology policy', *Science and Public Policy*, **14**(4), 182–8.

Porter, M. (1980). *Competitive Strategy: Techniques for Analysing Industries and Competitors*, The Free Press, New York.

Porter, M. (1983). 'The technological dimension of competitive strategy', *Research on Technological Innovation Management and Policy*, Vol. 1, pp. 1–33, JAI Press Inc.

Porter, M. (1985). 'Technology and competitive advantage', *The Journal of Business Studies*, Winter, pp. 60–78.

Schumpeter, J.A. (1970). *Capitalism, Socialism and Democracy*, 4th edn., Unwin University Books, London.

Senker, J.M. (1988). *A Taste for Innovation: British Supermarkets' Influence on Food Manufacturers*, Horton Publishing, Bradford.

5

Footfalls of the Future: the Emergence of Strategic Frames and Formulae

Ray Loveridge

Aston University

One of the more important dilemmas facing the teacher of strategic management is that confronting any would-be student of social phenomena. It is how to provide a determinate analytical structure on which to base prediction while allowing scope for human agency in interpreting reality and creating options (Halevy 1928). It is, for example, difficult to offer a credible account of the act of invention or of entrepreneurship in terms of the *post-hoc* rationalization of history and comparative static analysis of contextual structure. Equally, prediction gives way to plasticity, process and a tendency to solipsism in more existential explanations of executive behaviour. In much of the recent literature, there has been an attempt to resolve this dilemma through a recognition of historically recurring patterns of behaviour and belief as indicating the existence of a stable, underlying, structure of cognitive perception (Weber, 1924). In this chapter, a reconstruction of the histories of two companies, Lucas Industries, U.K. and Robert Bosch, F.R.G. is used to demonstrate the usefulness of a normatively rational analysis of market behaviour and market boundaries in providing an understanding of the interpretative frames and formulae used by strategists in both shaping and responding to their business context.

The Strategic Management of Technological Innovation. Edited by R. Loveridge and M. Pitt
(c)1990 by John Wiley & Sons Ltd

The rational analysis of strategic frames

The existence of a stable set of personally held values and orientations by which individual behaviour is structured and, therefore, predictable for others, is implicit in most forms of social analysis. For Schultz (1967) as for Weick (1979), the framing process is one of a constant re-enactment and confirmation of common sense classificatory schemata used to handle the intrinsic uncertainty of existence. For others such as Johnson (1984), the recipe has more clearly become an artefact that can be divided into the strategic frame and strategic formula. The first consists of 'beliefs about the organization's external environment and the organization's competitive position within it (1984, p.4). The second are the perceived 'best practices' in relation to strategic modes of operation and trading. They are detectable within the language of explanation used by executives and by their justification for past actions and future aspirations. They are present in the analogies and metaphors used to illustrate these constructs. In many ways, then, they accord with what some observers have described as 'ideologies' (Miles and Snow, 1978), others as 'theories of action' (Argyris and Schon, 1978), and yet others as 'ruling myth' (Hedberg and Jonsson, 1977). Johnson (1984) uses the term 'cognitive paradigm' interchangeably with that of 'frame and formula' (Johnson, 1987).

These are similarities in definition which emphasize continuity and consistency in behaviour and expressions of belief. They suggest that regularity and rationality of organized (though not necessarily bureaucratic) behaviour and of institutionalized sets of expectations within an organized context. Economists and other contextual determinists (structure–conduct–performance), generally assumed that the 'illiberality' or amount of threat present in their shared environment drives chief executives located in different organizations to concur on similar strategic recipes or formulae (Child, 1972). This may occur as a result of the elimination of deviants through natural selection (Smith, 1776; Hannan and Freeman, 1988) or through various modes of contextual learning. Recent analyses seem to have reverted to classical, supply-led, theories of political economy in their descriptions of contest for scarce resources arising from the survival needs of entrepreneurs (Pfeffer and Salancik, 1978). Uncertainty as to the actions of both competitors and collaborators is seen to produce concern for control over a strategic domain extending beyond the formal boundaries of organization and for resources that may prove contingent to the exercise of such control.

In spite of claims to the contrary, the new political economy seems not far removed from earlier contingency approaches to strategic analysis. These placed an emphasis on the ability of any given individual or group to absorb uncertainty central to organizational workflows in explaining their power or relative dependency on others (Hickson *et al.*, 1971). Perhaps the most succinct statement of this older approach to contextual determinism is that of Perrow (1970), who reduced the contextual parameters of organizational activity to two. The first of these was the exceptionality or variation in inputs of information and the second was that of inanalysability or uncodability. Increases along either dimension made the routinization of tasks and, therefore, the remote monitoring of performance, progressively more difficult for senior managers. A similar choice of variables underlies Williamson's later (1975, 1985) analysis of inter- and intra-organizational relationships.

Starting from more explicit assumptions of the nature of the market context than are allowed to contingency theorists, Williamson offers a typology of 'governance structures' based on the executives' desire to reduce the transaction costs (TC) of controlling the supply of an idiosyncratic resource or 'impacted information'. Given this degree of asset specificity, of bounded rationality and of an all-pervasive opportunism, corporate strategists will seek to 'organize transactions so as to economize on bounded rationality whilst simultaneously safeguarding themselves against the hazards of opportunism' (Williamson, 1985, p. 32).

However, in the development of TC analysis by David Teece (1981, 1986), the strategist appears to reverse his or her stance from defender to prospector. Seeking to gain competitive advantage, the executive sets out to create, protect and to appropriate idiosyncratic knowledge represented in technological innovation. Opportunism in the market-place is no longer, as is the case with Williamson, one-sided; purchasers as well as suppliers attempt to appropriate added value. Furthermore, there is a movement away from the over-simplified didactic relationship between a single buyer and seller in a vertical process of added value. In a more openly negotiated context, the function of hierarchy (and, more particularly, the multinational enterprise) becomes that of transferring a bundle of proprietary knowledge from one productive location to another within the boundaries provided by the firms' legal status. Economies of scope are thereby obtained through its utilization in different contexts and for different purposes (Teece, 1981). Synergies between different knowledge disciplines develop, often in a way that is specific to the firm and contained within the tacit knowledge or experiential interaction of individual employees within the socio-technical system of the organization (Shannon and Weaver, 1949; Kay, 1982).

The creation of such asset specificity has been related to the origins of the organization by Rumelt (1987) and by Casson (1982). It is through the differential nature of the competence or know-how of the entrepreneur that he or she gains the sponsorship of a financial institution and access to scarce resources (Schumpeter 1911). In this sense, the entrepreneur becomes an agent of the institutions whose sponsorship and/or recognition is required to gain both capital and custom (Alchian and Demsetz, 1972). If successful in increasing the number and/or strength of demands for his/her services then, entrepreneurial rents should become available for the qualitative development as well as quantitative growth in hierarchical control.

For both Rumelt and Casson, entrepreneurial assets extend beyond the choice of physical plant and equipment to a specific knowledge or know-how within one of four broad categories; (i) product invention (or process innovation that significantly affects the product package),(ii) administrative techniques, (iii) engineering techniques, (iv) agency or networking know-how (or 'know-who', as Casson describes it). This form of knowledge can evidently be transformed into a qualitatively different form from that described by Johnson as constituting the *strategic* formula. For entrepreneurial assets to provide a basis for organizational growth some amount of codification in the form of operational techniques transferable over cohorts of new recruits must be possible.

For ease of transfer and for greater control over the expanding activities of organizations by the entrepreneur it is likely that *operational* formulae will become codified in a more routine manner than is possible in the exercise of the strategic allocation of resources. This is because the strategic frame evolves in an inter-organizational context which is not only a negotiable one in a more formal sense than is normally true of 'internal' transactions but also subject to movements in interdependent networks of suppliers, customers and competitors that lie outside of the control of the entrepreneur (Hayek, 1960). The operational frame is likely to be more limited in scope, though commonalities in shared market position and/or operational knowledge can lead to wider professional and trade identities (Webb and Webb, 1897; Loveridge, 1972).

Empirical studies suggest two characteristics of the long-lasting innovative firm; (i) that its continued success is associated with a cumulative learning in the area in which it gained an early comparative advantage, (ii) that the acquisition of or access to the other areas of competence cited above constitute a necessary set of 'complementary assets' for the continued exploitation of the areas of initial competitive advantage (Teece, 1986; Pavitt, 1988).

The first of these characteristics has been explained by evolutionary economists in the mould of Nelson and Winter (1982) as relating to a process of incremental learning in which success reinforces the continued usage of particular search and design procedures and their foci on similar critical dimensions of the product or process. Development of the firm's performance along these dimensions can be plotted as a 'technological trajectory'. The importance of this trajectory is discovered in the experienced effects that it has on the *ex ante* beliefs possessed by designer/producers about what is feasible and worth trying.

Georghiou *et al.* (1986) see the analogy of a trajectory as relating to one of a matching technological corridor reflecting the role of the user-client needs in determining what is an acceptable form of innovation and, as von Hippel (1982) suggests, playing a major role in the development of complex components or intermediate products. In this manner, both the original client and the producer contribute towards the development of a 'robust' design giving rise either to several generations of derived outputs or to a family of products on similar principles (Rothwell, 1977). In this sense, innovation becomes cognitively 'routinized' in a way that is somewhat different from either the Schumpeterian (1950) view of the standardized administration of R&D and also from the normal use of the term in respect to the imposition of administrative control over operational processes. Essentially it refers to process of personal 'internalization' and the unconscious acceptance of certain problem solving procedures.

The second characteristic of successful innovation requires the availability of complementary assets. This crucially depends upon their codability and the degree to which the code is comprehended on a localized or on a generalized basis (Boisot, 1983). Movement between different functional or operational areas of the firm or interaction with outside contractors can demand an act of information translation between codes. As Winter (1987) points out the transfer of codified information is accomplished only through a capability to comprehend and act upon the information received accompanied by a willingness to do so. Information flows can, for example, be affected by a sense of insecurity throughout the period of trial and error experienced in the development of any truly innovative product or technique particularly under conditions of competitive threat. This can, according to Teece (1986), be related to the strength and mode of protection obtained through what he describes as the 'appropriability regime' provided by the State and/or other sponsoring agencies. Faced with the opportunistic nature of the market-place and the difficulties attending the construction of trusting relations in the world of TC economics, proprietorial rights over the innovation may be required.

In this atmosphere, it should hardly be surprising to discover, empirically, that synergy within the multi-product firm and across collaborative ventures appears to be somewhat hard to create, particularly in the Anglo-American context (Pavitt, 1988; Hitt *et al.*, 1988). The reasons relate to the apparent intra-organization segmentation of knowledge and interests formed within rather than across product divisions and functional specializations. Similar blocks to information flows appear as significant between organizations that have grown by a process of merger acquisition. This causes Pavitt and other observers to distinguish between groups that have grown through acquisition (traders) and those whose knowledge base has evolved incrementally from an original entrepreneurial asset (investors).

It might appear that knowledge assets that have been generated through a process of intra-organizational learning are not easily transactable in the open market. It might appear also that these descriptions apply to strategies that typify national styles and, perhaps, that these strategic responses relate to the national institutional contexts in which executive styles develop. In the following case histories it is hoped to demonstrate the reflexive nature of this process.

THE ENTREPRENEURIAL SHAPING OF FRAMES AND FORMULAE

Lucas Industries is one of the largest British companies, being 74th in its sales turnover in 1986 with a value of £1.62 billions. It is currently divided into three divisions; Automotive, Aerospace and Industrial Products. These products consist of electro-mechanical and electronic controls, fuel injection and combustion systems, and brakes for a range of land, sea and aerospace vehicles and industrial equipment. Central to its survival and growth since the formation of the founding company in 1875 has been its ability to maintain its contribution to a range of finished vehicles whilst moving from one source of generic energy to another; from oil and acetylene gas, through petrol and electricity to chemical combustion and integrated circuits. Over much of this period Lucas has dominated its home and Commonwealth markets and maintained a global position in motor components less only than AC/Delco (General Motors), Bendix (U.S.A.) and Bosch (F.R.G.). In the last decade it has attempted consciously to shift its manufacturing base into the Aerospace industry. It has done so while also attempting a major change in its operational formulae towards more flexible modes of operation.

The company was formed by a silversmith turned electro-plater, turned oil-salesman who, in mid-life purchased the patent on a novel form of marine lamp. Joseph Lucas redesigned the lamp with the help of a sheet-metal worker who was to become his first foreman. The new design enabled its construction in a sequence of specialized tasks before final assembly by low-skilled women operatives. As the business grew the range of vehicle lamps was extended and sold through catalogues and agencies established in the USA, Paris and Italy. The founder's son Harry took the company into the manufacture of bicycle components during the cycling boom of the 1890s and patented an electro-accumulator in 1888. After purchasing his first car in 1899 Harry Lucas formed a small group around himself to design lighting sets and later mechanical starters for motor vehicles.

From an initial workforce of six the number of permanent staff grew to 58 in 1880 with Joseph Lucas's sister supervising women staff and conducting quality control on both products and purchases. The first gas-powered lathes were introduced and by 1891 a purpose-built factory formed the setting for a flow-line moving production through six storeys. By that time it housed 300 employees and by the outbreak of World War One it had been expanded to contain 750 employees. In 1898 the firm changed from a partnership to a quoted company. As the company grew, so did the power of the foremen. However, by the early 1900s work organization took on many features of scientific management including timed piecework and stock control. Much of the Lucas's knowledge derived from personal trips by the chief executive to American factories.

A third generation of the founding family in the person of Oliver Lucas had by now served an apprenticeship in his father's development group. It is at this point that the German firm of Robert Bosch became significant to Lucas's development: at that time its sales were some five times greater than those of Lucas and so it was to remain save for short periods after two world wars.

In 1914, 99 per cent of the supply of magnetos for British motor vehicles and aeroplanes were manufactured in the Stuttgart factory of Robert Bosch. This company had grown from a workshop established by a journeyman mechanic in 1886 to a factory employing 5000 in 1914. At 25 years old Bosch had attended Technical High School and then University for one semester before leaving to find experience by working for Edison in USA and Siemens in Britain. At the request of a customer he invented the first low-voltage magneto ignition and within five years had gone into batch production. Eleven years later by adding his design for spark plugs Bosch began to supply complete ignition sets to the French motor manufacturer de Dion-Bouton, thus providing the significant breakthrough that enabled the development of a high speed mobile engine. By 1900 the system was being fitted to the first airship and later to the first aeroplanes. The first full scale factory at Stuttgart was built in 1901 and in 1909 it was accompanied by a second at Springfield, Ohio and by a repair shop in Yokohama, Japan. By 1914, 88 per cent of the Stuttgart output was exported and the Bosch development team worked closely with auto and aircraft manufacturers in Germany, France and the USA.

From the start the two companies appear to provide natural stereotypes in terms of their strategic formulae. The Lucas's were self-taught Birmingham engineers who developed a facility for adapting and patenting relatively low-tech component products and producing them cheaply in large batches. Bosch, who was to head his company until 1942, was an inventor, who shared with Daimler and Benz and other engineers in the Neckar valley at the end of the last century, an enthusiasm for the modern. Both founders were patriarchs in their treatment of labour and their families. But the language of Joseph and Harry Lucas conveys frugality, self-sufficiency and the need for employer sanctions through payment by results and the loss of half a day's wages for lateness. Until the 1920s Lucas directors would give magic-lantern lectures on subjects such as their Mediterranean holidays for the improvement of the workers. Bosch, like Ford, was a pioneer of high day wages, shorter hours, the six and a half day week and holiday contributions.

The contrast in their operational stance towards labour was not disassociated from their attitudes to suppliers and customers. The Lucas' displayed an early reliance on bank overdrafts and self-finance with a culling of their capital when a public company was floated in 1897. Local metal manufacturers Chamberlain and Wiggin joined the Board; such local engineering connections were to fill outside directorships for nearly a century. Bosch obtained his early finance from the Landerbank and from customers. By 1910 he felt able to present his old university with one million marks for research and in 1916 twenty million marks went to communal works in Stuttgart. When in 1917 he formed a joint stock company the directorship appointments represented the reciprocity that existed between the company and Deutsche Bank and principal customers. The strategic frames of the companies founders were thus shaped by the institutional context in which their founder's assets were recognized and sponsored toward particular developmental trajectories.

The emergence of an articulated frame and formula

In 1915 the British Government appealed to Lucas, one of the few large electrical manufacturers in the country, to manufacture magnetos. After consulting with his scientific consultant, Harry Lucas bought out one of the few British suppliers of magnetos, Thomson Bennett, a firm manufacturing a copy of the original Bosch design. By the end of the war Lucas employed 4000 people, had almost quadrupled its turnover and had expanded its range to aero-engine parts. It had also begun a relationship with Herbert Morris that was to provide the basis for its later growth as the primary supplier of electro-mechanical components to the British car industry.

During the 1920s Lucas derived the concept of supplying complete sets of original equipment for cars beginning with Morris, made up of magneto, dynamo and starter designs adapted from French Hoskiss, American Delco Remy and Bendix and German Bosch concepts. Bicycle components had already reduced to just over a tenth of the level of autosales within Lucas by 1925 and the company set out to acquire innovations through licence. In 1927 a Testing and Development Department was set up to monitor and 'reverse engineer' product improvements.

Over the period of growth and decline in British economic fortunes between the wars Lucas set out to acquire most of its smaller British competitors. Of these by far the most important were CAV, Rotax, Rists and Girling. The first company was taken over in 1926 and by virtue of its collaborative association with Bosch provided Lucas with access to the fuel injection systems for diesel engines invented by the latter in the following year. These were later to provide a vital bridge to gas-turbine technology and to petrol injection for motor vehicles. Rotax was acquired in the same deal and, through the prior acquisition of licences to produce Bendix starter drives and the later acquisition of the British aviation interests of the American company, became Lucas's main access to the aircraft supply chain. Rists too provided both access to the manufacture of wiring harnesses for aircraft and the integrating circuitry required to link all Lucas sub-systems. Girlings, a late acquisition in 1942, represented the local competition for Lucas's brake technology licensed from Bendix.

Throughout the economic depression Lucas 'had money to give away at the end of the year' (*Official History*; Nockolds, Vol. I, 1976, p. 204), indeed in 1929 it provided sufficient credit to keep the Rover company, one of its main customers, in business. It had obtained a virtual monopoly of the supply of lighting, starting and ignition sets as original equipment for motor-cars. It had on-going relations with the Bendix Corporation and Robert Bosch that provided an exchange of licences on new designs which was not entirely one way since Lucas were adept at improving original designs for large scale production. (In this sense they provided the classical 'maturing' influence of followers into a young market.)

In spite of constant pressure from Lucas for more Government protection for British component manufacture the company had secured a virtual global cartel and access to the pool of technical knowledge represented by a diversity of arrangements with principal specialist suppliers in the USA, Germany and France. For the most part these arrangements were personally negotiated by Oliver Lucas who shared the Managing Directorship (then the chief executive post) with Peter Bennett, the former chief executive of the acquired magneto company. Although contracts were agreed in detail by a specialized department their strategic direction was dictated by Oliver Lucas in his prospecting role and by Bennett in his direction of internal operations.

Up to 1936 work organization had drawn its inspiration from annual trips to the U.S.A. by teams of Lucas engineers. Both Oliver Lucas and Robert Bosch were admirers of Henry Ford and his innovations in production engineering. In the case of Bosch Fordist practices were introduced over a long period of expansion and diversification. Like the Bendix Corporation, Bosch chose to diversify more widely in seeking markets for electrical components. The supply of telecommunications equipment provided a main alternative outlet and it was early in the development of television. In all of these acquired or 'spin-off' enterprises

Bosch attempted to establish a participative style of management which was, nevertheless, functionally specialized and based on formal principles. By contrast managerial responsibilities in Lucas remained ambiguous and 'flexible'.

Both companies were faced with a highly fragmented market of car assemblers, and an even more divided chain of complementary component suppliers. But whereas in the German case the coordination of investment took place on the Industrial Boards of the Deutsche, Commerz and Dresden Banks, no such external coordinating mechanism existed for the largely self-financing component suppliers in the U.K. (Scott, 1988). By the 1960s almost all British suppliers had been acquired by engineering groups who exercised little more than a loose budgetary control over their operations. Most often they followed (and follow) a horizontal specialization in one stage of the supply chain with little vertical integration of processes. Hence the ability of Lucas to form common fronts with other domestic suppliers of non-competitive components against their common customers in vehicle assembly was negligible. The latter were used to playing off one supplier against another in the competition for orders. They were unused to entering into long-term collaborative relations in the design of a component or finished vehicle.

Assemblers, particularly Herbert Austin, resented Lucas's attempts to enter such monopoly relations with them. By contrast a major technological provider like Bosch was a highly respected member of their bank's Industrial Board and could use this forum to create a favourable climate for collaboration in their ventures. This was particularly the case when the German Government was preparing for war and provided an additional incentive for collaboration. (Bosch himself appears to have been against the Nazis though his technology provided a crucial part of their military effort.)

Lucas and Bennett nevertheless persisted in their attempts to establish a formally centralized system of control in spite of their uncertain trading environment. The introduction of the Bedeaux System in 1934 had been well prepared by a Work Study Department (established in 1927) but its effect on the work-force was traumatic and led to the first ever strike and parade of 5000 workers through Birmingham. The new system represented a major shift of management responsibility too; as one of the Lucas graduate trainees who was later to become vice-chairman of the Group, Arthur Siddall, described it:

'The technology, instead of being organized on the shopfloor, was decided outside the shop, who determined how things should be made and what should be used for making them. The previous system was based on the shop foremen or superintendents. They were manufacturing management, men of all work— some of them were damn good . . .' (quoted by Nockold, Vol. I, p. 290).

The reinforcement of the frame and formula

It is doubtful whether the Lucas System, as it came to be called, ever worked in practice entirely as it should. It provided one basis upon which the Board sought to monitor the performance of their diverse operating companies up to 1982. The company's management became Government advisers on production engineering during and after World War Two. When peace came 'normality' (to use a word that occurred frequently in interviews) meant resumption of centralized operating procedures and renewed pressure on the remaining British car manufacturers to accept standardized components. Standardization was eventually accepted by the 'Big Six' assemblers under pressure from the post-war Government to rationalize resources, lapsed in the 1950s, was restored on the back of Lucas's offer of an overall price discount of 5 per cent in 1957, but was quickly eroded by the assemblers' adoption of differentiated body styles and equipment in the mid-1960s.

The premature death of Oliver Lucas in 1948 brought an end to the highly individual deals struck with car assemblers. The succession of an accountant, Bertram Waring, as chief executive in 1951, was the first of a sequence of appointments of former personal assistants to Oliver Lucas as his successors as Chief Executive. All of them appear to have earned early recognition from their ability to organize his disjointed and peripatetic existence.

In the sellers' market that prevailed until the 1970s the formula laid down by the former regime was continued. However the formal functional responsibilities from the operating units to the centre were tightened and at the centre authority was divided between an Executive Board, and a Management Board consisting of the M.Ds of each of the sixteen subsidiaries. Investment decisions remained effectively with the Group Chairman. Functional lines of control were relaxed in 1966, but a build-up of central design capability came about through the establishment of a new machine design department and a new R&D facility. In 1974, after further changes in C.E.O., Bernard Scott attempted a further change towards an Executive Board consisting of functional and subsidiary directors and a Main Board consisting of functional and non-executive directors. It was apparent by then that the company's operations were becoming progressively isolated from the centre as directors of subsidiaries were 'too busy' to attend long-term policy planning meetings.

Bosch had also prospered with the astounding growth of its largest customer Volkswagen, and had also had a change in leadership since the death of its founder in 1942. Under the terms of his will the ownership of the company had to pass into the hands of a controlling trust fund to be established within thirty years of his death; dividends were to be spent on communal projects. This transfer of ownership took place in 1964 in fact.

The Foundation was run by seven trustees whose membership reflected the historical relationship between the company and German industrial banks, and more particularly with Deutsche Bank. Like all large German companies its post-war operational decision making had been affected by the dual structure of the Supervisory Board and Works Council. Quite clearly the degree of collaboration between management and labour rested, and rests on a long established spirit of enlightened company paternalism which entered many areas of out-of-work activity.

The Group elected to diversify into household goods in the 1950s and 1960s by modifying its industrial hand tools and producing new ranges of domestic products particularly white goods. Since the early 1960s these moves have derived from a Central Planning and Control Department who produce a three year rolling plan. This is regarded as the accumulation of a dialogue between the, then, 70 profit centres comprising separate divisions of the Groups' global activities. While the sheer communicative activity involved can be seen to have enormous cost implications, both the public and private statements of employees suggest the symbolic power of missionary statements made from the centre in Bosch compared with those of the expanded Lucas Group.

The inadequacy of the formula and displacement of the frame

By the mid-1950s the formal and informal cartel that had grown up in the global market for electro-mechanical components was breaking down. Outdated licences were not renewed across markets whose boundaries were reduced by GATT and, later, by the E.E.C. The Lucas Board became aware of the need for increased R&D and for product diversification. In both respects they failed to produce an effective alternative to prevailing practice.

Perhaps this is most evident in its initial success and subsequent abandonment of inventions in micro-circuitry. Beginning with the purchase of a licence for the production of semi-conductors from Bell Telephone in 1957 the Group formed a new Semi-Conductor Division of the electricals company. By 1961 development by a project team in the Birmingham Centre at the original Lucas Electricals factory had already 'pulled viable silicon crystals' according to participants interviewed by the author and by 1963 the largest batch production of chips in Europe had been set up in a new plant established in the suburbs. Regular visits to the Bell laboratories took place between 1957 and 1964 and Lucas itself patented and granted licences on transistorized devices over that period. Throughout the 1960s work went on with integrated circuits and devices were fitted experimentally to racing cars. British car manufacturers remained reluctant to move into electronically controlled engines.

At length, after Mercedes had adopted Bosch's much later design, the

Rover SDI and Jaguar saloons were fitted with Lucas devices. The results were calamitous for both supplier and user, since the new systems were subject to constant breakdown. The reasons seem to have been many but included the failure of the users to understand and collaborate in the design of the processes. Perhaps more important was the apparent failure of communication and cooperation within the groups responsible for the development of the devices. The early work had been done by a team of development engineers isolated in the suburbs—it was later passed to a combined group consisting of some of the original creators plus reliability engineers brought together at the centre. It seems that the latter, steeped in a traditional approach to quality control, and with little knowledge of solid state physics, were looking for quite inapplicable measures of reliability in tests done on the line while the former had failed utterly to appreciate the operating conditions that their devices had to work under. The newly formed central R&D department seemed to play little role in either the design or implementation of the new invention or in its later diffusion. Its work was largely confined to basic materials research and, at that time, its personnel were divided by academic specializations.

After an expensive initial attempt to automate chip production (at least ten years ahead of its time) and the failure of the British Government to legislate on 'lean burn' engine systems of the kind that had been developed by Lucas, the Board were reluctant to invest further in fundamentally new products. Over the same period a number of marketing disasters had also followed from ventures into medical scanning equipment, heat pumps and electric vehicles. The Board chose to licence out a major invention in ceramic materials, Sylon. Instead they invested heavily in the production of fuel injection and brakes in Germany and France and in the incremental development of these products whose immediate future seemed assured.

In the late 1960s the Group acquired additional aerospace component capacity during the Government inspired rationalization of the industry. The Lucas Board had never seemed successful in establishing any effective dialogue with its existing aerospace businesses. The Board's membership was, until recently, dominated by men who had been promoted from a life-times' experience in one or several of the automotive companies. By background they were immersed in the culture of large batch production engineering and of the technologically stable and relatively predictable forms of automotive products. It was as if successive Boards had acquired its aircraft interests as a result of persuasion and against their perceived interests. In 1942 Lucas had been invited by Rover to design the engine control systems for what became the Gloster/Whittle jet and after the war had been persuaded by men from the Ministry to remain in the gas-turbine supply business. By 1971 the newly formed Lucas Aerospace component company was a major supplier to the British industry and in particular to Rolls-Royce. The bankruptcy of that company in 1974 brought widespread plant closures and revealed the extent of Lucas's dependency on this aerospace sub-assembler.

Widespread redundancies brought into being a well organized shop-stewards movement. Their Proposals for an Alternative Corporate Plan largely reiterated many of Lucas's failed marketing proposals but brought unwelcome publicity to the company at a difficult time. By this time local bargaining with shop stewards throughout the main automotive group of companies also served both to create and to reveal disparities in the operation of the Lucas System (Loveridge et al., 1985). For a time Lucas management attempted to return to their war-time use of Joint Productivity Committees as a form of employee participation. The then Labour Government was supporting the European Community initiative designed to impose the German Works Council System throughout Europe. In the U.K. this initiative disappeared with the fall of the Labour Administration in 1979 and was later rejected by the European Assembly.

The relocation of the frame and formula

In 1981 Lucas made its first ever loss. Over the previous decade it had been losing sales rapidly in the automotive sector as its major remaining British customer had lost its share of the domestic market. Over the same period the output of British component manufacturers had fallen as a whole by a greater amount because of the importation of ready-to-assemble cars by U.K. based foreign assemblers (particularly GM; Vauxhall). In 1982 and 1983 the Aerospace and CAV fuel injection businesses also fell into loss. For a time the Group's Continental plants paid back the investment of the 1970s by helping to finance the redundancy pay-off to 30 000 British employees.

The chosen path to recovery was one suggested by the decentralization of the unwieldy Aerospace Division in the 1970s. At that time each business unit in Aerospace had been asked to choose as its own bench-mark the best of its competitors. It was then expected to achieve an equal performance to that of the latter or one that would keep it in business. This had the effect of refocusing the operational frame of local management and employees away from the centre in both marketing terms and possible modes of collective representation. In the words of the then Chief Executive 'whoever owns the problem manages the problem' (van de Vliet, 1986).

This strategy was now refined and adopted for the Group in 1983 under the title of Competitive Achievement Plan (CAP). Having determined the strategic control formula it was necessary to consider an operational formula. This was provided through the agency of John Parnaby, a former Professor of Production Engineering and advocate of the Japanese system of 'just-in-time' (J.I.T) management. Senior management were evidently aware of the effect that such systems were having on American management thought and indeed some subsidiaries had already experimented with devolved production systems, including the former home of the newly appointed Managing Director, Tony Gill. The appointment of one of Parnaby's former colleagues to a chair at Warwick University was also affecting the direction of change at both remaining major British car assemblers (Austin/Rover and Jaguar). Parnaby was offered a five year appointment on the Executive Board (afterwards renewed) in order to introduce the new system.

He chose to do it through a programme of executive training complemented by a consultancy service provided by an autonomous company; Lucas Systems. Each factory was to provide their own project team to which a systems consultant would be attached. After a pilot study in a small plant the Board actually chose a central Birmingham plant for a major restructuring according to the devolved design of the J.I.T systems. It was a plant with a poor reputation with customers and a record of conflict in industrial relations. Other units sent their emissaries to Japan to see the J.I.T system at work and invested a great deal of energy in bringing about a complete change in their workshops. Well over half of 150 businesses had adopted such systems by 1988. In some cases these units have been among the 35 that were subsequently closed or sold to competitors in spite of their success in increasing their productive performance.

These have all been in what the present Chief Executive, Tony Gill, describes as 'commodities'. They include the lighting business which was, only a decade ago, the location of one of the most significant innovations in process technology and, of course, the root from which the enterprise sprang 120 years ago. Decisions to close or to open businesses rest with the highly simplified divisional structure of boards representing automotives, aerospace and industrial products, with a strong direction given by the main board. Product R&D is seen to be largely located at the business unit where it is intended that it should be 'market led'. The automotives has inherited the former central R&D facility which now operates as an autonomous profit centre. The site on which it stands is progressively given over to problem-solving activity linked to other business units. Aerospace and industrial products are in the process of establishing their own central research facilities.

The cellular system of production or 'mini-factory' of the J.I.T. system entails that work passes between autonomous groups of fifteen or twenty operatives carrying out their own ordering and scheduling. Within cells all jobs are interchangeable with everyone, even the new grade of 'manufacturing craftsman', intended to clean the floor or otherwise facilitate production. The internal system of passing work in panniers or Kan-Bans is directly related to a value-chain extending beyond the factory to the suppliers and customers. Hence the ability of each cell to complete its tasks depends on the arrival and despatch of the product 'just-in-time'. The success of the internal system therefore depends on the certainty of delivery and quality within the whole value chain.

There was little to assure Lucas and other British sub-assemblers and assemblers of this certainty in the previous history of either the automotive or aerospace supply chains. Yet at the time of adoption the risk of J.I.T. was clearly discounted against the temporary power of employers to reshape the labour market in the early part of the decade. Lucas, like other British employers, has taken the opportunity to introduce recruitment techniques aimed to produce more committed employees, to restructure grading systems in order to promote task/skill flexibilities and has unilaterally abrogated centralized procedures for bargaining with unions. This has not been accomplished without conflict, particularly among skilled groups of employees in aerospace.

It cannot be said that either the labour or product markets are such as to guarantee the success of the system first developed by Toyota in a very different context. Since the mid-1980s the immigration of foreign assemblers has been accompanied by a return to the highest ever volume of vehicle sales in the U.K. that had been reached over a decade previously. Over the early 1980s the Lucas Board determined to move away from dependency on three major customers; Austin/Rover in cars; Massey Ferguson in commercial engines and Rolls-Royce in aero-engines. It was a time when these customers were attempting to restructure their own internal organization and to reform their relations with interdependent suppliers. Lucas Automotive was successful in gaining orders from the burgeoning Peugeot/Citroen group and from the newly arrived Nissan Company. The Group's main strategic thrust has, however been to acquire a number of aerospace suppliers in the U.S.A. through short-term loans and new bond issues that took advantage of the favourable exchange rates to gain highly idiosyncratic assets at bargain prices. In this way it has taken advantage of its known reputation in the after-sales market for Rolls-Royce and British Aerospace products to acquire a place as principal agents to American assemblers.

For Lucas, as for other specialist suppliers such as Bosch, the movement into CAD/CAM and other forms of integrated information technology by their major customers presents evident investment problems. In the absence of agreement on computer protocols and compatible software many final assemblers have adopted their own systems. This had meant that for contractors the value of their technological capability has to be discounted against its technical incompatibility with that of their customers and own suppliers. Lucas elected to pursue a modest investment in integrated manufacture (C.I.M.), focusing instead on the social-reorganization of production units. (Even so an experimental automated factory had been in operation on a New Town site for several years before closure in 1989.) Its policy has fuelled the resentment of some customers at Lucas's refusal to match their investment in networked CAD/CAM systems and at their suppliers growing independence from them as a result of supplying domestically based foreign competitors (see Clark and DeBresson in this volume). This resentment sometimes surfaces as it did when Lucas implied publicly that it supplied Nissan with superior components to those sold to Austin/Rover. It is also present in the recent complaints of assemblers to the Parliamentary Select Committee. Better collaboration along the supply chain may however be emerging from expensive mistakes made by assemblers in C.I.M. Rolls-Royce Engines, for example, has recently employed Lucas Systems as one of several consultants in reorganizing its manufacturing processes.

Bosch has also moved towards cellular manufacturing systems and has adopted technically advanced modes of C.I.M. much more widely than have Lucas auto plants. Its long-term R&D relationships with the major German manufacturers provides a more secure base upon which to undertake such investment. On the other hand the implicit boundaries drawn around its product market through its relationship with Siemens over the last century are now being challenged by the latter's aggressive marketing of its fully electronic vehicle control systems.

Thus competition between specialist suppliers becomes more real. Bosch has chosen to launch a new product through its first British production facility in order to compete with Lucas for the business of the Japanese immigrants to Europe. But for all specialist component companies the emergence of joint ventures and collaboration between their customers may actually challenge their existence as suppliers. In recent times partnerships between Japanese and American or British assemblers have brought with them the threat that the former will insist on factoring the agreement through its own Keiretsu network. For the most part Japanese assemblers have created or helped to establish native suppliers joined to them through common share ownership and collaborative 'clubs'. The use of dedicated IT networks within such Keiretsu have served to strengthen an already existing personal and proprietorial relationship. At the same time suppliers such as Nippon Denso, the electronics components supplier to Toyota, have become world leaders as a result of their ability to sell outside of their own group.

The Japanization of Lucas—transformation or transition?

In many respects, the actions initiated by the Lucas Board over the 1980s appear to exemplify the changes in strategy and operational paradigms said to have taken place among Western industrialists. These changes have been described in various ways, but all add up to a cognitive transformation in the manner in which markets, technologies and therefore, organizational structures and styles are perceived by senior executives and operational management. In most holistic explanations, the change is seen to be operationally led and to involve a movement from the highly centralized and functionally specialized bureaucracy represented in the Fordist model to the devolved, integrated modes of flexible specialization (Sabel, 1982; Perez, 1983). But in the breakdown of institutional boundaries that has accompanied a change in generic technologies, the long-term investment of resources is seen to demand much more risk-spreading through joint-ventures and alliances across industrial networks and vertically within supply chains. The 'firm' has therefore to acquire the flexibility of an agent mobilizing resources along dynamic networks (Miles and Snow, 1986).

In describing these events, the author has been aware that, over the latter part of the decade, senior management have been submitted to programmed learning in these concepts as well as to the advice of academic consultants couched in a similar discourse. To the extent that changes at Lucas and Bosch have been presented in interviews in these ideological terms, it becomes increasingly difficult to distinguish between 'expounded theory' and' 'theory in use'. A comparison with historical periods of contextual change in the company's history provides a useful control.

Quite clearly, the influence of the Japanese, or more particularly the Toyota exemplar has acted as a prototype for workplace changes at both Lucas and Bosch (see Turnbull, 1986; Vliet, 1986 for fuller accounts). The role of a 'frame-breaking' agent in bringing about organizational transformations has been emphasized by many academic observers (for example, Johnson, 1984; Tushman and Anderson, 1986). In the case of the Lucas Board, the crisis of confidence that led to the engagement of an outside expert, Professor Parnaby, was accompanied by a conflict over succession at chief executive level and the short-lived emergence of two managing directors. These events have almost direct parallels in the change in strategic direction signified by the appointment of Bennett as joint managing director in 1921 (Nockolds, 1976). The emergence of the CAP disaggregation of group interests, which provided the strategic frame within which to implement the workplace changes, was conceived out of the experiences of the Group Board in devolving control over their acquisitions in the Aerospace Division during the 1970s. Above all, the language in which both changes were brought forward was one already familiar to a Board schooled in the Lucas tradition of 'total systems management' and cost appraisal.

It seems difficult to overestimate the degree to which senior management came to use the language of scientific management as a legitimating ideology over a half a century. (Interviews conducted by the author testify to this as well as official records.) It was crucial to the rapid diffusion of radical change in the 1980s that its formula was couched within the same conceptual paradigm and familiar discourse of 'total systems' engineering, however far the rules of application have been turned on their head. (Within the formal model propagated by Lucas Systems 'the market' is still chiefly represented only by the establishment of competitor bench marking.)

In terms of the contextual imperatives that actually shaped operational frames at the level of business unit over the period from the late 1920s to the late 1970s, there can be little doubt that a significant gap prevailed between the local practices and those implicit within the procedures laid down centrally. The type of performance measures used and the systems employed for data collection disguised or subsumed large differences in local customs. Businesses acquired over the period retained quite different socio-technical systems geared to the wide variety of processes and markets represented by the 'auto-components industry', as well as a similar range of conditions in 'aerospace components'.

In some ways, the policy of disaggregation adopted in the 1980s by Lucas, although not by Bosch, might be regarded as a belated recognition of the 'objective' conditions within which it has been operating for over a century. The new Group structure could be seen to resemble a well established model in both the auto and aerospace industries. The evolution of motor vehicle components supply in the U.K. is much better represented by the loose holding structure of major engineering firms like GKN, IMI, Auto-Engineering, etc. These groups never *devolved* operational responsibility to their business units; for the most part, they left it in the hands of the local management at the time of acquisition (Loveridge, 1982). Lucas stands out in the business environment of West Midland engineering in their attempt to set and to monitor operating standards centrally. (Other examples are found in the more integrated process-based companies such as Cadbury, Dunlop and Triplex (Pilkington).)

Unlike most other British component suppliers, Lucas executives saw themselves as being able to negotiate on equal terms with vehicle assemblers. They therefore believed that they could maintain a stability of demand that could be matched by the routinized production of standard products. While other engineering groups spread their risks over a number of small businesses which often competed for the assemblers' business, Lucas maintained a strategic check on all market relationships from the centre (see Peters and Waterman, 1982 on loose/tight links). Like Bosch, the architects of the Lucas strategy believed they could help to bring about a rationalization of the demands of their domestic vehicle producers for differentiated components. For a time in the 1950s and early 1960s they succeeded.

This at a period when a high mass market was emerging for domestic machines in Europe as a whole. High profits justified the Lucas traditional policy of high dividend distributions. For Bosch it was an opportunity to gain economies of both scope, from its R&D asset, and scale, from its production of small motors, by expanding its direct access to consumer markets. In doing so it gained leverage in the chain of dependencies by which it was linked to car assemblers. But in general the segmented nature of vehicle markets (both cars and aeroplanes) and the differentiated niche strategies of assemblers made it difficult to achieve the Fordist dream that had been shared by both Oliver Lucas and Robert Bosch.

Ideals and imperatives

When Oliver Lucas and Arthur Bennett assumed the leadership of the firm in 1921, they inherited a tradition built on the adaptation or reverse engineering of others' designs, in the tradition of Birmingham engineering. Their standard of production engineering was however significantly higher than that of their local competitors. Lucas was already outstandingly different from other Midlands engineering firms both in size and extent to which their design expertise was based on a relatively esoteric body of knowledge. The recognition accorded them by H.M. Government during the First World War was to extend to their increased involvement in aircraft production. At that time, this national recognition was largely based on their ability to adopt and to adapt the Bosch magneto designs, under Bennett's guidance, at a time of national crisis. It is perhaps not surprising that the companies chosen as their exemplars by Bennett and Oliver Lucas, were both headed by executives who were personally leaders in their field of invention as well as skilled in mass production, namely Bendix and Bosch. But in doing so they also aspired to an operational frame and formula that was quite different from that which existed within their own operating context.

Up to the 1960s, the Lucas Board were justified in seeing their organization as part of an elite group of electrical sub-assembly manufacturers which dominated the world production of complex machines from motor-cars to power plants and included the burgeoning markets for domestic goods. It was a world in which cartels allowed the exchange of new inventions through licences to operate within protected markets considered by each participant to be too expensive to enter. These companies were quite unlike the sub-contractors described in either the Fordist (regulation) model of Aglietta (1975) or that of the TC model of Williamson (1975). Both approaches discount the enormous historical strength of sub-assembly manufacturers and their contribution to invention and innovation (Rosenberg, 1979).

The extent to which final users may be unable, or regard it as uneconomic, to resolve their dependency on expert supply chains through vertical integration is rarely acknowledged (Pavitt, 1988). In the case of the motor industry, large specialist suppliers continue to exist alongside the internally integrated supply chains of Ford, GM, Fiat and the major Japanese Keiretzu groups. The latter depend on external expertise in resolving the multi-disciplinary problems presented by the design of technically complex sub-assemblies providing this expertise remains at least competitive with that which has been achieved, so far, within internal relationships. According to the public statements of executives within the latter groups the existence of external competition at all levels of the supply chain (and not simply in other monolithic competitors) is considered a necessary part of their mode of maintaining control within their *internal* hierarchy. One might feel that this indicated a certain complementarity between 'markets' and hierarchies within the strategic frames and formulae of these executives?

The exemplars adopted by Lucas and Bennett in the 1920s were companies such as Bosch and Bendix founded and led by an inventor possessing academic training and an ability to bridge the conceptual gap between research and its implementation within their personal understanding and style of action. The route taken in the expansion of their organizational activity by these exemplary companies was dictated by the incremental development of core inventions (the magneto, spark-plug, fuel injection pump and telephone exchange by Bosch; the carburettor and generator by Bendix).

Lucas achieved its early success through the adaptation of others' inventions and the acquisition of firms such as Thompson-Bennett, CAV and Girling. Even today, its most successful products such as Girling anti-locking brakes have been developed through locally discovered changes in design involving tacit skills developed in the workshop. While Heads of R&D have occupied Board positions in Lucas since the 1960s and have appeared in both the official history and in the author's interviews as figures of mythical significance, their role seems somewhat ambiguous. They appear as heroes but only rarely as champions of innovations which have had strategic impact on the course of the company's development. Lucas continues to spend more than most U.K firms on R&D (on average, 8 per cent of sales value per annum), but the degree to which central initiatives have historically led to significant in-house innovation seem relatively small.

To a large extent, this seems to have related to the dominance of production engineering in local and corporate discourse. The incident involving development and production engineers in a confrontation over the implementation of micro-circuitry production is, perhaps, an archetype of an inter-disciplinary failure to communicate. But, equally, the imperatives of the situation might be seen to have been seen as the customers' requirements for a cheap, reliable system delivered to their specifications in batches to be determined by the extremely variable needs of the ultimate consumer. In this environment, necessity was indeed often the mother of invention. Even while Bennett was attempting to remove 'conceptive' control from the foreman's activity through the introduction of a routinized production system the need for constant adaptation often made these formal arrangements dysfunctional. This malfunctioning was multiplied by layers of administration coordinative mechanisms. In the mid-1970s, having weathered the immediate crises that led to the nationalization of British Leyland and Rolls-Royce (aero-engines), Lucas business unit executives were still 'too busy' to attend central strategy meetings because of the turbulence present within their relations with others in their particular supply chains.

Sponsorship and recognition

Over the post-war period, the frame and formula established between the wars was carried on by the dynastic protégés, the former PAs of Oliver Lucas. They had helped to establish a policy of financial rectitude which included a high level of self-financed investment and generously distributed profits. External directorships continued to be dominated by local executives from other Midlands firms whose financial policies were very similar. While Lucas's replacement and servicing sales expanded beyond the markets provided by British assemblers, its dependence on the latter diminished very little in the overall growth in trade. These were relationships sustained by professional purchasing agents and engineers, but still resting on highly personalized relationships with senior executives. These relationships with the motor industry provided the main comparators for the Board, probably to the continued detriment of their position in the aerospace market.

Their status in aerospace had to a surprising extent been created by the recognition and by the blandishments offered by H.M. Government. To a large degree this stemmed from their long established relationships with Government buying departments and the preference of the latter for stable partnerships. These user/designer relations have evidently been more successful in creating the long-term trust required in the design of complex sub-assemblies than those existing in the motor industry. For this reason, the otherwise conservative Board have seemed able to risk much greater investments in single projects than it has in motor vehicles where its relations with both Government and assemblers has been extremely variable. The Government's sponsorship has also been more apparent in suggesting entry into new markets. Moreover in two wars and under the post-war Labour Administration, Lucas became the model for productivity campaigns sponsored by H.M. Government and assisted Government departments in educational campaigns in production management. One of the major techniques employed were trips by British businessmen to exemplar American companies chosen by Lucas engineers (today Mecca has moved to Japan).

Until recently, the banks do not appear to have played any long-term role in Lucas's strategy except as a continuous source of short-term finance. Under the more adventurous financial strategy of leveraged borrowings adopted in the late 1980s, Lucas has created a treasury management strategy based on a global finance market. It remains a radically different strategy from the one adopted by Bosch from its origins. Like most 'modern' firms in Germany, Bosch grew rapidly in the early years of the century with the help of liquidity supplied by the regional banks, through orders secured both through its foreign subsidiaries and through the build-up of military expenditure at home. The risks taken in, for example, building American and Japanese plants, were covered both by the security provided by the banks and the unique value of an invention protected world-wide by patent and exploited extensively through licences. The continued cross-linkages created by the appointment of engineers from major electrical engineering firms as bank directors helped to ensure the coordinated development of this specialized sector as a whole. In determining the paths along which to diversify the assets created by his inventions, Robert Bosch was able to exercise far more control over the firms' inter-organizational context than has ever been available through the institutionalized arenas in which Lucas has negotiated.

The current attempt by the Lucas Board to reshape its identity depends crucially on a reputation already achieved in key niche markets for high value-added motor components and in its high after-sales servicing of aero-engines in the U.S.A. There has been a determined effort to shake off its previous nineteenth-century identity as 'King of the Road'—the manufacturer of cycle lamps. In this endeavour it has secured recognition in joint ventures with several leading Japanese and American companies and the sponsorship of foreign as well as domestic sources of finance. There seems a strong possibility that the structure that emerges will provide greater implicit control from the centre than the former centralized system of formal administration. This is likely to derive from its reputational power in raising capital from a world-wide finance market and feeding it to increasingly autonomous subsidiaries. Meanwhile, the reality of this autonomy is being tested by increased movement of these latter business units into new and hitherto untouched areas of supply to consumer durable goods. A continuation of this movement might avail the Group of economies of scope so far unrealized and neglected, possibly owing to its lack of marketing expertise at the centre. In doing so, it should be assisted rather than hindered by the very rapid erosion of market boundaries across all of the formerly segmented specialist suppliers of electrical mechanical (now mechatronic) components to assemblers of complex machines. In this situation, the formerly nationally specific orchestrators of industrial change, such as Deutsche Bank, appear to be losing their place on the rostrum.

THE SOCIAL EMBEDDEDNESS OF INFORMATION EXCHANGE

It is clear that the entrepreneurial investments spoken of by structural theorists are as much cognitive or psychological as financial and material. The founding of a new organization is often fraught with risk for those involved. The imprint of these early experiences and of responses to the contingencies of its birth are often to be found into maturity (Stinchcombe, 1965). Similarly, the routines of design and production that are central to the analysis of evolutionary economists are perceived by them to be internalized precepts of action rather than externally imposed administrative regulations. There is, then, a convergence among strategic analysts on what Rumelt (1979) describes as 'framing' theories of management behaviour (although the epistemological points of departure and methodological approaches among social scientists differ quite widely).

Both interpretative and positivistic theorists see managers as using heuristics or 'rules of thumb' to simplify a complex reality (Schenk, 1988). Indeed, both Johnson (1984, 1987) and Spender (1980 suggest that there is a high degree of concurrence between what they variously describe as the 'paradigm' or as the 'formula' used by boards of directors within industries, that is, by those sharing contingently similar contexts. Much the same assumption is made by Child and Smith in this volume, when they suggest that sectoral frames of reference develop around a 'cognitive arena', within which there are shared transactional networks. Material and social boundaries to action can be considered as 'objective conditions' within any given period or state of knowledge. Shearman and Burrell (1988) suggest that this concurrence is part of a learning process across a set of firms that come to be recognized as constituting an 'industry' by other sponsoring or regulatory institutions within the economy. The means of gaining recognition for this collective state and for protecting its boundaries against potential contenders seems, therefore, to be part of the strategic problematic and not an operational given.

There are, however considerable commonalities in the use of terms such as 'exemplars' and 'comparators' in the work of framing theorists. These suggest that concrete references or sets of other players develop a strategic significance in the minds of chief executives. As Merton and Rossi (1957) and more recently, Urry (1973) point out, the reflexive nature of face-to-face interactions within some groups make their importance quite different from references having only abstract significance. Experiential references can have a more immediate impact on the formulae or modes of response adopted by the actor in that the social meaning of this information can be most evidently negotiated within personal interactions with others. But as Urry (1973) suggests, the reference orientations of any individual are likely to be multiple, temporally related to on-going events and historically dependent one on another. While this latter research cautions against the extraction of elements within the cognitive frame, it is clear that the choice of external referents and their inter-relationship one with the other can display an internalized ideology or 'theory in use' (Argyris and Schon, 1978). The executive can generally be seen to believe that his/her organization has a legitimate right to claim the support and recognition of others and that (s)he is justified in adopting certain paths to achieve those ends. For this reason, it is likely that some players in the 'strategic set' will take on the function of legitimizing exemplars and others of justifying comparators.

But significant audiences will also include those who offer more abstract qualities of sponsorship and recognition in the struggle to appropriate or to create unique assets, as well as those that exercise immediate influence on their day-to-day usage. The difficulty for the researcher is evidently that of attributing value to distal references such as 'what might stockholders think' in constructing a view of the strategist's frame. As Argyris and Schon (1978) remind us, ideological statements can reflect both instrumentally expounded statements of belief (such as Annual Reports) as well as the internalized values that actually drive executive action.

The assets themselves are contained in the substantive procedures by which the organization adds value for its market. By and large, these are located at the operational level of the organization and are often insulated from the longer-term allocative or strategic decision-making processes. Many of the tensions within organizational life derive from the isolation of the strategic frame from that used in the performance of operational tasks. It is usually the case that the reference groups used by strategists will differ in kind and significance from those used by operatives or by operations management.

This tension may be taken as one instance of formally codified information being perceived as having more or less value in different situations according to the perceived need for its applications and the perceived capability of the actors to make use of it. Boisot (1982) sees this evaluation as taking place within a 'cultural space'. In general, the process by which meaning (and therefore value) is attached to information by reference to an ongoing situational knowledge has been ignored by economists and other 'rational' theorists.

This is a significant omission in their analysis for many reasons; three will suffice. Firstly, tacit or situationally specific knowledge is almost always required for the delivery of a service, however codified the substantive elements of a task. Secondly, the basis of market contests lies in a threat to survival that induces an equal and opposite tendency to appropriate and close off opportunities at all levels of internal as well as external markets Williamson's concept of 'atmosphere' offers little insight into the negotiated process of social integration represented in 'relational transactions'. Thirdly, the advance of knowledge within industrialized societies had led to the increased use of specialized and esoteric codes. Hence, control over expert groups has increasingly depended on the managerial monitoring and assessment of *outputs* rather than a knowledge of *inputs* to the value-added process. Coordination has become increasingly difficult in the face of a horizontal tension that exists between specialized corpuses of 'expert' knowledge. Buttressed by language and discourse these separate frames cannot easily be overcome at any one cognitive level of interaction (Bernstein, 1975; Peters and Waterman, 1982).

A crisis that engages all players at several levels of cognition in a manner that is overlaid with emotion is likely to create the conditions under which both frames and formulae are reconfigured (see Pitt in this volume). The potential threat contained in a change in competence-destroying technologies of a generic nature such as micro-electronics may create the conditions under which such a cognitive restructuring can occur (Tushman and Anderson, 1986) but only when it is recognized and handled strategically. The economic crises experienced in many industries over the last two decade have been accompanied by apparent changes in both the strategic configurations perceived by leaders in their sectors and in their prevailing operational modes. While there has been no lack of academic prognostication, much seems to be based on the assumption of a universal and holistic transformation of frames and formulae (Aglietta, 1979; sDosi, 1982; Abernathy *et al.*, 1983).

These explanations tend to subsume the complex learning process by which changes in operational formulae acquire strategic significance and lead to a reconstruction of the executive's from of reference.

CONCLUSIONS

The mode of analysis proposed by this study is one in which the strategist is seen as acting within a recognizable interpretative logic. But it is one that must be approached through a contextual explanations of strategy. This approach might well be described as postulating a mediated rationality in which a collectively shared view of the rational nature of intra- and inter-organizational relations is mediated by the personal frame of reference of the Chief Executive. This frame imparts meaning to the processes and outcomes of an on-going stream of transactions in a way that can be captured only imperfectly by broad economistic phrases and terms such as 'investment' and 'asset', and then only by identifying the nature of the latter for the actor. Instead of reverting to ideographic evidence alone, the author has sought patterns in the recorded history of institutions in order to establish the existence of important reference groups used as normative exemplars and comparators by strategists. Crucial to the internalization of these references by the main actors has been their endorsement by audiences responsible for sponsoring and recognizing the formation of institutional boundaries and their subsequent maintenance. Here the structural values and ideologies present in different patterns of national regulation might also be seen to shape the events and metamorphoses that represent the respective histories of Bosch and Lucas.

In this evolving stream the significance of invention and innovation can be seen to have widely different meanings for organizational development. Equally the stage of the development of organization gives greater or lesser significance to these disjunctures in knowledge. In the early days of the firm, the establishment of a collective identity built around the creation, protection and exploitation of a unique 'asset' can prevail into the period of bureaucratization that accompanies quantitative growth (Child and Kieser, 1976). The history of Bosch and Lucas and of other members of their former global strategic set, suggest that this focus taken together with the extent and manner in which the entrepreneur choses and acquires complementary 'assets' is quite crucial to the later technological performance of the organization.

REFERENCES

Abernathy, W.J., Clarke, K.B., and Kantrow, A.M. (1983). *Industrial Renaissance*, Basic Books, New York.
Abravanel, H. (1983). 'Mediating myth in the service of organisational ideology', in L.R. Pondy *et al*. (eds), *Organizational Symbolism*, JAI Press, New York.
Aglietta, M. (1975; 1979 edition). *A Theory of Capitalist Regulation*, New Left Books, London.
Alchian, A.A., and Demsetz, H. (1972). 'Production, information costs and economic organizations', *American Economic Review*, **62**, December, 777–95.
Argyris, C., and Schon, D.A. (1978). *Organizational Learning: A Theory of Action Perspective*, Addison-Wesley, Reading, Massachusetts.
Barnett, C. (1986). *The Audit of War*, Macmillan, London.
Bernstein, B. (1982). *Class, Codes and Control*, Routledge & Kegan Paul.
Boisot, M. (1983). 'Convergence revisited: the codification and diffusion of knowledge in a British and Japanese firm', *Journal of Management Studies*, **1**, 159–90.
Casson, M. (1982). *The Entrepreneur: an Economic Theory*, Martin Robertson, Oxford.

Child, J. (1972). 'Organization structure, environment and performance: the role of strategic choice', *Sociology*, **6**, 1, 1–22.

Child, J., and Kieser, A. (1976). 'The development of organisations over time', in W.H. Starbuck (ed.), *Handbook of Organizational Design*, Vol. II, Elsevier, Amsterdam.

Child, J., and Smith, C. (1987). 'The context and process of organizational transformation—cadbury Limited in its sector', *Journal of Management Studies*, **24**, 6, November.

Dosi, G. (1982). 'Technical paradigms and technological trajectories—a suggested interpretation of the determinants and directions of technical change', *Research Policy*, **11**, 3.

Geoghiou, L., Metcalfe, J.A., Gibbons, M., Ray, T., and Evans, J. (1986). *Post Innovative Performance*, Macmillan, London.

Greeman, C., Clark, J., and Soete, L. (1982). *Unemployment and Technological Innovation*, Pinter, London.

Halevy, (1928; 1953 edition). *The Growth of Philosophical Radicalism*, Faber, London.

Hannan, M.T., and Freeman, J. (1988). 'Density dependence in the growth of organizational population', in G.R. Carroll, *Ecological Models of Organizations*, Ballinger, Cambridge, Mass.

Hayek, F.A. (1960). *The Constitution of Liberty*, Routledge & Kegan Paul, London.

Hedberg, B., and Jonsson, S. (1977). 'Strategy making as a discontinous process', *International Studies of Management and Organization*, **VII**, 89–109.

Hickson, D.J., Hinings, R., Lee, C.A., Sneck, R.E., and Pennings, J.H. (1971). 'A strategic contingencies theory of intra-organizational power', *Administrative Science Quarterly*, **16**(2), June, 216–29.

Hitt, M.A., Ireland, R.D., and Goryunov, I.Y. (1988). 'The context of innovation: investment in R&D and firm performance', in U.E. Gattiker, and L. Larwood, *Managing Technological Development*, de Gruyter, Berlin.

Johnson, G. (1984). 'Managing strategic change—a frames and formulae approach', presented to the Strategic Management Society Conference, Philadelphia, USA, October.

Johnson, G. (1987). *Strategic Change and the Management Process*, Basil Blackwell, Oxford.

Kay, M. (1982). *The Evolving Firm: Strategy and Structure in Industrial Organization*, Macmillan, London.

Loveridge, R. (1972). 'Occupational change and the development of interest groups among white collar workers in the UK: a long term model', *British Journal of Industrial Relations*, **X**, 3, 340–65.

Loveridge, R. (1980). 'Strategic frames and formulae in decision making events', *British Academy of Management Workshop*, University of Bradford, 5 January.

Loveridge, R. (1982). 'Business strategy and community culture', in P. Dunkerley, and G. Salaman (eds), *The International Yearbook of Organizational Studies*, Routledge & Kegan Paul, London.

Loveridge, R., Lloyd, P., and Broad, G. (1985). Workplace control and co-determination, *Department of Employment Monograph Series*, No.30, London.

Merton, R.K., and Kitt, A.S. (1950). 'Contributions to the theory of reference group behaviour', in R.K. Merton, and P.F. Lazarfeld (eds), *Continuities in Social Research*, Free Press, Glencoe.

Miles, R., and Snow, C. (1978). *Organizational Strategy, Structure and Process*, McGraw-Hill, New York.

Miles, R., and Snow, C. (1986). 'Organizations: new concepts for new forms', *California Management Review*, **28**(3), 62–73.

Mowery, D.C. (1987). *Alliance Politics and Economics*, Ballinger, Cambridge, Mass.

Nelson, R., and Winter, S. (1982). *An Evolutionary Theory of Economic Change*, Harvard University Press, Cambridge, Mass.

Nockolds, H. (1976). *Lucas: the First Hundred Years* (2 vols), David and Charles, Newton Abbot.

Patel, P., and Pavitt, K. (1987). 'The technological activities of the UK: a fresh look', in A. Silberston (ed.), *Technology and Economic Progress*, Macmillan, London.

Pavitt, K. (1988). 'Strategic management in the innovating firm', *ESRC Designated Research Centre Discussion Paper No. 61*, Science Policy Research Unit, University of Sussex.

Pavitt, K., Robson, M., and Townsend, J. (1989). 'Technological accumulation, diversification and organization in UK companies, 1945–1983', *Management Science*, **35**, 1, January, 81–99.

Perez, C. (1983). 'Structural change and assimilation of new technologies in the social economic and social systems', *Futures*, October, 357–75.

Perrow, C. (1970). *Organizational Analysis—a Sociological View*, Tavistock, London.

Peters, T.J., and Waterman, R.H. (Jr) (1982). *In Search of Excellence*, Harper & Row, New York.

Pettigrew, A. (1987). *The Awakening Giant—Continuity and Change in ICI*, Blackwell, Oxford.

Pfeffer, J., and Salancik, G.R. (1978). *The External Control of Organizations: a resource dependence perspective*, Harper and Row, New York.

Rosenberg, N. (1979). 'Technological Interdependence in the American Economy', *Technology and Culture*, **20**, 28 September.

Rothwell, R. (1977). 'The characteristics of successful Innovators and Technically Progressive Firms', *R&D Management*, **7**, 191–206.

Rumelt, R.P. (1979). 'Evaluation of strategy: theory and models', in D.E. Schendel, and C.W. Hofer (eds), *Strategic Management*, Little, Brown, Boston.

Rumelt, R.P. (1987). 'Theory, strategy and entrepreneurship', in D.J. Teece (ed.), *The Competitive Challenge*, Ballinger, Cambridge, Mass.

Sabel, C.F. (1982). *Work and Politics: the Division of Labour in Industry*, Cambridge University Press, Cambridge.

Schultz, A. (1967). *The Phenomenology of the Social World*, North Western University Press, Evanston.

Schumpeter, J. (1911) (1934 edition). *Theory of Economic Development*, Oxford University Press, Oxford.

Schumpeter, J.A. (1950). *Capitalism, Socialism and Democracy*, Harper & Row, New York.

Schwenk, C.R. (1988). 'The cognitive perspective on strategic decision making', *Journal of Management Studies*, **25**, January.

Scott, J. (1988). 'Intercorporate structures in Western Europe: a comparative historical analysis', in M.S. Mizruchi and M. Shwartz (eds), *Intercorporate Relations—the Structural Analysis of Business*, Cambridge University Press, Cambridge.

Shannon, C.E., and Weaver, W. (1949). *The Mathematical Theory of Communication*, University of Illinois Press, Chicago.

Sharp, M. (1989). 'Corporate strategies and collaboration—the case of ESPRIT and European electronics', in M. Dodgson (ed,), *Technology Strategy and the Firm: Management and Public Policy*, Longman, London.

Shearman, C., and Burrell, G. (1988). 'The structures of industrial development', *Journal of Management Studies*, **24**, 4, July, 325–45.

Smith, A. (1776) (1965 edition). *An Inquiry into the Wealth of the Nature and Courses of the Wealth of Nations*, Modern Library, New York.

Spender, J-C. (1980). 'Strategy making in business', unpublished Doctoral Thesis, School of Business, Manchester University, U.K.

Stinchcombe, A.L. (1965). 'Social structure and organizations', in J.G. March (ed.), *Handbook of Organizations*, Rand McNally, Chicago.

Teece, D.J. (1981). 'The market for know-how and the efficient international transfer of technology', *The Annals of the Amercian Academy of Political and Social Science*.

Teece, D.J. (1986). 'Profiting from technological innovation: implications for integration, collaboration, licensing and public policy', *Research Policy*, **15**, 285–305.

Turnbull, P. (1986). 'The 'Japanisation' of production and industrial relations at Lucas Electrical', *Industrial Relations Journal*, **17**(3), 193–206.

Tushman, M.L., and Anderson, P. 'Technological discontinuities and organizational environments', *Administrative Science Quarterly*, **31**, 439–65.

Urry, J. (1973). *Reference Groups and the Theory of Revolution*, Routledge & Kegan Paul, London.

Vliet, A. (1986). 'Where Lucas sees the light', *Management Today*, June.

von Hippel, E. (1982). 'Appropriability of innovation benefit as a prediction of the source of innovation', *Research Policy*, **11**, 95–115.

Webb, S., and Webb, B. (1897). *Industrial Democracy*, Longmans, Green & Co, London.

Weber, M. (1949 edition). *The Methodology of the Social Sciences*, Free Press, Glencoe, Illinois.

Weick, K. (1979). *The Social Psychology of Organizing*, Addison-Wesley, Reading, Illinois.

Whipp, R., and Clark, P. (1986). *Innovation and the Auto Industry*, Frances Pinter, London.

Williamson, O.E. (1975). *Markets and Hierarchies: Analysis and Anti Trust Implications*, Free Press, New York.

Williamson, O.E. (1979). 'Transaction-cost economies: the governance of contractual relations', *Journal of Law and Economics*, October.

Williamson, O.E. (1985). *The Economic Institutions of Capitalism: Firms, Markets, Relational Contracting*, Free Press, New York.

Winter, S.G. (1987). 'Knowledge and competence as strategic assets' in Teece, D.J. (1981).

Part 2

Structures, Boundaries and Alliances

6

The Strategic Analysis of Inter-organizational Relations in Biotechnology

John Freeman and **Stephen R. Barley**

Cornell University

Whatever else it might be, an organizational strategy is first and foremost a scheme for confronting constraints and exploiting opportunities posed by an organization's environment. Consequently, much of the literature on organizational strategy draws heavily on the work of environmentally oriented organizational theorists. To the degree that organizational theory informs work on organizational strategy, one might expect the latter to reflect problems associated with the former. Similarly, one might argue that developments in organizational theory should prove useful for strategy research.

Taking these propositions as starting points, we attempt to show how current conceptions of organizational strategy reflect shortcomings associated with traditional theories of organizational environments. By drawing on organizational ecology and network theory we propose what we believe is a more advantageous framework for conceptualizing strategic action within an organizational field. Finally we illustrate our notions with

The Strategic Management of Technological Innovation. Edited by R. Loveridge and M. Pitt
© 1990 by John Wiley & Sons Ltd

examples drawn from the domain of commercial biotechnology, the area on which our research currently centres.

ORGANIZATIONS AND ENVIRONMENTS: THEORETICAL BACKGROUND

Until the mid-1960s most organizational theorists focused on either the study of bureaucracy (Merton *et al.*, 1952; Gouldner, 1954; Blau, 1955) or decision making (Simon, 1947; March and Simon, 1958; Cyert and March, 1963). Thus, with few exceptions (Selznick, 1944), researchers of the era attended almost exclusively to processes and structures internal to organizations. However, by the mid-1970s, organization studies experienced a paradigmatic *reorientation* as an increasing number of scholars began to explore how organizations relate to their environments (Emery and Trist, 1965; Lawrence and Lorsch, 1967; Thompson 1967; Terreberry, 1968; Perrow, 1970, 1972; Pfeffer, 1972; Pfeffer and Salancik, 1978, Aldrich, 1979). The reorientation largely reflected the influence of general systems theory, whose diffusion into social science led organizational theorists to conceptualize organizations as 'open' systems (Scott, 1981). If organizations were open systems, they could no longer be adequately described as self-sufficient entities concerned solely with the management of internal affairs, since open systems face the additional dilemma of acquiring resources and disposing of products.

Research on organizational environments subsequently developed in two directions. The first stream of thought as *contingency theory*, attempted to correlate different organizational structures with different environmental attributes; for instance, variability and predictability (Perrow, 1967; Lawrence and Lorsch, 1967). Optimal structures were often prescriptively matched to varying degrees of environmental turbulence or complexity. Debate centred on whether a strongly deterministic view of adaptation was warranted.

Whereas determinists argued that environments dictated organizational structures in a direct and inevitable manner (Hickson *et al.*, 1971), voluntarists claimed that structures arose from management's decisions about how to respond to environmental demands (Child, 1972). Hence, the voluntarists implied that the relationship between environment and structure could be directly manipulated. Voluntarism held considerable sway in business schools where it eventually gave rise to a strategically oriented literature on organization design. Experts argued that by assessing environmental attributes, managers can implement strategies by building structures, thus organizations became tools for achieving strategic objectives. But regardless of whether organizations were held to adapt mechanically or voluntarily, contingency theories routinely portrayed the environment abstractly and analytically.

The second stream of thought, *resource dependency theory*, drew heavily on theories of social exchange (Blau, 1964; Emerson, 1972a, 1972b) to define the environment as the set of organizations with whom a focal organization exchanged goods and services (Evan, 1966). Resource dependency theory postulated that because organizations specialize in a limited range of activities, they are rarely self-sufficient and, therefore, turn to other organizations for the resources they required. Because most resources are finite and scarce, organizations were said to reduce uncertainty by establishing relatively stable ties with the organizations on whom they depend (Cook, 1977; Pfeffer and Salancik, 1978). However, such arrangements entail costs. Depending on a resource's scarcity, its criticality, and the availability of alternate suppliers, organizations may gain or lose power by engaging in exchange. Resource dependency also drew heavily on theories of decision-making under conditions of bounded rationality (Simon, 1947; March and Simon, 1958; Cyert and March, 1963; Thompson, 1967). This led to the 'strategic contingency' view of power. Like contingency theory, resource dependency theory rapidly sired a strategic literature that focused on how organizations might reduce their own dependencies while enhancing the dependency of their exchange partners.

Although contingency and resource dependency theory conceptualized organizational environments and mechanisms of adaptation somewhat differently, they showed surprisingly similar *limitations*. First, both approaches promulgated a firm-centred perspective that focused on the relationship between individual organizations and their immediate environments. As a result, researchers gave little attention to whether organizations were constrained by distal events or by relationships among organizations with whom they had no ties. Yet studies of industrial concentration and political change have repeatedly shown that organizations are often severely constrained by distant events and by organizations with whom they have no direct contact. Powerful social actors are often able to influence environments in systemic ways when viewed from the perspective of remote organizations that are presumed to be unaffected. For example, changes in the U.S. Department of Defence's internal accounting procedures clearly alter practices and relations among contractors and subcontractors.

Second, researchers in each tradition usually ignored the fact that organizations and their environments *change*. In fact most investigations of environmental attributes and organizational dependencies were designed synchronically, so researchers had little choice but to draw static conclusions. But environmental conditions rarely remain stable over long periods of time.

Third, contingency and resource dependency theory both offered a relatively *undifferentiated* vision of organizations and their environments. Most studies employed sampling strategies based on a limited number of analytic traits which, in turn, encouraged researchers to treat organizations drawn from different populations as homogeneous. Moreover, by classifying environments in terms of abstract but presumably generalizable dimensions, researchers failed to recognize that concepts like uncertainty have meaning only in concrete contexts (Barley, 1986). The upshot was to rely on a set of narrow typologies that underestimated the heterogeneity of environments and organizational forms (McKelvey and Aldrich, 1983).

Finally, both theories wittingly or unwittingly emphasized an organization's unbridled *ability to adapt*. Neither showed much concern for organizational births or deaths, much less the rise or demise of whole populations of organizations. Paradoxically, the view of political processes developed in the resource dependence tradition underlies much current ecological reasoning on the inertial properties of organizations (Hannan and Freeman, 1987).

Strategy research rooted in contingency and resource dependency theory mirrors these shortcomings. Most strategic analysis centres on single organizations whose freedom to act is assumed to be unfettered by any larger system of organizational relations and inter-dependencies. Environments are routinely assessed in terms of abstract attributes, such as complexity or munificence, whose meaning is held to be similar across all situations. Even organizations themselves are often assigned to one of a small number of categories based on *a priori* typologies of strategic action. As a result, environments and strategies are both treated as having limited variety. Moreover, general typologies of strategic action are presumed to cut across diverse organizational populations and to apply regardless of the contextual particulars of a population's milieu. Finally, the strategic literature not only emphasizes an organization's ability to adapt, its orientation is decidedly voluntaristic: implicitly or explicitly, organizational action is viewed as the will of a small group of managers.

Even though top managers frequently articulate views of an organization's strategy and may even spend considerable time plotting tactics, one may question the degree to which their statements and plans account for an organization's trajectory. The *link* between business strategy as formulated by top management and the actions taken by an organization may be as tenuous as the long-discredited link between official goals and observed behaviour. However, as Mintzberg (1987) notes, it is possible to speak of strategy without assuming cognitive deliberation or conscious choice. Although most of the strategy literature focuses on consciously devised ideas about how to do business, on what Mintzberg terms 'strategy as plan', strategy may also be understood as patterned consistencies which may or may not be intended. This approach opens the door for an analysis of organizational strategy and environment that avoids the shortcomings of contingency and resource dependency theory: an approach in which it is *irrelevant* whether strategies represent intentions or merely the sediments of actions and events. From this point of view, strategy is nothing more than an organization's characteristic way of relating to its environment.

In recent years, population ecologists and students of interorganizational networks have attempted to redress a number of the shortcomings associated with earlier theories of organizational environments (Hannan and Freeman, 1977). Although the two approaches developed independently, both conceptualize environmental dynamics at a level of analysis beyond that of the individual organization. Each concerns itself with the heterogeneity of organizational forms, and both employ the notion that organizational forms are *isomorphic* with the structure of environmental resources. However, because population ecology extends a theory of environmental attributes while network theory retains exchange theory's emphasis on relationships, each addresses different problems and is subject to different limitations. Nevertheless, because the strengths and weaknesses of the two are complementary, much can be gained by attempting to combine them. One particularly fruitful point of synthesis for investigations of organizational strategy concerns the study of organizational niches.

POPULATION ECOLOGY: NICHES AS RESOURCE SPACES

In bioecology, the concept of the niche expresses a population's (or a species') role in a community, the population's 'way of earning a living'. Modern niche theory traces its emergence to Hutchinson's (1957, 1978) abstract geometric definition. Hutchinson defined a niche as the set of environmental conditions within which a population can reproduce itself. Because growth rates usually reflect numerous aspects of the environment, the relevant environment can be construed as an N-dimensional space in which each dimension corresponds to an environmental condition, for instance, average rainfall or average diurnal temperature fluctuation. Hutchinson defined a population's '*fundamental niche*' as the hypervolume formed by the set of points for which the population's growth rate is not negative. In other words, the fundamental niche consists of the set of all environmental states in which the population can grow or, at least, sustain its numbers. Hutchinson also contrasted the population's fundamental niche with its '*realized niche*', the set of environments in which a population can maintain itself given the existence of competing populations. A population's realized niche is therefore generally narrower than its fundamental niche.

Population ecologists have applied the bioecologist's concept to the study of organizational populations (Freeman and Hannan, 1989). An organizational niche is that combination of resources and constraints within which a particular organizational form can arise and persist. The notion stresses the isomorphism of population and environment by way of a fundamental duality: niches define forms, but forms also define niches. By framing the relation between organizations and environments in terms of niches, population ecology seeks to address several limitations of earlier environmental theories.

First, even though theorists have long recognized that organizations require a variety of resources, previous theories offered little guidance about which resources to analyse. In contrast to earlier approaches, organizational ecologists have demanded a more differentiated system for classifying environments and organizational forms. For instance, instead of describing resources in terms of analytic dimensions, ecologists suggest that researchers enumerate the actual resources on which particular organizations depend. Moreover, ecologists assert that researchers cannot enumerate resources without substantial knowledge of the population's social, economic and political context since arrays of resources are likely to be specific to populations. Nevertheless, the ecologists suggest a criterion for choosing from an array those resources most worthy of analysis. Analytic effort should focus on resources that have significant implications for the survival of the form of which the organization is but an individual case. The long-run advantage of this approach is not only its greater sensitivity to the possibility that environmental dynamics may vary among types of organizations, but the promise that contextual sensitivity may eventually lead to a more stable system for classifying organizations.

Second, while previous theories have construed resources in terms of scarcity and munificence, they have rarely recognized that the meaning of such terms is *relative to the context* in which an organization operates. Notions such as scarcity and munificence imply a comparison between the available quantity of a resource and the quantity that is required. The amount of a resource an organization uses has direct implications both for its structure and with whom it will establish relations of exchange. For instance, organizations that employ unskilled workers on a sporadic basis are likely to hire labour through a contractor because it is not worth their while to establish a personnel office. On the other hand, firms that require many highly trained and mobile employees are likely to develop an elaborate recruiting operation and to establish ties with a number of geographically dispersed organizations that train such individuals.

Although both firms may experience labour scarcity with equal severity, their different niches and levels of resource usage entail different strategies and structures.

Finally, earlier environmental theorists treated variability as a static concept: rates of change were classed simply as high or low. In fact, however, almost all resources display cyclical patterns of availability. The crucial difference is the *periodicity* of the cycles. A so-called stable environment may only appear to be stable because its cycles of change are protracted. The capacity to ride out a niche's resource fluctuations is one of most important limits on a population. As with levels of resource usage, frequencies of resource fluctuation have implications for the strategies and structures that best suit life in a niche.

Population ecologists usually summarize niche structures by using *fitness functions*: rules that relate environmental conditions to the growth rates of populations. A fitness function implies a *carrying capacity*, the maximum number of organizations a niche can support. Propositions about fitness can therefore be restated as propositions about carrying capacities. Specifically, the structure of a niche can be defined in terms of how environmental conditions affect carrying capacities. From this vantage point, two populations are distinct if, and only if, their carrying capacities are affected differently by the same environmental conditions or similarly by different conditions. By implication, no two populations can occupy the same niche.

Although the concept of a fundamental niche is characterized by the survival or growth of an isolated population, few populations are actually isolated. Consequently, niche theory must address dynamics that occur when populations interact. From an ecological perspective, two populations interact if, and only if, the presence of one affects the growth rate of the other. Interaction is termed 'competition' when two populations lower each other's growth rate. 'Predator–prey' interactions occur when one population's growth lowers the growth of the second, while growth of the second spawns growth of the first. (For example, as the number of mice in habitat increases, so does the number of owls. But as the population of owls rises, the number of mice declines.) Symbiosis or commensalism refers to the case in which each population positively influences the other's growth rate.

When populations interact, the expansion (or contraction) of one population alters the niche of the other. For instance, if two populations compete, the presence of one reduces the set of environments in which the other can sustain itself. By equating competition with niche overlap, ecologists have been able to relate observations of realized niches to dynamic models of population growth and decline. Competition between populations can be parsimoniously expressed as a system of Lotka–

Voltera equations:* However, most ecologists have not used the approach because competition is usually indirect and hence, difficult to observe. Instead, they have relied on the close relation between competition theory and niche theory to obtain indirect estimates of competition from the overlap of niches defined in terms of observed utilization of resources. A population's utilization profile summarizes its level of dependence on a continuous resource or on categories of a discrete resource. The width of the population's niche is the variance of its resource utilization. For example, construction firms that bid only on contracts for renovations of residential housing have low variance in their utilization of one resource, size of contract. They therefore have a narrow niche when compared to a population of firms that bid on residential renovations as well as many other types of projects.

A study by McPherson (1983) illustrates how ecologists may combine the notion of carrying capacities with utilization profiles to analyse the niche structures of a community of organizations. Using data on people's membership in a set of voluntary associations such as labour unions, sports clubs, and youth groups, McPherson characterized each organization's realized niche as a space defined by such dimensions as the age, occupation, sex and education of its members. Competition between kinds of associations was construed as the degree to which realized niches overlapped. Assuming aggregate equilibrium, McPherson used these estimates to solve Lotka–Voltera equations for the intrinsic carrying capacities that would exist in the absence of competition. Finally, McPherson showed that by inserting estimates of the intrinsic carrying capacities into the model, he could verify each population's observed size. McPherson noted that the result confirmed the validity of the typology of associations he used in his analysis.

Although we believe that relations between organizations and environments can be studied using niche theory, excepting McPherson's work, recent ecological studies have suffered from a number of critical weaknesses. First, researchers have operationalized niches too simply, using at most two resource parameters. For instance, in their study of niches in the restaurant industry, Freeman and Hannan (1983) defined

---

* Lotka–Voltera equations take the following form:

$$\frac{dx_1}{dt} = r^1 X_1 \left(\frac{K_1 - X_1 - a_{12}X_2}{K_1} \right)$$

and

$$\frac{dX_2}{dt} = r_2 X_2 \left(\frac{K_2 - X_2 - a_{21}X_1}{K_2} \right)$$

where X_i is the size of population i; k_i is the environment's carrying capacity for population i; r_i is the natural rate of interest for population i; and a_{ij} is a competition coefficient measuring the effect of an increase in population j on population.

niches entirely in terms of cuisine and market. Fully specified niches are likely to require multivariate specifications. A second weakness has been the continuing practice of defining environments in abstract terms, such as grain and variability. Such characterizations are not only difficult to relate to the substantive issues that interest most analysts, but they are insensitive to the contextual particulars that make niches unique. These abstract conceptualizations need to be supplemented with more concrete environmental characteristics.

Finally, most work on organizational populations has concerned organizational demography, the dynamics of a single population, or the interplay between two populations. Consequently, niche theory's promise of a community level ecology of organizations has been left unexplored. For example, in their studies of national labour unions and semiconductor manufacturing firms, Hannan and Freeman (1987, 1988, 1989) attempted to document how shifting resources and exogenous events affect the life chances of organizational populations. In particular, they have shown how demographic processes (such as the rate of immigration), economic changes (e.g., booms and busts in the national economy), changes in legislation (passage of the Wagner Act) and technical innovations affect a population's birth and mortality rate. But these variables were not conceptualized in terms of niches since such an approach presumes that one has identified important resources in advance. In his study of the niches occupied by newspaper publishing companies, Carroll (1987) went further by showing how political and demographic factors generate temporal variations in resource availability. However, his empirical analysis focused only on sources of revenue and the issue of competition between specialists and generalists. Carroll opted for a simplified conceptualization of niche to make his analysis more tractable. Barnett and Carroll (1987) provided a more detailed description of the niches of various forms of telephone companies and are clearly moving in the direction we advocate.

SOCIAL NETWORK THEORY: NICHES AS PATTERNED RELATIONSHIPS

Like population ecologists, students of inter-organizational networks have treated environments as distributions of resources that influence strategies and structures. However, since network studies of interorganizational relations are rooted in notions of exchange, they have avoided several of population ecology's limitations. As we have seen, population ecologists usually construe environments in abstract terms such as variability and grain. Network theorists generally define environments more concretely and contextually. In network theory organizational populations differ from biotic populations in that organizations rarely exploit their resource space directly. Instead, organizations acquire resources through transactions with those controlling the resources. Relationships with other organizations can therefore be considered proxies for resources, if not resources in their own right, since they constitute the channels by which goods and services are distributed. Hence, the organization's environment is decidedly concrete: it consists of other organizations and the relationships that bind the organizations in a coherent system. This definition pushes network analysts to more contextually sensitive definitions of a resource space. Of primary importance to network analysts are the patterned flows of specific goods and services among a particular set of actors.

We have also seen that population ecologists often define resource spaces unidimensionally. In contrast, network analysts routinely operationalize environments as flows of multiple resources. Organizations may turn to some organizations for financing, to others for personnel, and to still others for information, raw materials, and political support. Network analysts display and analyse such relations as separate sociograms or *adjacency matrices* whose cells contain data on the presence and intensity of a particular type of tie between two organizations. When analysed simultaneously across numerous organizations, these multiple relations define a complex system, or network, that inscribes the structure of an organizational field.

An organizational field can be understood as a community of organizations with some functional interest in common, for instance criminal justice, mental health, or the manufacture and sale of computers (Knoke and Rogers, 1979). Since such networks are often composed of a variety of different organizations, the notion of an organizational field moves environmental and strategic theory beyond the level of the individual population to the study of relationships among populations. As such, network theory offers an opportunity to study the niches that comprise an organizational community in the relatively precise and discrete terms of a unique pattern of relations. To date, however, few network theorists have attempted to study the structure of an organizational field.

Most previous studies of inter-organizational networks can be grouped into two broad streams by the type of organizations they investigate. One stream has focused primarily on non-profit, community organizations (Laumann and Pappi, 1976; Galaskiewicz, 1979; Laumann et al., 1978; Knoke and Rogers, 1979; Boje and Whetten, 1981), the other on profit-making firms (Mariolis, 1975; Burt, 1979, 1980a, 1980b; Mizruchi, 1981; Galaskiewicz and Wasserman, 1981; Gogel and Koenig, 1981; Mintz and Schwartz, 1984). Community studies are notable for their sensitivity to the fact that organizations may be tied to each other in numerous ways. However, almost by definition, community studies have examined small sets of heterogeneous organizations confined to geographically bounded environments; they have rarely included multiple representatives of the same population.

In contrast, network studies of profit-making firms generally involve large numbers of organizations and include at least several members of the same industry, if not the same population. Since most analysts have drawn their samples from lists of the economy's largest firms, it is unlikely that any sample constituted a coherent organizational field since few members of such samples have a common functional interest. Studies of networks among profit-making firms have also focused almost exclusively on one type of tie: the interlocking directorate. Hence, we know a considerable amount about the structure of networks formed by interlocking boards, but next to nothing about networks of firms formed by other types of ties.

Both streams of research have generally employed graph theoretic constructs such as cliques, centrality, and density. While these concepts are useful for describing either total networks or relations among specific actors within the network, they do not allow one to identify a set of organizations with *identical patterns* of resource consumption. The latter is crucial for operationalizing the notion of niche in network terms. Researchers may find the notion of structural equivalence and the techniques of block-modelling useful for identifying populations and niches in terms of patterned relationships.

In strict terms, two actors are structurally equivalent if both have an identical pattern of ties to all other actors in a network (Lorrain and White, 1971). However, if one wishes to construct equivalence classes from actual data, such a criterion is often too stringent. Most analysts have used blockmodelling techniques to group actors into equivalence classes on the basis of the general similarity of their relations to other actors in the network (Breiger, Boorman, and Arabie, 1977; White, Boorman, and Breiger, 1976; Burt, 1976; Sailer, 1978). It is also possible to model networks by using contextually meaningful categories for grouping actors into clusters (Barley, 1988). At the individual level of analysis, blocks or equivalence classes have been equated with the notion of a role (White, Boorman, and Breiger, 1976; Burt, 1976b; 1980b). Although few studies have blockmodeled interorganizational networks (Knoke and Rogers, 1979; Burt, 1983; DiMaggio, 1986; Fennel, Ross, and Warnecke, 1987), if niches are construed as specific patterns of resource ties, then at the organizational level of analysis structurally equivalent blocks should contain organizations that occupy the same niche. In other words, niches are to organizations what roles are to individuals.

The elegance of this approach resides in its ability to define organizational populations in terms sensitive to the conceptual duality of niche and form (DiMaggio, 1986). If a niche is a set of resources that sustain a specific form and if forms map on to unique populations (Hannan and Freeman,

1988), then *populations* must be sets of organizations with an *identical pattern of resource consumption*. If an N-dimensional resource space consists of the relations through which inputs are acquired and outputs are disposed, then populations can be viewed as all organizations that evidence a similar vector of relationships. Thus, populations are structurally equivalent sets of organizations derived by blocking flows of assets, goods, services, information, personnel, and other resources pertinent to an organizational field. So niches are the content of the ties that define a block, while forms are the patterns or *gestalts* inscribed by those ties.

It follows that the difference between one form of organization and another is a strategic difference in so far as the *strategy is implicit in the form*. Strategy resides in the pattern of resource exchanges that define an organizations membership in a different block or niche. For instance, in choosing to purchase a retail franchise, an entrepreneur chooses a particular market segment, a technology for delivering a product or service, a mode of advertising, and a motif for architecture and interior design. A business such as a McDonald's restaurant involves a prepackaged strategy just as it involves a prepackaged product. Such franchises entail legal obligations to serve food in a specified way and to buy ingredients from specified suppliers. An advantage of such an arrangement is that risk of early demise or the liability of newness (Stinchcombe, 1965; Freeman, Carroll, and Hannan, 1983) is greatly reduced because the organization can be shoe-horned into a well organized niche or system of inter-organizational relations.

The network approach to identifying niches presumes that an organization's method of securing resources is sufficiently stable over time that the organization's strategy can be recognized. If an organization's method of securing resources is unstable, one cannot predict how future transactions are likely to be patterned. Without such predictability the idea of a strategy would make little sense. This is not to say that transactional patterns are fixed, simply that if there is no observable pattern, consciously derived plans are unlikely to bear fruit since all plans must be laid against some background of consistency.

Any industry or organizational community is, therefore, likely to contain a variety of organizational forms defined by different patterns of resource exchange that represent different organizational strategies. A *strategic analysis* of such a community would seek to identify niches, the patterned ways by which resources are acquired. The analysis would also seek to determine whether particular strategies or niches are differentially associated with such organizational outcomes as profitability, merger, acquisition, and disbanding. To appreciate the details of such an approach, consider the outline of a network oriented analysis of the niche structure of commercial biotechnology in the United States.

COMMERCIAL BIOTECHNOLOGY

Definitions

Broadly construed, biotechnology includes techniques as old as Western civilization itself: for instance, the cultivation of micro-organisms for brewing and the intentional cross-breeding of plants and animals. In recent years, biotechnology has come to mean techniques for directly manipulating the genetic code of plants or animals, and the use of biogenetically engineered micro-organisms in the manufacture (or degradation) of materials of economic value. The term 'biotechnology' usually refers to recombinant DNA (rDNA) cell fusion and related technologies, as well as advanced bioprocess engineering.

Firms from a number of traditional industries have been involved in the commercial development of biotechnology. For instance, applied research using rDNA and cell fusion technology is currently being conducted by firms in the pharmaceutical, chemical, agricultural, and energy industries. These technologies are also being exploited by a number of newly established firms that specialize in genetic engineering and cannot be clearly subsumed under a standard industrial code. Other firms have been established exclusively to fund biotechnology ventures and still others have formed to provide necessary equipment and supplies. So it is more accurate to speak of a community of organizations involved in biotechnology than of a biotechnology industry. We use the term 'community' to indicate that these organizations form a whole, held together by commensalistic and symbiotic ties (Hawley, 1950, 1986). The community encompasses all public or private organizations that pursue, produce, sponsor, fund, or regulate research involving recombinant DNA, cell fusion, or monoclonal antibody technologies and the products manufactured by processes derived from them.

Within the United States, the biotechnology community includes at least nine categories of organizations:

1. *New biotechnology firms* (NBFs) established to pursue applied R&D in areas of commercial promise (e.g., Biogen, Genentech, Cetus, Centocor, etc.);
2. *University departments* of microbiology and related disciplines carrying on basic and applied research;
3. *Research institutes* partially or fully engaged in biotechnical research (e.g., The Whitehead Institute, The National Institutes of Health);
4. *Established corporations* in the chemical, pharmaceutical, agricultural, and other industries that have either begun their own biotechnical R&D operations or who are involved in funding arrangements with NBFs,

university departments, or research institutes (e.g., Shearing-Plough, Monsanto, Shell Oil, Kodak, Corning Glass, etc.);
5. *Venture capital firms* that have provided a substantial amount of funding to NBFs;
6. *Regulatory bodies* having jurisdiction over the products and processes of biotechnical research (e.g., Food and Drug Administration, U.S. Department of Agriculture, etc.);
7. *Industrial associations* established to further the aims of this community (e.g., The Industrial Biotechnology Association, The Association of Biotechnology Companies);
8. *Scientific bodies* that fund conferences, sponsor research, disseminate relevant information (i.e., the National Academy of Science);
9. *Suppliers* of equipment and biological reagents for rDNA and cell fusion research or bioprocess engineering.

These categories are not proper organizational populations. Instead, each is likely to contain several distinct populations defined by differential patterns of inter-organizational ties and resource consumption. The objective of a strategic analysis of the niches formed by inter-organizational ties is to differentiate between structurally equivalent sets of organizations nominally included in broad *a priori* categories. These categories are likely to contain the majority of American organizations engaged in the commercialization of biotechnology (Office of Technology Assessment, 1984; U.S. Department of Commerce, 1984; Olsen, 1986).

The biotechnology community

DNA's basic structure and means of replication was reasonably well deciphered by the late 1960s. But the techniques then available for manipulating genetic sequences made any attempt to alter an organism's genetic code look rather distant (Yoxen, 1983). With the occurrence of several breakthroughs in the early 1970s, genetic engineering began to appear not only viable, but imminent. In 1973, Herbert Boyer of the University of California at San Francisco and Stanley Cohen of Stanford demonstrated a technique for splicing a gene from one organism into the genetic structure of another so that the second expressed proteins characteristic of the first. Two years later, Cesar Milstein and Georges Kohler of the British Medical Research Council Laboratory of Molecular Biology successfully fused cells from a mouse myeloma with cells derived from mouse B-lymphocytes to create a hybridoma, a self-replicating antibody-producing line of cells grown *in vitro*. These two research programmes respectively proved the feasibility of recombinant DNA and monoclonal antibody technology.

In 1974, a number of scientists involved in rDNA research called for a moratorium on further research until the technology's risks could be more adequately assessed, leading to the Asilomar Conference of 1975 at which most top American molecular biologists ratified the notion of a voluntary moratorium and the formation of what became the Recombinant DNA Advisory Committee. Although the moratorium was never formally rescinded, its gradual waning can be traced to the U.S. Patent Office's decision in 1976 to grant Stanford and the University of California patent rights to Boyer and Cohen's technique for using plasmids to transfer foreign DNA to a micro-organism. The decision triggered commercial interest in using genetically engineered micro-organisms for manufacturing a variety of marketable products (Yoxen, 1983).

At first glance, the biotechnology community has much in common with the semiconductor and computer industries. Both are characterized by rapid technical change, small innovative firms, sizeable expenditures on R&D, massive infusions of venture capital, and rapid growth (OTA, 1984; Oakey, 1984). However, there are important differences between the two technical communities. Whereas small computer firms have generally been founded by engineers who previously worked for larger companies in the microelectronics industry (Brittain and Freeman, 1980; Rogers and Larson, 1984), most NBFs have been established by scientists associated with universities. While engineers often possess knowledge of production as well as R&D, academics typically have no production experience. Furthermore, the immediate goal of many new microelectronics firms is the production of a device whose prototype has already been developed. In contrast, biotechnology firms are generally founded to pursue applied, and in some cases basic, research in molecular biology. Many NBFs lack production facilities, and even if an NBF intends to market a product its founders are likely to lack the experience necessary for bringing products to market (Dubinskas, 1985).

The nature of R&D and the product life cycle also differ. The development phase of a typical biotechnology product is relatively long (OTA, 1984), exacerbated by the fact that biotechnology's most promising products, therapeutic drugs and diagnostic assays for humans and animals, are subject to stringent regulatory procedures without parallel in microelectronics. Another important difference is that microelectronics products, particularly in the industry's early years, were sold as components of more complex systems. Many biotechnology products are aimed at end-users or professionals employed by them.

Established electronics corporations are likely to possess competencies and facilities that allow them to enter quickly markets opened by smaller firms. In contrast, established corporations have only recently begun to develop expertise in biotechnology so that command of the science still resides largely with academic scientists and smaller firms. Markets for many of biotechnology's products are also likely to be small and hence not amenable to the economics of scale characteristic of mass production.

Finally, unlike microelectronics, the biotechnology community's fate is linked to an uncertain legal environment and shifting public opinion. Biotechnical research poses hazards that are not present in the development of electronic products. Even though we do not know whether the risk of developing and accidentally releasing a dangerous organism is as serious as sometimes portrayed, the fear of such a possibility has engendered more regulation of research than has ever occurred in microelectronics. Surprise court rulings, congressional action and public outcries will probably continue to have dramatic implications for the structure of the biotechnology community as well as for the nature of the ties that form among its members.

Because the early phases of biotechnical product life cycles are typically protracted, even the oldest NBFs have yet to show more than trivial returns on sales (OTA, 1984). During the early 1980s, venture capitalists were the primary source of funding for biotechnology start-ups in return for a significant proportion of a firm's equity. Since then, the high costs of research and development, clinical trials and marketing have in most cases exceeded the amount of venture capital available to biotechnology firms and which has actually declined (OTA, 1984). Most NBFs have been unable to obtain debt financing since they lack sufficient collateral. Consequently, NBFs have been forced to finance their R&D by establishing ties to other organizations with commercial interest in biotechnology.

According to the Office of Technology Assessment (1984, p. 270), biotechnology's reliance on contract research is without parallel in any commercial area except perhaps for small defence contractors. Joint ventures as well as research, licensing, manufacturing, marketing and product development agreements between NBFs and established firms have also been prevalent. Under the typical R&D licensing agreement an established firm funds an NBF's development of a product and acquires exclusive licence to market the product, while the NBF retains patent rights. The R&D limited partnership is the most recent trend in financing commercial biotechnology: an NBF usually assumes the role of general partner and hence, all liability. The limited partners buy a share of the NBF's future profits or losses. Investors may be corporations, pension funds, mutual funds, or private individuals. The limited partners provide funding in return for equity in a specific product or product line. Unlike other forms of equity financing, limited partners do not participate in the management of the firm by sitting on the firm's board or voting as stockholders.

THE STRUCTURE OF RESOURCE NETWORKS

The brief, but dynamic history of commercial biotechnology raises a number of substantive issues regarding the evolving ecological structure of the biotechnology community. In particular, much speculation exists as to why NBFs have formed particular types of ties with established corporations (OTA, 1984; U.S. Department of Commerce, 1984), for example, how organizations respond strategically to technological or environmental pressures. Thus, firms without production facilities may seek licensing agreements more frequently than firms with such facilities. NBFs may be less likely to seek licensing agreements when their products promise to have a limited market for which large scale production facilities are unnecessary. Biotechnology's institutional and regulatory milieu may affect relations among organizations. For example, since the Food and Drug Administration's requirements for assuring the safety of therapeutic drugs are more stringent than those for *in vitro* assays, NBFs working on therapeutic products, such as interferon, may be more likely to establish ties with established firms than NBFs specializing in diagnostic products, since the former need to defray the heavier costs of clinical trials.

While such conjectures point to the strategic importance of interorganizational ties, their orientation is reminiscent of traditional contingency and resource dependency theory. Specifically, each hypothesis focuses on the situation faced by an individual firm while overlooking the importance of the firm's position within the biotechnology community as a whole. In contrast, an organization's ecological niche and, hence, its strategy are likely to be better defined by the larger *gestalt* formed by the firm's global pattern of ties. Thus, if one is interested in identifying niches within an organizational community, one must understand how different patterns of exchange define different positions within the network. To illustrate this ecological approach consider the structures inscribed by the relations in which Genentech and Centocor, two U.S. biotechnology firms engage. Both firms are well-known NBFs. Genentech specializes in recombinant DNA research while Centocor develops monoclonal antibodies.

Genentech

Founded in 1976 by Robert Swanson and Herbert Boyer to take advantage of the Cohen-Boyer patent, Genentech is widely considered one of the most successful NBFs in the United States. The company's initial work on somatostatin was conducted in Stanford's laboratories by Stanford personnel using Genentech's venture capital. Initial funding was provided by various venture capital firms. After its phenomenally successful public stock offering in 1980, Genentech relied less heavily on venture capital and entered cooperative agreements with a number of multinational corporations. As Figure 1 illustrates, despite the fact that Genentech received little money from direct sales, its revenues grew rapidly in a relatively short period of time.

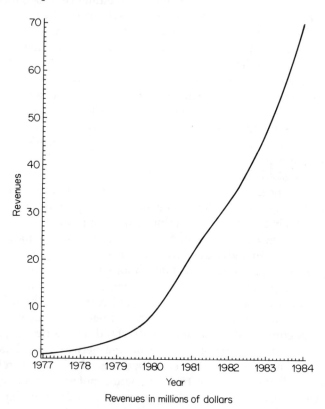

Revenues in millions of dollars

Figure 1. Genentech revenues

Although Genentech was originally founded as a research specialist, its objective is now reputedly to become a major contender in the pharmaceutical industry (Olsen, 1986; Kenney, 1986). To this end, the company has sought to develop products and capabilities that can serve as the basis for vertical integration. Genentech apparently plans to expand by manufacturing in its own facilities and ultimately to sell products under its own label (Daly, 1985, pp. 73–5). The firm has attempted to rely exclusively on its own technology and has resisted arrangements that would require it to surrender control of its technical expertise (Bioscan, 1987, p. 210).

Technologies that do not fit neatly with Genentech's emphasis on human and animal therapeutics are spun off into joint ventures. This strategy suggests that Genentech will eventually have to begin competing with a number of its current partners and associates.

Because Genentech concentrates on therapeutics, its development costs greatly exceed the costs of basic research. Although venture capital and research contracts formed the cornerstone of Genentech's early funding, the cost of clinical trials and the drawn-out process of seeking patent protection forced it into an elaborate set of relationships with other companies, a pattern rare in other high-tech industries such as semiconductors and computers.

Figure 2 portrays schematically a number of the ties that exist, or have existed, between Genentech and other organizations. The figure indicates the kind of relationship which organizations have engaged in with Genentech and the direction of resource flows. Types of agreements or relationships representing *sources* of funding, information, or technology are listed on the left-hand side of the figure with arrows pointing toward Genentech. Relations or agreements through which Genentech *provides* funds, information or technology to other organizations are listed on the right-hand side of the figure with arrows pointing toward the recipients. Relations that entail reciprocal or *mutual exchanges* of resources are depicted by double-headed arrows and are located on the figure's vertical axis.

In addition to venture capital, Genentech has received direct infusions of capital via two mechanisms: equity investments and R&D limited partnerships. A number of firms, including those who provided early venture capital, have acquired blocks of Genentech's stock. The size of these holdings are relatively small when compared to corporate equity in some biotechnology firms (none exceeds 9 per cent).

Since its inception, Genentech has formed three R&D limited partnerships. Genentech Clinical Partners I was established in 1982 to fund the development of human growth hormone (hGH) and gamma interferon (IFN). The fund netted Genentech 56 million dollars. Clinical Partners II raised $34 million in 1983 for financing Genentech's research on tissue plasmagen activator (tPA). In 1984 Genentech raised $32 million via Clinical Partners III to fund development of its tumour necrosis factor (TNF). In 1987, Genentech bought out the limited partners who had invested in Clinical Partners I and II.

An NBF may also seek *non-financial* resources by entering into agreements with established organizations. In 1985, Genentech and Alfa Laval signed a supply agreement for Alfa Laval to provide Genentech with fermentation equipment. Genentech has also licensed technologies from several patent holding organizations. For instance, Genentech has licensed a drug delivery system from Exovir, a glyocoprotein remodelling technology

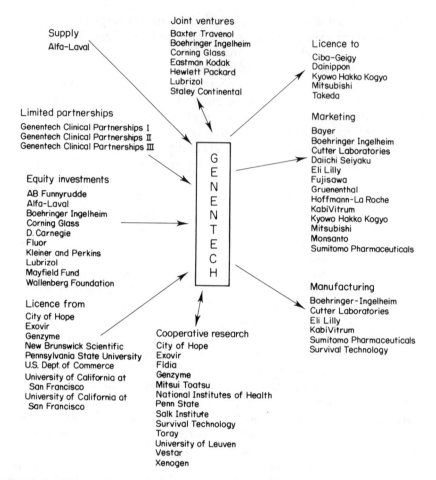

Figure 2.   Genentech network. (From Bioscan, 1988; Daley, 1985; Kenney, 1986; Olsen, 1986)

from Genzyme, and a technology for DNA synthesis from New Brunswick Scientific.

Over the years, Genentech has involved itself in a number of joint ventures aimed at developing specific products tangentially related to therapeutics. Boehringer Ingelheim and Genentech are partners in two joint ventures: Genentech Canada and the Research Institute for Molecular Pathology in Vienna. From 1983 to 1987 Hewlett Packard and Genentech cooperated through HP-Genenchem, a joint venture founded for developing hardware and software for use in biotechnology. Travenol-Genentech Diagnostics was formed to develop, manufacture and market human diagnostics. It was dissolved in 1986 an its products are now being marketed by one of Baxter-Travenol's subsidiaries. Genencor is a 1982 joint venture between Corning Glass and Genentech to manufacture and market industrial enzymes. Staley Continental and Eastman Kodak later joined Genencor as additional partners. Finally, GLC Associates which focuses on developing bioengineered vitamin C is Genentech's joint venture with Lubrizol.

Cooperative research and joint development agreements differ from joint ventures in that they do not entail the formation of a separate corporate entity. Instead, two organizations jointly pursue research or development on a specific product or process. Genentech has either research or development agreements with a number of organizations. In most cases, Genentech's partner has special skills, knowledge or access to technologies which Genentech lacks. In return, Genentech provides funds. Consequently, as Figure 2 reveals, many of Genentech's partners in research have also licensed their substances or techniques to Genentech.

Because most NBFs lack large scale production facilities and have limited marketing expertise, agreements with other companies provide the NBF with an avenue for distributing its products as well as raising additional funds. For example, Genentech has licensed bovine interferon to Ciba-Geigy, TNF to Dainippon, tPA to Kyowo Hakko Kogyo, tPA and human serum albumin to Mitsubishi, and alpha and beta IFN to Takeda. Under a licensing agreement a firm sells its knowledge or techniques for a lump sum payment and, in the case of patented inventions, royalties. The purchaser may develop, manufacture, market or suppress the product as it sees fit. By licensing a technology, an NBF gives up considerable control over the future of its product or technique.

NBFs retain greater control by negotiating, marketing or manufacturing agreements. In a marketing agreement the NBF retains its right to develop and produce a product but sells its right to distribute the product in certain markets. As Figure 2 indicates, Genentech has relied heavily on marketing agreements, mostly with foreign firms who have acquired the right to sell the product only in certain foreign markets. In a manufacturing agreement one firm typically agrees to manufacture a product for another. For instance, Genentech has allowed Cutter Laboratories to manufacture Factor VIII (a protein involved in the clotting of blood and useful for the treatment of haemophiliacs); and Eli Lilly its human insulin. Boehringer Ingelheim manufactures Genentech's tPA for use outside the United States; Survival Technology makes tPA for use in its own automatic intramuscular injection systems for heart attack victims; and both KabiVitrum and Sumitomo Pharmaceutical make Genentech's human growth hormone for foreign markets. As Figure 2 suggests, Genentech has frequently offered its exchange partners manufacturing as well as marketing rights.

Centocor

In contrast to Genentech which focuses on human and animal therapeutics produced via the techniques of recombinant DNA, Centocor has concentrated on developing monoclonal antibodies for *in vitro* and *in vivo* diagnostic test kits and medical imaging. Centocor was founded in 1979 by Michael Wall and Hubert Schoemaker at the University of Pennsylvania for developing assays useful in the diagnosis of cancer. Unlike Genentech, Centocor is reputed to have no plans for becoming an integrated pharmaceutical company. Instead, it aims to develop a variety of products useful in the health care industry and then to market them through the industry's existing channels (see Olsen, 1986). Centocor funds and then licenses promising technologies from universities, research institutes and other organizations. Its own personnel further develop the products and perform the clinical trials necessary for FDA approval. Once the product is approved, Centocor places the product on the market by using the resources of established firms. Because it deals primarily in monoclonal antibodies for use in diagnostic test kits, it faces neither the stringent regulatory requirements nor the difficult manufacturing problems associated with the rDNA therapeutics of Genentech.

Figure 3 portrays Centocor's organizational ties following the same conventions used in Figure 2. Like Genentech, in its early days Centocor relied on venture capital and then, in 1982, raised considerable additional capital through a public stock offering. With the exception of the FMC corporation, most of the organizations that hold significant blocks of Centocor's stock are venture capital firms. In contrast, corporations account for most of Genentech's large investors. In recent years, Centocor has also relied heavily on R&D limited partnerships to raise capital for specific projects.

Centocor developed one of its products by entering a supply agreement with Celltech Limited which provides Centocor with the MAbs that Centocor uses in Myoscint, a radioisotopically-labelled antibody for diagnostics following a myocardial infarct. This supply agreement differs significantly from Genentech's single supply agreement. Whereas Genentech's supplier provides hardware, Centocor's supplies a bioengineered substance that Centocor uses as an ingredient in its own products.

Centocor also relies more heavily than Genentech on licensing agreements. Genentech licenses products from seven organizations, Centocor has licensing agreements with eighteen. All of these agreements cover bioengineered substances developed by researchers outside Centocor and, with the exception of Polycell, all are with universities, research hospitals or research institutes. Of Genentech's seven licensing agreements, three are

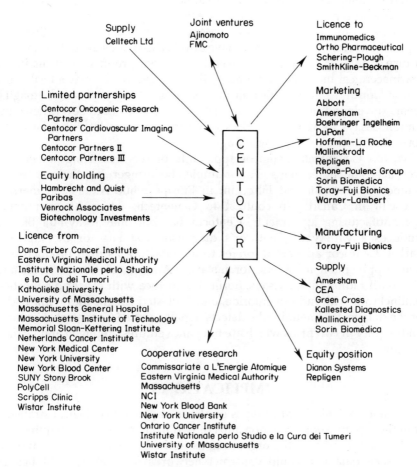

Figure 3. Centocor network. (From Bioscan, 1988; Daley, 1985; Kenney, 1986; Olsen, 1986)

with corporations for techniques or hardware necessary for its research and development process. The different nature of Genentech and Centocor's licensing and supply agreements is consistent with their reputed strategies: Genentech aims to rely heavily on in-house R&D; Centocor to develop products that originate elsewhere.

Like Genentech, Centocor has several joint ventures (though less than Genentech) and a number of cooperative research and joint development agreements. In 1981, Centocor established a joint venture with the FMC Corporation to develop human antibodies and to work in immunoregulation. Centocor acquired FMC's interest in Immunorex in 1986. In 1987 it set up a joint venture with Ajinomoto for marketing

Panorex, Centocor's MAb for the treatment of pancreatic cancer. All of the numerous partners with whom Centocor has cooperative research or joint development agreements are non-profit, research organizations. As Figure 3 indicates, many of these have also licensed substances to Centocor. In turn, Centocor has licenced several of its own products to firms in the pharmaceutical industry. Immunomedics and Ortho Pharmaceuticals hold licenses for and pay royalties on their sale of a variety of cancer imaging agents developed by Centocor. In 1984, Centocor licensed several enzyme immunoassays to SmithKline Beckman and two years later it licensed its fibrin diagnostics to Schering-Plough.

Yet, like Genentech, Centocor apparently prefers marketing agreements over licensing agreements. For example, Boehringer Ingelheim markets Centocor's Myoscint and Fibriscint in Europe while Amersham markets Myoscint in Canada. In contrast to Genentech, which has a number of manufacturing agreements, Centocor has but one: Toray-Fuji Bionics produces a variety of Centocor's diagnostic test kits for the Japanese market. Centocor appears to prefer to manufacture its own products and then supply them to others: for instance, Amersham, Green Cross, and CEA with radioimmunoassays, Sorin Biodmedica with cancer diagnostics, Mallinckrodt with indium chloride, and Kallested Diagnostics with MAb's for the diagnosis of neural tube defects. Finally, unlike Genentech, Centocor holds equity in two firms which are not joint ventures: Repligen and Dianon Systems.

IMPLICATIONS

One cannot reliably identify types of niches or general ecological strategies from the organizational relationships of two firms. One would require data on ties among most of the organizations in the biotechnology community. With such data one could construct networks for each type of tie: for joint ventures, licensing agreements, marketing agreements and so forth. These networks could then be blocked and analysed simultaneously or individually using blockmodelling programs like CONCOR or contextually meaningful clusters of organizations such as venture capital firms, established pharmaceutical companies, foreign firms, NBFs involved in rDNA and NBFs producing MAbs. This approach takes advantage of data on numerous firms and would yield a positional rather than a relational analysis (Burt, 1980aa,b). Our arguments regarding niche structures assume a *positional* analysis while our data on Genentech and Centocor are *relational*. We make such a leap for the purposes of illustration. Currently we are collecting data which will enable us to conduct a positional analysis of the ties among all American organizations involved in commercial biotechnology. Nevertheless, by analysing and contrasting Genentech's and Centocor's 'ego' networks as portrayed in Figures 2 and 3, one can appreciate the type of information that might be gleaned from a full-scale positional analysis of interorganizational ties among members of the biotechnology community.

For instance, a comparison of the two figures indicates that many more organizations licence products to Centocor than to Genentech, a majority of which are non-profit research organizations. If these firms are postulated to form a block in a blockmodel, we may conclude (i) that members of this block are more likely to have relations with 'Centocor-like' firms than with 'Genentech-like' firms and (ii) that these relations entail a flow of research support from the NBF to the research organization (cooperative research) in return for a flow of licensed substances from the research organizations (licence from).

Separate blockmodels of the various relations might also indicate that 'Genentech-like' firms are more likely than 'Centocor-like' firms to enter manufacturing agreements with multinationals and that these relations are more likely to be 'multiplex'. That is, 'Centocor-like' firms are less likely to have several agreements with the same firm. Centocor has marketing, supply, or manufacturing agreements with, or licences its technology to, eighteen different firms. Only four of these firms (22%) have more than one type of agreement with Centocor. In contrast, of the seventeen firms with whom Genentech has similar relations, seven (41 per cent) are partners in multiple agreements. All else being equal, one might argue that multiple agreements signal greater dependence on a multinational corporation than do single agreements.

In a similar vein, a variety of multinationals hold substantial equity in Genentech, while mostly venture capital firms hold large blocks of Centocor's stock. One might conjecture that multinationals which hold equity in an NBF are likely to have other ties to the NBF. Of the other five multinationals with equity in Genentech, three have at least one other relation with it. Logically, these firms should have greater influence over Genentech than should the venture capital firms which have only one tie with Centocor. Thus, all else being equal, outside firms are likely to have more influence over Genentech than over Centocor.

Finally, if foreign and domestic multinationals are posited as two distinct blocks then 'Genentech-like' and 'Centocor-like' firms would have another significant difference. Thirteen of the seventeen firms with whom Genentech has marketing or manufacturing agreements or to whom it licences or supplies its products are foreign multinationals, not U.S. corporations. In contrast, while Centocor engages in similar relations with eighteen firms, only eight are foreign multinationals. This pattern might be interpreted as indicating that Genentech is more worried than Centocor about retaining the domestic market for itself.

At this stage, how might one compare the niches and strategies of 'Centocor-like' and 'Genentech-like' firms in terms of their position in a network of interorganizational relations? 'Centocor-like' firms could be characterized as *'Brokers'*: members of a niche whose specialty is acquiring technologies and products from research organizations and then redistributing them to multinationals for marketing or use in other products. The status of broker is consistent with indifference to foreign partners in marketing, licensing and supply agreements. 'Genentech'like' firms could be characterized as *'original developers and would-be producers'*. Unlike brokers, they licence fewer substances from research organizations while using a variety of agreements to supply their own products to a variety of multinationals. Their aspiration to become producers is indicated by their preference for entering marketing, manufacturing and licensing agreements with foreign rather than domestic firms.

This analysis corresponds to the strategies that Genentech and Centocor report for themselves. However, it is possible that other firms might evidence similar patterns of ties without articulating similar strategies. Lack of congruence is not a problem for the network analysis of niches, since whether or not a firm *intends* its position within a network is largely irrelevant to the analysis.

To conclude, we emphasize that our comments are conjectures offered purely for illustrative purposes. Within the entire network of relations among organizations in the biotechnolgy community the positions we have called a 'broker' or 'original developer and would-be producer' might not appear. Nevertheless, we are confident that the general approach we have outlined would yield positions analytically similar to those we have suggested, even if they differ in their details. In short, niches are likely to be defined by differential patterns of resource flows among organizations. The *structural* aspects of those niches can be understood to *encode* an organizational *strategy*, be it latent or manifest. As such, niche explication should be regarded as a potentially powerful means of strategic organizational analysis, particularly when studying dynamic, fast-evolving communities based on specialized forms of technological innovation.

ACKNOWLEDGEMENTS

This research was supported by National Science Foundation Grant SES-8811489.

REFERENCES

Aldrich, H. (1979). *Organizations and Environments*, Prentice-Hall, Englewood Cliffs.
Barley, S.R. (1986). 'Technology as an occasion for structuring: evidence from

observations of CT scanners and the social order of radiology departments', *Administrative Science Quarterly*, **31**, 78–108.

Barley, S.R. (1988). 'Actions, institutions, and technical change in radiology: toward a role-based theory of technology and the social organization of work', unpublished manuscript. School of Industrial and Labor Relations, Cornell University.

Barnett, W.P., and Carroll, G.R. (1986). 'Competition and mutualism among early telephone companies', *Administrative Science Quarterly*, **32**, 400–21.

Bioscan: The Biotechnology Corporate Directory Service (1988). Oryx Press, Phoenix, AZ.

Blau, P.M. (1955). *The Dynamics of Bureaucracy*, University of Chicago Press, Chicago, IL.

Blau, P.M. (1964). *Exchange and Power in Social Life*, John Wiley, New York.

Boje, D.M., and Whetten, D.A. (1981). 'Effects of organizational strategies and contextual constraints on centrality and attributions of influence in interorganizational networks', *Administrative Science Quarterly*, **26**, 378-95.

Breiger, R.L., Boorman, S.A., and Arabie, P. (1975). 'An algorithm for clustering relational data with applications to social network analysis and comparison with multidimensional scaling', *Journal of Mathematical Psychology*, **12**, 328–83.

Brittain, J.W., and Freeman, J.H. (1980). 'Organizational proliferation and density-dependent selection', in J. Kimberly and R. Miles (eds), *Organizational Life Cycles*, pp. 291–338, Jossey-Bass, San Francisco.

Burt, R.S. (1976). 'Positions in networks', *Social Forces*, **55**, 93–122.

Burt, R.S. (1979). 'A structural theory of interlocking corporate directorates', *Social Networks*, **1**, 415–35.

Burt, R.S. (1980a). 'Cooptive corporate actor networks; a reconsideration of interlocking directorates involving American manufacturing', *Administrative Science Quarterly*, **25**, 557–82.

Burt, R.S. (1980b). 'Models of network structure', *Annual Review of Sociology*, **6**, 79–141.

Burt, R.S. (1983). *Corporate Profits and Cooptation*, Academic Press, New York.

Carroll, G.R. (1987). *Publish and Perish: The Organizational Ecology of Newspaper Industries*, JAI Press, Greenwich, CT.

Child, J. (1972). 'Organization structure, environment, and performance—the role of strategic choice', *Sociology*, **6**, 1–22.

Cook, K.S. (1977). 'Exchange and power in networks of interorganizational relations', *Sociological Quarterly*, **18**, 62–82.

Cyert, R.M., and March, J.G. (1963). *A Behavioral Theory of the Firm*, Prentice-Hall, Englewood Cliffs, NJ.

Daly, P. (1985). *The Biotechnology Business: A Strategic Analysis*, Francis Pinter, London.

DiMaggio, P. (1986). 'Structural analysis of organizational fields: a blockmodel approach', in B.M. Staw and L.L. Cummings (eds), *Research in Organizational Behavior*, Volume 8, pp. 335–70, Greenwich, CN: JAI Press.

Dubinskas, F.A. (1985). 'The culture chasm: scientists and managers in genetic-engineering firms', *Technology Review*, **88**, 24–30.

Emerson, R.M. (1972a). 'Exchange theory, part I: a psychological basis for social exchange', in J. Berger *et al.* (eds), *Sociological Theories in Progress*, Vol.2, pp. 38-57, Houghton Mifflin, Boston.

Emerson, R.M. (1972b). 'Exchange theory, part II: exchange relations, exchange networks, and groups as exchange systems', in J. Berger *et al.* (eds), *Sociological Theories in Progress*, pp. 58–87, Houghton Mifflin, Boston.

Emery, R.E., and Trist, E.L. (1965). 'The causal texture of organizational environments', *Human Relations*, **18**, 21–32.

Evan, W.M. (1966). 'The organization set: toward a theory of interorganizational relations', in J.D. Thompson (ed.), *Approaches to Organizational Design*, pp. 175–91, University of Pittsburgh Press, Pittsburgh, PA.

Fennell, M.L., Ross, C.O., and Warnecke, R.B. (1987). 'Organizational environment and network structure', in N. DiTomaso and S.B. Bacharach (eds), *Research in the Sociology of Organizations*, Volume 5, pp. 311–40, JAI Press, Greenwich, CN.

Freeman, J.H., and HYannan, M.T. (1983). 'Niche width and the dynamics of organizational populations', *American Journal of Sociology*, **88**, 1116–45.

Freeman, J.H., Carroll, G.R., and Hannan, M.T. (1983). 'The liability of newness: age dependence in organizational death rates', *American Sociological Review*, **48**, 692–710.

Galaskiewicz, J. (1979). *Exchange Networks and Community Relations*, Sage, Beverly Hills, CA.

Galaskiewicz, J., and Wasserman, S. (1981). 'A dynamic study of change in a regional corporate network', *American Sociological Review*, **46**, 475–84.

Gogel, R., and Koenig, T. (1981). 'Commercial banks, interlocking directorates and economic power: an analysis of the primary metals industry', *Social Problems*, **29**, 117-28.

Gouldner, A. (1954). *Patterns of Industrial Bureaucracy*, Free Press, Glencoe, IL.

Hannan, M.T., and Freeman, J. (1977). 'The population ecology of organizations', *American Journal of Sociology*, **82**, 929–64.

Hannan, M.T., and Freeman, J. (1987). 'The ecology of organizational founding: American labor unions, 1836-1985', *American Journal of Sociology*, **92**, 910–43.

Hannan, M.T., and Freeman, J. (1988). 'The ecology of organizational mortality: national labor unions, 1836–1985', *American Journal of Sociology*, **94**, 25–52.

Hannan, M.T., and Freeman, J. (1989). *Organizational Ecology*, Harvard University Press, Cambridge, Mass.

Hawley, A. (1950). *Human Ecology*, Ronald, New York.

Hawley, A. (1986). *Human Ecology: A Theoretical Essay*, University of Cicago Press, Chicago.

Hickson, D.J., Hinings, C.R., Lee, C.A., Scheck, R.E., and Pennings, J.M. (1971). 'A strategic contingencies theory of interorganizational power', *Administrative Science Quarterly*, **16**, 216–29.

Hutchinson, G.E. (1957). 'Concluding remarks', *Cold Spring Harbor Symposium on Quantitative Biology*, **22**, 415–27.

Hutchinson, G.E. (1978). *An Introduction to Population Ecology*, Yale University Press, New Haven, CN.

Kenney, M. (1986). *Bio-technology: The University–Industrial Complex*, Yale University Press, New Haven, CN.

Knoke, D., and Rogers, D.L. (1979). 'A block model analysis of interorganizational networks', *Sociology and Social Research*, **64**, 28–52.

Laumann, E.O., Galaskiewicz, J., and Marsden, P.V. (1978). 'Community structure as interorganizational linkages', in A. Inkeles (ed.), *Annual Review of Sociology*, Vol. 4, pp. 445–84, Annual Reviews, Palo Alto, CA.

Laumann, E.O., and Pappi, F. (1976). *Networks of Collective Action: A Perspective on Community Influence Systems*, Academic Press, New York.

Lawrence, P.R., and Lorsch, J.W. (1967). *Organization and Environment: Managing Differentiation and Integration*, Richard D. Irwin, Homewood, IL.

Lorrain, F.P., and White, W.C. (1971). 'Structural equivalence of individuals in social networks', *Journal of Mathematical Sociology*, **1**, 49–80.

March, J.G., and Simon, H.A. (1958). *Organizations*, Wiley, New York.

Mariolis, P. (1975). 'Interlocking directorates and control of corporations: the theory of bank control', *Social Science Quarterly*, **56**, 425–39.

Mckelvey, B., and Aldrich, H. (1983). 'Populations, natural selection, and applied organizational science', *Administrative Science Quarterly*, **28**, 101–28.

McPherson, M. (1983). 'An ecology of affiliation', *American Sociological Review*, **48**, 519–32.

Merton, R.K., Gray, A.P., Hockey, B., and Selvin, H.P. (1952). *Reader in Bureaucracy*, Free Press, Glencoe, IL.

Mintz, B., and Schwartz, M. (1984). *Bank Hegemony: Corporate Networks and Intercorporate Power*, University of Chicago Press, Chicago.

Mintzberg, H. (1987). 'The strategy concept I: five P's for strategy', *California Management Review*, **30**, 11–24.

Mizruchi, M.S. (1981). *The Structure of the American Corporate Network*: 1904–1974, Sage, Beverly Hills, CA.

Oakey, R. (1984). *High Technology Small Firms: Innovation and Regional Development in Britain and the United States*, Saint Martins Press, New York.

Office of Technology Assessment (1984). *Commercial Biotechnology: An International Analysis*, U.S. Government Printing Office, Washington D.C.

Olsen, S. (1986). *Biotechnology: An Industry Comes of Age*, National Academy Press, Washington, D.C.

Perrow, C. (1967). 'A framework for the comparative analysis of organizations', *American Sociological Review*, **32**, 194–208.

Perrow, C. (1970). *Organizational Analysis: A Sociological View*, Brooks/Cole Publishing Company, Monterrey, CA.

Perrow, C. (1972). *Complex Organizations: A Critical Essay*, Scott, Foresman, Glenview, IL.

Pfeffer, J. (1972). 'Merger as a response to organizational interdependence', *Administrative Science Quarterly*, **17**, 328–94.

Pfeffer, J., and Salancik, G.R. (1978). *The External Control of Organizations: A Resource Dependence Perspective*, Harper & Row, New York.

Rogers, E.M., and Larsen, J.K. (1984). *Silicon Valley Fever: Growth of High-Technology Culture*, Basic Books, New York.

Sailer, L.D. (1978). 'Structural equivalence: meaning and definition, computation, and application', *Social Networks*, **1**, 73–90.

Selznick, P. (1944). *TVA and the Grass Roots*, University of California Press, Berkeley.

Simon, H.A. (1947). *Administrative Behavior*, Macmillan, New York.

Scott, W.R. (1981). *Organizations: Rational, Natural, and Open Systems*, Prentice-Hall, Englewood Cliffs, NJ.

Stinchcombe, A.L. (1965). 'Social structure and organizations', in J.G. March (ed.), *Handbook of Organizations*, Rand McNally, Chicago.

Terreberry, S. (1968). 'The evolution of organizational environments', *Administrative Science Quarterly*, **12**, 590-613.

Thompson, J.D. (1967). *Organizations in Action*, McGraw-Hill, Glasgow.

U.S. Department of Commerce, International Trade Administration (1984). *High Technology Industries: Profiles and Outlooks—Biotechnology*, Government Printing Office, Washington, D.C.

White, H.C., Boorman, S.A., and Breiger, R.L. (1976). 'Social structure from multiple networks. I. Blockmodels of roles and positions', *American Journal of Sociology*, **81**, 730-79.

Yoxen, E. (1983). *The Gene Business: Who Should Control Biotechnology*, Harper & Row, New York.

7

The Mutual Organization: a New Form of Cooperation in a High Technology Industry*

Christian Koenig

ESSEC

and

Raymond-Alain Thiétart

University of Paris IX-Dauphine and ESSEC

The link between technology and the structure of organizations has long been the subject of studies and theories. In practice, the rapid development of strategic alliances, particularly in high-technology, high-risk projects, raises new strategic and managerial issues, yet joint ventures and partnerships of various sorts spring up and show promising performance. This chapter examines the characteristics of a form of cooperative agreement

* An earlier version of this chapter appeared in *Technology in Society*, vol. 10, 1988, Special Issue on Macro-Engineering, edited by Mel Horwitch

The Strategic Management of Technological Innovation. Edited by R. Loveridge and M. Pitt
©1990 by John Wiley & Sons Ltd

we call the mutual organization, assesses the role of corporate strategies, technology, and managerial structure therein, and draws lessons concerning conditions for its success.

A TRANSACTION-COST APPROACH TO ORGANIZATIONAL FORMS

Many institutional arrangements are available for running economic activities. Among others, these include arm's length transactions and the vertically integrated firm. Economists have long thought that markets and firms are alternatives for organizing economic activity. But each of these extremes of the spectrum of organizational forms involves costs: costs of setting up, running, and monitoring an arm's-length market transaction on the one hand; costs of running the organization on the other hand. The problem for organizers is to design a governance structure that strikes an adequate balance between the costs of 'organizing' and the costs of 'running the economic system' (Arrow, 1969, 1974; Coase, 1937; Williamson, 1975).

Williamson (1975, 1979, 1981, 1985) has developed an analytical framework of transaction-cost economics to help establish why this balance may shift in favour of market mechanisms or the integrated firm. A transaction occurs when a good or service is transferred across a technologically separable interface. Generally speaking, the firm is said to economize on transaction costs when it internalizes transactions. But this does not suffice to predict which type of governance structure should prevail in general: not only does the firm itself incur costs but the institutional arrangement has to match the attributes of the transaction.

If the transaction is a one-shot enterprise, is highly certain in its terms, and does not require specific resources (say, a special machine or specially trained personnel), then the integrated firm enjoys no advantage relative to an arm's-length market transaction. Actually, the *frequency* with which transactions recur, the *uncertainty* to which they are subject and the degree to which they are supported by durable and specific investments all draw the balance in favour of the integrated firm.

The frequency argument states that, all other things being equal, if a transaction recurs often and if the terms of the transaction vary little, both parties would be better off entering some form of long-term contract or integration in order to save on the cost of renegotiating or rewriting the contract every time a transaction occurs.

The uncertainty argument is more difficult to grasp. When there is so much uncertainty surrounding a transaction that no contingent-claims contracts can be written, then the integrated hierarchy is the preferred governance structure (Williamson, 1975). A firm in a nascent and therefore risky industry might want to reduce risks by contracting with an existing distribution channel rather than commit resources to the creation of its own channels. But if the product is so new or so different that distribution channels either do not exist or are inappropriate, we are back to Williamson's interpretation: market transactions are not possible, and the firm is induced to integrate. However volatile the market, integration will also be preferred when other firms already foreclose procurement or distribution channels or when trade secrets are at stake in the transaction (Harrigan, 1983). For instance, Airbus had to set up its own sales force to contact customers since all the other commercial aircraft manufacturers were already forward integrated. When a market is volatile, it is risky to extend the degree of vertical integration or the number of vertically integrated stages from procurement to distribution, unless uncertainty is such that otherwise no transaction would occur.

Similarly, the balance shifts in favour of integration when distributors of a differentiated good have sufficient power or are not easily controlled, and can actually enhance or 'debase' the quality of the good sold to them by the producer, who then cannot fully 'appropriate the rents' that would normally accrue to it. Or the producer can incur costs that it would avoid if it could exercise quality controls over distributors. This attribute of transactions is referred to as distribution- or demand-externality (Williamson, 1979, 1981) or product degradation risk (Walker, 1988).

The third transaction attribute to consider is asset specificity. This means that when a transaction requires that each of the two contracting parties invest in durable, specific assets, they should avoid short-term contracts and prefer integration. Among such durable specific assets are special tools, machinery, plants, specifically trained manpower, and specific R&D programmes.

In addition, two characteristics of human behaviour have to be taken into account when organizing economic activity: bounded rationality and opportunism. Bounded rationality refers to the cognitive and computational limits of human actors that constrain them in formulating and solving complex problems, processing information and anticipating the consequences of their decisions (Simon, 1957, 1979). Opportunism, or *moral hazard*, refers to the potential unreliability of human actors, in the sense that they are willing to modify, *ex-ante* or *ex-post*, the terms of the transaction in their own interest. This problem can be mitigated if parties are so numerous that competition among them reduces their ability to act opportunistically, but it becomes serious when those involved are few in number. In this *small number* situation, as Williamson (1975, p. 27) puts it, 'it is in the interest of each party to seek terms most favourable to him, which encourages opportunistic representations and haggling'.

To summarize, transaction costs economics identifies two sets of problems:

- Comparing the cost of planning, adapting and monitoring task completion under alternative governance structures and then matching governance structures with the attributes of the transaction.
- Organizing economic activity in a way that economizes on bounded rationality and safeguards transactions against opportunism.

Thus, when transactions are repetitive, occur in a context of high uncertainty and when the contracting parties have invested in durable, transaction-specific assets, the integrated firm, with its hierarchical decomposition between strategic and operational decisions (to economize on bounded rationality) and its system of incentives and controls (to attenuate opportunism), seems to offer the appropriate institutional arrangement for economic activities.

This conceptual framework is very useful when trying to understand which activities are internalized in a firm and which are not; but it tends to focus on two extreme forms of institutional arrangements: spot market transactions and total internalization. As pointed out by Thorelli (1986) it leaves out a variety of network arrangements in which two or more organizations are involved in long-term relationships. Network arrangement is an organizational form intermediate between markets and hierarchies that has undergone considerable development in our economies, particularly in the form of joint ventures, consortia, and other cooperative structures such as those prevailing in the European aerospace industry (Contractor and Lorange, 1988; Ghemawat, Porter and Rawlinson, 1986). The transaction costs framework also leaves out micro-organizational issues, such as power coordination, culture or decision making, which will be addressed later.

The transaction cost framework has been challenged in the literature of interorganizational cooperation, based on social exchange theory or resource dependence (Aldrich, 1976; Johanson and Mattson, 1987; Perrow, 1986; Pfeffer and Salancik, 1978; Schmidt and Kochran, 1977; see also Granovetter, 1985). Yet, we believe that a revised transaction cost framework is helpful in explaining the emergence of new organizational forms, including networks and what we shall call the mutual organization. As Robins (1987) states: 'the transaction cost paradigm provides tools for examining the economic environment and identifying competitive advantages of specific organizational configurations'. Williamson (1985) and many other authors (Child, 1987; Jarillo, 1988; Teece, 1987, 1988) more recently identified numerous forms which fall between the two extremes of arm's-length market transaction and integration.

THE MUTUAL ORGANIZATION AS A NEW FORM OF ALLIANCE

We now turn to those institutional arrangements intermediate between markets and hierarchies. At one extreme (see Figure 1) we can picture a prime contractor and a number of independent sub-contractors who are engaged in one-shot contracts that, typically, are not renewed when the terms of the contract have been met. In the context of the aerospace industry, this is still a pervasive arrangement in which a plane manufacturer allocates tasks on various so-called 'systems' (flight controls, avionics and the like) to specialized firms for the duration of the project. At the other we have the vertically integrated firm, as used to be the case in the aircraft industry when European firms, imitating their large U.S. counterparts, each manufactured complete planes (although they also subcontracted subsystems), despite the fact that economies of scale did not warrant this extent of integration.

In some industries, such as construction, there is a recurring relationship between the prime contractor and a group of sub-contractors. Although each project, say a new house, involves new contracts between the general contractor and special trade sub-contractors, so that there are no advantages in favour of integration, the preferred arrangement is usually intermediate between market transactions (i.e., contracting with different sub-contractors for each project) and integration. Eccles (1982) called this intermediate arrangement the *quasi-firm*: 'Through a continuing association, both parties can benefit from the somewhat idiosyncratic investment of learning to work together' and they can also reduce uncertainty about future business relationships. The quasi-firm is a coordinated contracting mode which relates a prime contractor as principal and a group of sub-contractors as agents in a recurring relationship. The quasi-firm is a *network* since two or more organizations are involved in a long-term relationship (Thorelli, 1986). In this network each party bears risks relative to its own activity.

In other forms of networks all members are joint risk-takers (e.g. the various forms of partnerships). Figure 1 illustrates various forms of alliances. In the typical joint-venture two or more firms put complementary assets together in a self-contained organization with production of goods or services of its own. In this case, the competitive success of the joint firm depends in part on its own environment and competitive conditions. Another form of alliance associates two unequal partners in a risk-sharing scheme in which the imbalance between the partners makes the alliance resemble a glorified sub-contracting arrangement rather than true partnership (Figure 2). The relationship of Boeing with Japanese firms on the Boeing 767 is an example of this, where Boeing designs and sells the plane but allocates work to risk-sharing partners who have no ownership rights (Moxon, Roehl, and Truitt, 1988). Another form of alliance consists in the 'exchange of hostages': each of two competing firms collaborates on a project developed by the other, a form of mutual check. Yet, when General Electric and Rolls-Royce collaborated on each other's new engine project, each simultaneously pursued development of new engines which competed with the engines developed in common. This *reciprocal deception* led to divorce.

The *mutual organization* is a network like the quasi-firm, but the parties are both principals and agents and 'learning to work together' presumably also prevails in this co-contracting mode. The major difference from quasi-firm lies in risk-allocation. If there were a sufficiently large number of potential members in the mutual organization there would be no need for such a partnership and a less committing quasi-firm would do. It is the small-number situation that contributes to the preference for joint risk-taking. Moreover, the mutual organization means that there is asset specificity in the transaction. Not only is there the specificity of learning how to work together but each member invests tangible or intangible assets specific to the mutual organization. This is true in the aerospace industry, so why do not all members merge to form a vertically integrated firm?

In one way, the mutual organization resembles the integrated firm in that the activities of the members are integrated into one organization which could implement the hierarchical decomposition principle propounded by Williamson for economizing on bounded rationality and safeguarding against opportunism. But there are major differences. The mutual organization is a network of principals and agents and problems of bounded rationality and opportunism appear among them, whether or not they

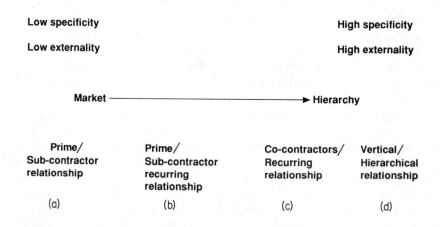

Figure 1

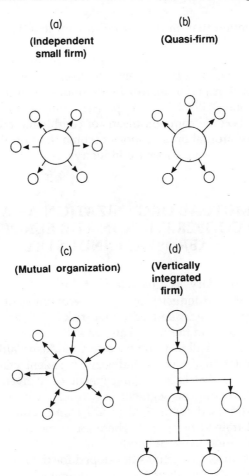

(a)
(Independent
small firm)

(b)
(Quasi-firm)

(c)
(Mutual organization)

(d)
(Vertically
integrated
firm)

Figure 2 From market to hierarchy

set up a hierarchy to operate their organization. There are problems of communication and coordination of principals and of members trying to appropriate the results of the mutual organization for their own profit. To the typical agency-relationship problems of the integrated firm (Pratt and Zeckhauser, 1985) are thus added problems of relationships among principals. The mutual organization also differs from the integrated firm in terms of asset specificity. When several firms form a mutual organization to operate a project or conduct an economic activity, they do not convey all their assets to the organization: transaction-specific assets are only a share of the total assets of each member. Thus, there is more asset specificity in the mutual organization than in the quasi-firm, but less than in the integrated firm.

Another obstacle to vertical integration is the opposition of each firm to a merger at the possible expense of loss of control. Members of a mutual organization prefer, almost by definition, a plural organization to an integrated one. Alliances, joint projects and ventures, consortia and networks have become common means of conducting economic activities. Why, then, have mutual organizations become the prevalent way of doing business in the European aerospace industry?

THE MUTUAL ORGANIZATION AS A FORM OF COOPERATION IN THE EUROPEAN AEROSPACE INDUSTRY

The aerospace industry is characterized by high asset specificity in exchange. It is also an industry with large economies of scale, experience effects and high technical and economic risks. All these factors assume particular importance in the European context.

For simplicity, we shall focus on the aircraft industry although conditions are usually similar in the space industry. A salient characteristic of the aircraft industry is that each major element of a given aircraft—or one might say, each production segment (e.g. fuselage, cockpit, tail, wings)—is specifically designed for a final product and requires costly, specific equipment and trained personnel. There are some economies of scope in aircraft manufacture since some production segments can serve on different airplane models: fuselage sections developed for the Boeing 707 were used on the 727 and the 737 and there is similar commonality between the Airbus A-300 and A-310. But this only confirms the high degree of asset specificity involved in producing and assembling all the major elements that constitute the final product.

The aerospace industry is a risky business: numerous technical failures have marked the history of this activity, but we want to emphasize the great commercial and economic risk. Commercial risk is related to development time from project launch to operation, which averages over five years for a commercial aircraft to over ten years for some space projects or advanced aircrafts. This duration makes accurate market predictions very hazardous. Development also entails high economic risks. Typical development costs for a commercial aircraft amount to the sales equivalent of 70 planes, i.e., $3.5 billion for the Airbus A-300. Moreover, the sales equivalent of 40 additional planes has to be spent over the life-cycle of a project for product improvements, totalling 110 unit sales equivalent. For an aircraft like the

Airbus A-300 the cumulative financial risks amount to $5 billion, which is more than any single European firm can bear (For a description of risk and rivalry in the aircraft industry see Newhouse, 1982.)

At the present time, no aircraft manufacturer has enough capital to face such a risk. However, in an industry where experience effects are important (the experience coefficient is about 0.8), Boeing manages to use resources drawn from successful programmes such as the 727 or 737 to help finance new programmes. Lacking this experience and adequate financial resources, European manufacturers are naturally induced to pool risks in a mutual organization.

This inducement is reinforced by economies of scale. These have to be considered in terms of the size of the market. It is generally considered that the break-even point for an aircraft programme is between 300 and 600 units, depending on whether it is a new aircraft or an offspring of another programme and on the potential to gain experience early. No single European country has a market large enough to make such a programme profitable. Joining together to gain access to a larger market does not solve the problem: the combined market of all the main participating countries in Airbus is not large enough either. However, pooling market access in a mutual organization may help get programmes started and gain early experience.

Another reason why the mutual organization has become the preferred mode of cooperation in European aerospace is that, faced with the economics briefly mentioned above, all manufacturers have developed risk-sharing schemes with partners. But cooperation may take several forms. The most common has been the prime contractor sub-contractor organization (quasi-firm) in which the dominant partner remains responsible for overall design and for selling the product while sub-contractors bear risks relative only to their own share of the work. This is still a common practice in military aircraft. To mention only one case in the commercial sector, Aeritalia (Italy) has long been a sub-contractor for Boeing, notably on the 747 and the 767.

Many European manufacturers and governments, however, have shown reluctance toward a cooperation system in which a national firm would be dominated by a foreign partner. Political opposition to integration or to extensive sub-contracting has been an additional inducement to forming mutual organizations in which the main sub-contractors are also partners. Whether or not this has been a judicious choice and a workable arrangement can be appreciated only in the light of experience. For this reason, we have looked at four case studies of cooperation in the European aerospace industry in order to illustrate the pitfalls of collaborative ventures, the performance of mutual organizations, and the managerial issues that arise in them.

Concorde: the false mutual organization

'Concorde was an entirely political aeroplane. The plane was to show that we were good Europeans.' Sir Richard Way, Permanent Secretary (Ministry of Aviation, 1963 quoted by May (1979).

The Concorde provides an interesting case for illustrating the conditions required to set-up an adequate mutual organization. Although a major technological achievement, the Concorde project was a financial and commercial disaster, with development costs above those forecast and with only 16 units produced.

It took several years for the two industrial partners, British Aircraft Corporation (BAC) and Sud Aviation (later Aerospatiale), to come up with a suitable project as the basis for a government agreement signed in November 1962 by Britain and France. The two firms did not agree on whether to build a short range (Sud Aviation) or a long-range (BAC) version of the Supersonic transport (SST). The inter-governmental agreement actually stated that equal attention should be devoted to both versions, but they themselves did not agree on the purpose of the whole project. Both had in mind the development of high-technology employment, but the goals of Gaullist France were to pre-empt U.S. firms in this seemingly lucrative and strategic business, regardless of cost, whereas the British saw in the agreement an opportunity to show the French they could be good Europeans and thus soften de Gaulle's views on Britain's application for EEC membership. Ironically, de Gaulle vetoed Britain's EEC application two months after the Concorde agreement, although he supported continuation of the project.

In the 1962 agreement the rules of the game for managing the project were also set. The basic objective was complete equality in managing the project and sharing the work, costs and sales proceeds. It was agreed that an integrated management be set up, a first in international cooperation. A Standing Committee of officials from the two countries, later called the Directing Committee, with alternating chairmanship, was to supervise the programme and exercise control through an Aircraft Committee of Directors and an Engine Committee of Directors, also with alternating chairmanships.

Staffed by high-level officials who had little time to spend on the project and for whom compromise was difficult, the Directing Committee delegated more and more work to the Administrative and Technical Sub-Committee which was responsible for subcontracts. For each contract the Sub-committee was presented with possible choices by BAC and Sud Aviation which had to accept tenders from a pre-established list. The Sub-committee then recommended a choice to the Directing Committee based on the 50–50 equality rule. Consequently, the final decisions were not always in line with the wishes of the design teams and from specification to placing a contract could take well over two years and often entailed duplicated costs (Costello and Hughes, 1976; Hochmuth, 1974; May, 1979). At the production level there was no real integrated management, but a replication of the government committees' dual structure. This resulted in power struggles, haggling and inability to compromise, so that more and more decisions moved upwards to the Directing Committee and then to government level.

Thus, the Concorde programme was plagued from the outset by a cumbersome dual structure that, according to Geoffrey Knight, a former BAC vice-chairman, 'had built into it rich sources of discord, delay and duplicated expenditure' (Knight, 1976). It was not until mid 1964 that Sud Aviation finally gave in and agreed to BAC's long-range version. But even then, all conflicts were not resolved. The airlines were pushing for a stretched version and the French were unyielding. Two fuselage length increases were finally accepted, causing additional costs and delays.

Dramatic cost escalations triggered worries and criticism, bringing the first serious crisis in the programme. When the Labour Party returned to power in Britain, Prime Minister Wilson was under pressure to negotiate an end to the project with the French. He was then reminded that the 1962 agreement did not include a break clause, an oddity originally backed by the British government, which thought its French counterpart might wriggle out soon after signing the agreement. De Gaulle's reaction was to order that all contacts at government or industry level be suspended. This crisis, which lasted ten weeks, had one positive impact however: informal cross-links were developed below the top level and this encouraged better cooperation between the firms (Hall, 1980).

Financial rumblings and growing tension between dissenters and supporters in both countries finally led to an inquiry in 1965 into cost controls. In 1966 the whole organization was reshuffled but there was still no integrated management with strategic power. Although cross-links had developed, coordination was still lacking and decisions moved up to the newly formed government committees, the Concorde Management Board (CMB) and above it the Directing Committee. With tighter control (by technical agencies of both governments) most conflicts were resolved and decisions made at top government level. Simultaneously, key people in most of the committees were being replaced. In early 1969, the CMB finally asked the firms to designate project managers who were to report to the chiefs of the Committee of Directors: Henri Ziegler, a prominent figure in the French aircraft industry who had recently become the head of Sud Aviation, and Geoffrey Knight, BAC's vice-chairman.

In addition, contract incentives (or disincentives) were changed. Until 1968 contracts were placed on a cost-plus basis; after that, a complex scheme was set up. A firm's profits decreased when there was cost escalation unless the cost increases could be credibly attributed to increases in the aircraft's performance. As a result, manufacturers had an incentive to suggest design modifications, at no risk to themselves (Hall, 1980).

In the last years of the programme, before the 1974 termination, tension increased between BAC and Sud Aviation. Sud Aviation merged in 1970 with Nord Aviation and SEREB (Ballistic missiles) to form SNIAS, known today as Aérospatiale; Ziegler reduced the importance of the Concorde project in the new firm and appointed a new chief engineer. In the meantime, Sir George Edwards tightened BAC's structure. But the two firms appeared again as competitors on a new front: SNIAS was launching Airbus with German manufacturers, while Sir George was (unsuccessfully) trying to get public funding for launching the BAC 311, a competitor to Airbus. Moreover, BAC and Sud Aviation (SNIAS) were competing for Concorde sales. Market projections had been grossly exaggerated. Airlines displayed little enthusiasm for a supersonic transport since they had already had to write-off fleets of propeller planes to replace them with jets in the late 1950s and early 1960s. Although the American SST decision of 1963 spurred the Concorde programme, jumbo-jet development looked more attractive to airlines. In this unfavourable context there was no integrated marketing management and the two firms were competing for market prospects all over the world.

The lessons to be drawn from the Concorde programme are numerous. Government bureaucratic involvement in management was never properly dealt with. Lack of goal congruence, inadequate incentives, fuzzy intermediate goals, lack of controls and information systems, and above all, a cumbersome dual structure all caused strategic power to drift away from manufacturers to government levels. Although international cooperation did function at lower management levels, the Concorde operation arguably never achieved the status of a mutual organization.

Airbus Industrie: the emergent mutual organization

'After Concorde, an engineer's dream built by politicians, comes Airbus, a businessman's dream built by engineers.' *L'Expansion Magazine*, 1980.

Having gained 30 per cent of the market for wide-body aircraft, Airbus Industrie is now a serious contender for the new generation of single-aisle planes. Although its financial performance is probably still mediocre, the commercial success of Airbus Industrie shows that mutual organization is a workable concept.

Encouraged by their respective governments to cooperate with European partners, Hawker-Siddeley, BAC, Bréguet and Nord Aviation started in the early 1960s, separately or in teams, on projects for a large capacity, short-to-medium-range aircraft. Airlines expressed a need for a new plane, called an 'airbus', although they did not agree on a precise configuration.

As in the case of the Concorde each manufacturer defended its own design. In particular, BAC put pressure on the British government to support an entirely British airbus designed for British European Airways (BEA, later to merge with BOAC to form British Airways). After much haggling, an essentially political compromise was struck whereby the French, British and German governments signed a 'memorandum of understanding' for Rolls-Royce leadership in the design of a new engine, the RB207, and for Sud Aviation design leadership on the airframe, working with Hawker-Siddeley and Deutsche Airbus (a joint venture of MBB and VFW-Fokker). Snecma of France and MTU of Germany also worked on the engine.

During the period 1967–9 Rolls-Royce chose to work on the RB211 engine for the Lockheed Tristar. Partly for this reason, partly because airline orders failed to materialize, the airbus project was scaled down to a 250-seat version, the A300B. The British government (under pressure from BAC) and Rolls-Royce resigned from the project in early 1969, but Hawker-Siddeley decided to continue working as an associated partner on the wings. In May 1969 France and Germany signed a new agreement for the A300B programme: Sud Aviation and Deutsche Airbus were to have an equal share in it and Hawker-Siddeley was to be a design consultant and sub-contractor, along with Fokker (a Dutch company) and CASA (a Spanish company).

A year and a half later Airbus Industrie was set up under French law as a 'groupement d'intérêt économique' (or GIE). For Airbus Industrie, the GIE offered several advantages. A GIE has no capital, makes no profits and is fiscally transparent. Profits, risks and taxes percolate down to members. Partners are joint risk-takers, each is responsible for securing adequate funding from its government and all are liable for the debts of the GIE out of their own assets. The GIE helped formalize cooperation without a merger of the firms involved. It also represented security to customers, to whom it offered a unique interface for sales and service. The GIE is able to receive new members, it avoids locking up large amounts of capital and, finally, it is a flexible organization able to react to changes in membership and in customer demands.

Between 1970 and 1972, the Dutch and Spanish governments joined the programmes, but only CASA became a member of Airbus Industrie, with a 4.2 per cent share. Aerospatiale (successor to Sud Aviation) and Deutsche Airbus each held 47.9 per cent. In 1978, after British Aerospace was created by merging BAC and Hawker-Siddeley, the British decided to join in and help develop the A310, a smaller version with a newly designed wing.

The organization of production has remained broadly the same for the A300B, the A310 and the new, single-aisle, 150-seat A320. It was originally master-minded by German engineer F. Kracht, who had long been working with French firms. Large completed sub-assemblies are flown in a specially designed plane (originally used by NASA for rocket transport) to the Toulouse Aérospatiale facilities for final assembly. To date, no major problems or delays have been encountered in this operation.

We now turn to the structure of Airbus Industrie. It is basically a coordination device and a conflict resolution mechanism. The Board of Control, which consists of representatives of the four member firms, is similar to the board of any corporation. It was headed for twenty years by the late Franz-Joseph Strauss, a key German political figure and a keen supporter of Airbus. The Board is assisted by an executive committee and a financial committee, which liaise between the Board and the management of Airbus Industrie. Management's decisions have to be approved by the Board of Control.

Governments supervise Airbus Industrie through an organ called the Executive Agency. Each member firm secures development funds from its own government. Royalties paid to the latter by Airbus Industrie itself were originally calculated so that governments would be reimbursed after the sale of the 360th aircraft (of the A300 and A310).

Originally, Airbus was a mostly French structure. Ziegler was its first executive president, succeeded by B. Lathiére, with R. Béteille, a renowned technician, as general manager. This early French dominance reflected its large share of the total work after Britain quit the project. After Deutsche Airbus had built up infrastructure and technical competence the Germans started asking for more power in the Airbus organization. This view was shared by British Aerospace when it became a member. It took several years for British Aerospace, Deutsche Airbus and CASA representatives to be appointed to top executive positions in the Airbus structure, while Airbus Industrie's own recruitment increased.

Most of Airbus Industrie's own personnel are hired for clerical work, but a number of executives, including a recently created Human Relations Management Officer, were recruited directly. A personnel statute, now in preparation, will make it necessary for employees of the member firms to choose after a few years between joining the Airbus workforce or returning to their mother company.

Airbus Industrie, however, may still be viewed as 'a minor child that cannot give orders to its parents', as one of its executives put it. It has nevertheless persuaded members to accept the same basic work allocation (integration, systems, cockpit–wings–fuselage–tail) for the A320 as for the A300 and A310. This was originally resisted by member firms who feared their own engineers would lose expertise in designing and making a complete aircraft. At the design stage all member firms still submit concurrent projects, Airbus acting as a final arbiter.

Various incentives exist to make members act in favour of the common good. Being both principal and agent, each member has a vested interest in minimizing costs so as to increase sales and profits. Also, since all accounting is done in US dollars, a decrease in the latter's value acts as an incentive to control costs. But there are disincentives as well. Procurements are made on a territorial basis, that is, purchases made in the country of a member are paid by the respective firm, whereas the costs of purchases made in a third country are shared according to membership shares. Although this procedure may ultimately be changed, it has long been an incentive to handle procurement outside the Airbus community. Finally, there is still a risk of member firms' competing with each other on other projects, which increases tensions and conflicts within the mutual organization.

Because of its GIE status, it is impossible to make a financial assessment of Airbus's performance. Nevertheless, with a foothold gained in the U.S. market with orders from Eastern, Pan Am, and Northwest Airlines, it demonstrates the promising commercial results of a mutual organization that has succeeded in keeping government at arm's length, though some problems of managing a mutual organization persist.

ESRO: a truly mutual organization

Hochmuth (1974) in his book on transnational organizations gives two other interesting illustrations of the 'do's' and 'don'ts' of a mutual organization. In the late 1950s there were discussions between government officials and scientists on the possibility of creating an organization comparable to CERN (Centre Européen de Recherche Nucléaire) in space science. Pierre Auger, then the head of CERN, was asked to coordinate the first efforts which in 1962 resulted in ESRO (European Space Research Organization). His experience at CERN undoubtedly helped him design the new organization: a director-general with real power, assisted by a council composed of two delegates (one scientist, one government representative) from each of the ten member countries.

The initial objective of this new organization was purely scientific: to run 400 experiments requiring 440 rockets and 34 satellites over a period of 8 years. To achieve this objective the director-general had a great deal of freedom to make decisions within broad guidelines and a three-year budget voted by the council. Decisions in the council were taken by majority vote (two thirds) except those concerning the budget which required unanimity. Appointment and dismissal of top personnel was delegated by the Council to the director-general.

ESRO was also a legal entity. It could acquire assets and design and construct facilities. Finally, in order to keep the ten countries committed to the space projects, the so-called 'fair return' principle was enforced: 80 per cent of each country's contribution had to be spent in the country itself. Thus, the industries of the member countries were to get a large share of their own country's financial input.

In order to carry out its mission ESRO needed industry cooperation and wanted to have a single contractor for each project to facilitate coordination and control. However, to keep specialists busy in each firm and in each country, the European organization awarded contracts to prime contractors, which in turn sub-contracted to other firms. Though this arrangement was reasonably satisfactory for ESRO, it was not the best solution for industry. Theoretically, 20 European firms could bid for each of the satellite projects. Since only one firm was awarded the contract this procedure wasted time and resources. It also created unnecessary red tape in the ESRO administration. As a result, contracting was difficult.

This difficulty, compounded by cost underestimation and the worsening European economic situation, resulted in a gradual shift of power within ESRO. The director-general eventually lost his influence and power to the Administration and Finance Committee of the council. Over time the decision-making process moved up the hierarchical ladder and led to a continuing increase in administrative costs. Finally, the bureaucratic process prevailed, and impaired strategic vision.

In the mid-1960s the ESRO Council realized the organization was paralysed and that a solution had to be found. Instead of seeking a political solution, ESRO tried to solve its problems by itself. Mr Bannier, a Dutchman, was appointed by the Council to audit the international organization and suggest policies to improve its functioning. His recommendations were as follows: restore real authority to the director-general, limit the number of problems submitted to the council and its committees, award a greater autonomy to ESTEC and ESOC (the technical arms of ESRO) and ease the budget rules.

At the time these changes were proposed, Professor Auger left the organization and was replaced by Professor Bondi who, with the help of a small advisory committee, started to reinforce ESRO's financial and political situation. He tried to get stronger commitment to the organization's projects from hesitant countries. He initiated a shift toward application satellites to provide a stronger base for ESRO's scientific activities. Finally he organized a better ESRO-industry relationship. Pushed by market forces and ESRO's new leadership, the industry reorganized itself around three consortia:

MESH, COSMOS, and STAR. Under this new structural arrangement, the working relationships between ESRO and the industry improved. ESRO had the necessary expertise and the decision-making power and dealt with one responsible prime contractor at a time. Even if costs ended up twice the initial estimates, ESRO achieved scientific success. This success may be attributed to the specific organizational characteristics of the venture: strong leadership, shared vision, a good ESRO-industry relationship, goal congruence, international institutionalization, and satisfactory crisis management.

ELDO: a stillborn mutual organization

ELDO (European Launcher Development Organization) presents a quite different story. The ambitious ESRO scientific programme led to the idea that Europe should have its own spacecraft. The British, with their Blue Streak rocket as a possible first stage for the launcher and the French, with de Gaulle's objective of European independence and the Emeraude rocket as a compatible second stage, were behind the idea.

A five-year programme with a budget of $196 million to build a European spacecraft capable of launching small satellites was agreed during the first international conference of January 1961, attended by 12 countries. After hard negotiations, work was allocated. The British got the first stage; the French the second stage; the Germans the third stage; the Italians the satellite; the Belgians the ground control; and the Dutch the telemetry.

A convention was signed in April 1962 and ratified in early 1964. Although it was apparently similar to ESRO's, there were striking differences. For example, supervision of the programme was undertaken at the country level. Also, ELDO could place contracts only if these were agreed by the government of the country where the work was to be carried out. The reasons for this procedure were historical. France and Great Britain already had large industrial and scientific teams working on space projects. These teams had to continue working and tasks allocated and financed, pending ratification of the convention; the convention merely formalized the procedure. The result was that national bureaucracies kept control over a technical capability that ELDO could never develop independently afterwards.

Another striking difference between ESRO and ELDO was the preponderance of 'politicians' and bureaucrats in the latter. At ESRO scientists were in charge, but ELDO had definite political undertones. For example, at the ELDO preparatory conferences national representatives were either from the ministry of economics or from the ministry of aviation. In the case of ESRO, representatives were scientists and delegates from governmental organizations concerned with science. Whereas the chief executive officer (i.e. the director-general) of ESRO was a renowned scientist (Pierre Auger), ELDO had a distinguished diplomat as secretary general (Ambassador Carrobio di Carrobio). The secretary general was the principal (but not the chief) executive officer. The result was that ELDO had a structure highly dependent on politicians and bureaucrats. For instance, governments transmitted the demands of their national industries to ELDO without any real evaluation of the proposed projects. Eventually they found themselves in a quasi-coalition with their national contractors against ELDO's global interests. This structure was an anomalous mix of national programmes in which government priorities were dominant and to make things even more difficult, the rule of unanimity on all substantial issues had been adopted.

Early in 1964 three events occurred. First, the budget was revised to $300 million from $196 million. Second, France challenged the project on the basis that the planned rocket was already obsolete. Third, Italy requested that a 'fair return' principle be adopted. Instead of discussing these issues at the council level an "intergovernmental conference' took place in January 1965. The procedure of addressing problems at the highest level was to remain a characteristic of ELDO's decision-making process. Another consequence of this 'consultation' was that no decision was taken and a period of uncertainty followed, leading to a slowing down of the work. At the beginning of 1966 two other events occurred. The British Minister of Aviation sent a memorandum to other nations questioning the usefulness of continuing ELDO. A few days later, ELDO announced that estimated costs were $425 million.

After several meetings at ministerial level and thanks to the successful firing of the first complete EUROPA I configuration (with the last two stages inert), decisions were made. These reallocated the countries' contributions; renewed commitment to the EUROPA I programme; introduced a new project for launching a 200 kg satellite; increased the budget to $626 million; introduced a 'just return' principle with a minimum of 80 per cent of the budget spent in the contributing country; created a 'Committee of Ministerial Alternatives' institutionalizing a decision-making structure above the council level; and established two management directorates responsible for the initial and supplementary programmes. The formation of the new directorates weakened the authority of the general secretariat. For example, the DFA (Director of Future Activities) was almost autonomous from the Secretariat. He had the rank of deputy secretary general, a separate budget, and reported directly to the ELDO Council. Even though these decisions gave renewed impetus to the organization, the crises that had led to them slowed the work pace and lowered morale.

On the industrial front the main actors decided to coordinate their efforts to prevent the collapse of ELDO. They created an Industrial Integrating Group whose role was to solve the problem of lack of coordination between ELDO and the different industrial partners. By mid-1967, SETIS was created. Its role was to advise ELDO on technical matters. Unfortunately, it had no financial or operational responsibility and was compelled to leave supervision of industrial contracts to the government technical units.

In 1968 a new crisis hit ELDO. Estimated costs had now more than tripled since the inception of the programme to $720 million. Britain warned that it would not participate after 1972. In late 1970 a fourth European Space Conference was held in Brussels. Britain again confirmed that she would withdraw from the programme after 1972. Belgium, France and Germany determined to go ahead by themselves, but constrained by their resources, had to find an organizational solution. A few years later their solution was the creation of a new European organization: the European Space Agency.

ELDO represents a case of a mutual organization that was never able to achieve precise objectives. First, no clear impetus from the top was ever given. Participants aimed for parochial goals without really caring about the overall objective of the organization. There was a definite lack of goal congruence. Second, the power structure was in the hands of politicians so that decisions rapidly reached the ministerial level. ELDO managers never had the last word since decisions were made outside the formal structure. Third, the industrial integrator could never build an operational and multinational organization that might have been used to coordinate all European space industry efforts. Finally, the succession of crises and escalating costs led to low morale, demotivation and demobilization of personnel.

OBSERVATIONS AND CONCLUSIONS

Airbus, Concorde, ESRO, and ELDO illustrate four different forms of mutual organization. Though at the macro-organizational level they seem to exhibit the same characteristics they are, in fact, very diverse at a micro-organizational level. Their diversity, however, narrows when the four organizations are compared two by two in terms of their relative success. For example, Airbus and ESRO, the two successful organizations are very comparable in terms of their organizational features; on the other hand, Concorde and ELDO, the two less successful ventures are similar with respect to their managerial design (see Table 1).

In the first two, Airbus and ESRO, professionals are in charge; they have full power to implement decisions and are the real champions of their project. The two organizations are relatively decentralized in managerial and administrative decision-making and have a relatively short and rational decision making process. Coordination amongst partners and integration of industrial firms are satisfactory. Roles are well defined and there is a shared view of what the organizational goals should be. At Airbus and ESRO there is a reinforcing mechanism that induces partners to stick together. The partners are shareholders and suppliers and the 'fair return' principle insures against an inequitable sharing of the resources. When there is a conflict it is often resolved within the organizational structure, allowing speed and relatively unbiased attention to the problems. In the same vein, insofar as control of operations is concerned, only end results and financial standards matter. The general managers are charged with controlling and implementing a corrective action if necessary. Finally, each organization is flexible and adaptable in responding to a changing environment.

When we turn to Concorde and ELDO the picture is different. Here the top management is highly involved in internal politics. Decisions are taken at ministerial level and general managers lack the power to give the organizations the required leadership. The impact of such an arrangement is that all administrative and strategic issues are dealt with at the apex of the structure. Centralization of decisions increases the time needed to analyse alternatives, make a choice and translate it into actions. Conflict resolution takes place more often at the highest level i.e. at the government level. There is also a definite lack of coordination among partners and with industry. Overall goals are generally not well shared by the national participants; divergent objectives prevail. Similarly, there is no real organizational identity; cultural values are blurred. There are few reinforcing mechanisms and incentives, or they are unsatisfactory and frequently constitute sources of conflict. Finally, both organizations are inflexible and parochial, preventing them from adapting to new situations easily.

Thus, analyses of the four organizations indicate that the more successful are those with what can be called adaptive forms of management; the less successful ones are of the stable kind. In a rapidly changing environment in which technological and economic forces are uncertain, it is likely that a more adaptable type of management leads to success as is the case with Airbus and ESRO. When organizations lack strategic vision, leadership, positive incentives and a rapid decision-making process, as in the case of Concorde and ELDO, they are bound to fail.

In addition, the mutual organization raises several managerial issues including, *ex-ante*, the right choice of partners who need to be complementary and synergistically reinforcing. The ownership mix and risk allocation among participants have to be such as to induce cooperations and achieve a fair balance among individual contributors. The cost of the transaction has to be matched by the benefits to each participant. Also, the objectives of the mutual organization need to be clarified and shared by its members to limit future haggling. Finally, to secure an efficient and smooth functioning organization, role definition is essential.

Opportunistic partner behaviour needs to be limited and monitored through balanced incentives and appropriate control mechanisms. The mutual organization must be granted enough power to make decisions almost by fiat. Generally, the decision process which, by its very nature, is long in this type of organization needs to be improved to prevent low morale and an absence of direction. The weak sense of loyalty observed in some mutual organizations is also an important issue. Organizational identity has to be developed as quickly as possible and grounded in specific cultural values. Career management is another problem. Member organizations should reward their employees' efforts when they are assigned to the new structure and give them a guarantee concerning full appreciation of their time and work. Finally, to prevent problems of competition between the mutual organization and individual partners, precise and equitable rules should be drawn.

These managerial issues and the recommendations that have been made are based on the study of the four cases discussed above. It is likely, however, that further analyses of new situations will improve our understanding of an increasingly important form of organization: the mutual organization. In particular, issues related to communication, incentives and the building of trust within this type of organization need further research. The industrial challenges associated with the management of large cooperative ventures require innovative organizational structures. Alliances among diverse partners seem to be one of the key factors for survival in an environment where risks are high and resources are scarce. The mutual organization, when appropriately managed, is one of the answers to these challenges.

REFERENCES

Airbus Industrie (1981). 'Legal organization of Airbus Industrie', Airbus Industrie Contracts Department, Toulouse.
Aldrich, H.E. (1976). 'Resource dependence and interorganizational relations', *Administration and Society*, **7**, 419–55.
Arrow, K.J. (1969). 'The organization of economic activity: issues pertinent in the choice of market versus non market allocations', in *The Analysis and Evaluation of Public Expenditure: the PPB System*, Vol. 1, US Joint Economic Committee, 91st Congress, 1st Session, US Government Printing Office.
Arrow, K.J. (1974). *The Limits of Organization*, Norton, New York.
Child, J. (1987). 'Information technology, organization and the response to strategic challenges', *California Management Review*, **30** (1), 33–50.

Table 1. Micro-organizational features of the four mutual organizations

Organization features	Airbus	ESRO	Concorde	ELDO
Top management group	Professional. Powerful. Responsible. Loyal. Strong leadership.	Professional Powerful. Purposeful. Loyal. Strong leadership.	Political. Management by committee. Fuzzy and changing power structure. Fuzzy and changing leadership.	Political decisions at the ministerial level. Split loyalty. Powerless.
Degree of decentralization	Decentralization of operating decisions. Relative decentralization of administrative functions.	Decentralization of administrative and operating functions. Specialists have freedom.	Centralization of administrative functions. De facto decentralization at operating level.	Centralization of administrative functions. De facto decentralization at operating level.
Decision making	Relatively short decision-making process. Rational means-ends analysis.	Relatively short decision-making process. Rational means-ends analysis.	Long decision-making process. Political and technological analysis.	Long decision-making process. No real planning.
Coordination	Good industrial coordination (manufacturing level). Integration of marketing and servicing functions.	Good research and industrial coordination. Good coordination between industrial partners.	No coordination between industrial partners. Informal coordination at the technical level. Dual organization for manufacturing and marketing.	No effective coordination with industry. No effective coordination among countries
Goal congruence	Commercial goal shared by all partners.	Scientific goal shared by all partners.	Technological goal congruence. Divergent upper goals.	No goal congruence between partners.

	Airbus proposes, partners decide, manufacture and take risks.	ESRO develops and designs, industry manufactures.	...ut work allocation is clearly defined. Partners discuss, decide and manufacture.	ELDO is responsible but has no power. Under supervision of their government, national firms design and manufacture.
Organizational culture and identity	French leadership (at the beginning) Growing institutionalization and loyalty. Partners 'transcend' themselves.	European organization. Scientific community is the key in building an organizational identity.	No formal organization. Shared engineering pride.	Blurred organizational identity. Parallel and parochial organizations.
Incentives	High opportunity Cost of abandoning project. Partners are shareholders and suppliers.	'Just-return' principle applied. Firms are not risk-takers.	Government agreement prevents them from resigning. Industrial partners are not risk-takers.	Imbalance between partners. 'Just-return' principle is violated.
Conflict resolution	Airbus Industrie is a matchmaker. Important conflicts are solved at the Executive Committee level.	Important conflicts are solved at the Council level. Problems are solved within the organization.	Partners must agree on everything. Important conflicts are solved at the ministerial level.	Partners must agree on everything. Lengthy conflict resolution process. Conflicts are solved at the government level.
Control mechanism	Financial and end results. Executive Committee and director-general are the ultimate controllers. Chairman of the council is a gatekeeper.	Financial and end results. Director-general is the ultimate controller. Control checks are not frequent.	Technological results. In principle: Committee of Directors decides; management board monitors.	Mostly financial. Financial committee is the ultimate controller. Frequent control checks.
Adaptation	Great flexibility. Airbus structure allows operational flexibility (Airbus manages product modification asked by customers).	Flexible and adaptable.	Some technological and design adaptability but impaired by decision-making problems. Partners stick to their position.	Low flexibility. Parochial decision-making. Technological conservatism. Heavy weight of national industries.

Coase, R.H. (1937). 'The nature of the firm', *Economica*, **4**, 386–405.

Contractor, F.J., and Lorange. P. (eds) (1988). *Cooperative Strategies in International Business*, Lexington Books, Lexington, Mass.

Costello, J., and Hughes, T. (1976). *The Concorde Conspiracy*, Charles Scribner, New York.

Eccles, R. (1982). 'Quasi-firms in the construction industry', *Journal of Economic Behaviour and Organization*, **2**, 335–57.

Ghemawat, P., Porter, M.E., and Rawlinson, R.A. (1986). 'Patterns of international coalition activity', in M.E. Porter (ed.), *Competition in Global Industries*, Harvard Business School Press, Boston, Mass.

Granovetter, M. (1985). 'Economic action and social structure: the problem of embeddedness', *American Journal of Sociology*, **91** (3), 481–510.

Hall, P. (1980). *Great Planning Disasters*, Weidenfeld & Nicolson, London.

Harrigan, K.R. (1983). *Strategies for Vertical Integration*, Lexington Books, Lexington, Mass.

Hochmuth, M.S. (1974). *Organizing the Transnational*, Sijthoff, Netherlands.

Horwitch, M. (1982). *Clipped Wings: the American SST Conflict*, MIT Press, Cambridge, MA.

Jarillo, J.C. (1988). 'On strategic networks', *Strategic Management Journal*, **9** (1), Jan.–Feb., 31–42.

Johanson, J., and Mattsson, L.G. (1987). 'Interorganizational relations in industrial systems: a network approach compared with the transaction-cost approach, *International Studies of Management and Organization*, **XVII** (1), 34–48.

Knight, G. (1976). *Concorde: The Inside Story*, Weidenfeld & Nicolson, London.

May, A. (1979). 'Concorde—bird of harmony or political albatross', *International Organization*, 481–508.

Moxon, R.W., Roehl, T.W., and Truitt, J.F. (1988). 'International cooperative ventures in the commercial aircraft industry: gains sure, but what's my share', in F.J. Contractor and P. Lorange (eds), *Cooperative Strategies in International Business*, Lexington Books, Lexington, Mass.

Newhouse, J. (1982). *The Sporty Game*, Alfred A. Knopf, New York.

Perrow, C. (1986). *Complex Organizations, A Critical Essay* (3rd edn.), Scott-Foresman, Glenview, IL.

Pfeffer, J., and Salancik, G.R. (1978). *The External Control of Organizations*, Harper & Row, New York.

Pratt, J.W., and Zeckhauser, R.J. (1985). *Principals and Agents: The Organization of Business*, Harvard Business School Press, Boston, MA.

Robins, J.A. (1987). 'Organizational economics: notes on the use of transaction-cost theory in the study of organizations, *Administrative Science Quarterly*, **32**, 68–86.

Schmidt, S.M., and Kochran, R.A. (1977). Interorganizational relationships: patterns and motivations', *Administrative Science Quarterly*, **22**, June, 220–34.

Simon, H. (1957). *Administrative Behaviour* (2nd edn), Macmillan, New York.

Simon, H. (1979). 'Rational decision making in business organizations', *American Economic Review*, **69**(4), September, 493–513.

Thorelli, H. (1986). 'Networks: between markets and hierarchies', *Strategic Management Journal*, **7** (1), 37–51.

Teece, D.J. (1987). 'Profiting from technological innovation: implications for integration, collaboration, licensing and public policy', in D. Teece (ed.), *The Competitive Challenge*, Ballinger, Cambridge, Mass.

Teece, D.J. (1988). 'Capturing value from technological innovation: integration, strategic partnering, and licensing decisions', *Interfaces*, **18** (3), 46–61.

Walker, G. (1988). 'Strategic sourcing, vertical integration, and transaction costs', *Interfaces*, **18** (3), 62–73.

Williamson, O.E. (1975). *Markets and Hierarchies*, The Free Press, New York.

Williamson, O.E. (1979). 'Transaction cost economics: the governance of contractual relations', *Journal of Law and Economics*, **XXII** (2), 233–61.

Williamson, O.E. (1981). 'The modern corporation: origins, evolution, attributes', *Journal of Economic Literature*, **XIX** (4), 1537–68.

Williamson, O.E. (1985). *The Economic Institutions of Capitalism*, The Free Press, New York.

8

The Changing Structures of R&D: from Centralization to Fragmentation

Richard Whittington

School of Industrial and Business Studies, University of Warwick

Industrial research and development (R&D) has been undergoing increasing fragmentation since the 1970s. Large central laboratories have been cut back; contract research has boomed; and small firms and small units have claimed a rising share of innovation. This chapter examines these trends as they have affected six central laboratories and eight independent R&D organizations over the last decade. Underlying the trends, it will be argued, is a general effort to intensify market pressures on R&D workers. Scientific and technical work is being increasingly 'externalized' (Smith, 1989) by a deliberate shift from internal facilities towards outside contractors operating on the open market. Where complete externalization is impossible, remaining in-house R&D facilities are indirectly 'exposed' to market pressures by transferring them to decentralized profit-centres, with research contracts allocated on a competitive basis (Whalley, 1986).

This chapter situates the growing externalization and exposure of R&D workers within a general crisis of 'organized capitalism' (Lash

The Strategic Management of Technological Innovation. Edited by R. Loveridge and M. Pitt
(c)1990 by John Wiley & Sons Ltd

and Urry, 1987). From the mid-1970s, large capitalist enterprises have faced an increasingly turbulent and competitive environment. The pace of technological change has rapidly accelerated; the economy has been subjected to cyclical fluctuations (1974–5 and 1979–81) of a severity unprecedented since the Second World War; and growing trade union organization has forced a crisis of control over the labour process (Shutt and Whittington, 1987). At the same time, markets have globalized, governments have deregulated and the 1980s takeover boom has put company managements under increased pressure to maximize performance. The result has been a reversal of the historical development of advanced capitalist economies towards organizing economic activity within ever larger, bureaucratic corporations. The U.K. in particular has seen falling industrial concentration, decentralization by large firms to peripheral plants and a burgeoning of the small firm sector (Shutt and Whittington, 1987). Under increased competitive pressure, large corporations have sought greater productivity and flexibility by fragmenting their operations into decentralized internal profit centres or delegating to external sub-contractors. Thus large corporations have responded to greater market pressures by turning market forces upon their own managers and workers. Traditionally centralized and cosseted, R&D facilities have been particular victims of this general substitution of market disciplines in place of bureaucratic control.

THE RISE AND RESTRUCTURING OF INDUSTRIAL R&D

The beginnings of industrial research are often dated from Thomas Edison's establishment of the Menlo Park laboratory in 1876, where scientific labour was applied in unprecedentedly systematic fashion to the problem of electric light (Hughes, 1985). By 1899, there were 139 laboratories in US industry, mostly employed in the routine science of testing and standard setting (Rosenberg, 1985). From around the turn of the century, a few very large laboratories began to emerge, dedicated to more fundamental research. Noble (1977) describes how General Electric founded its first research laboratory in 1901 under the management of a professor of physical chemistry from the Massachusetts Institute of Technology. This laboratory grew from only eight employees in 1901 to 555 by 1929, when one of its members won General Electric's first Nobel Prize. In the interwar period, such large, basic research laboratories remained confined to a handful of companies led by Bell, du Pont, Dow and General Electric itself. However, the Second World War transformed industrial R&D activity in the United States: between 1933 and 1945, the number of scientists employed in

American manufacturing rose from 11,000 to 46,000 (Chandler, 1985). There began what Graham (1985) describes as the 'Age of Big Science'. Led by the aerospace and electronics sectors and supported by high Government spending, large American corporations of the post-war period invested generously in large central laboratories that were deliberately detached from manufacturing sites and directed towards research for the medium to long term.

Industrial research in the United Kingdom followed a similar pattern, but perhaps with less pace and less commitment. The numbers of scientific staff directly employed by British manufacturing were proportionately lower during the interwar period than in the United States and companies were reluctant to invest in facilities of their own, preferring to rely on outside consultants and research associations (Mowery, 1986). This failure to internalize R&D had adverse consequences. According to Mowery (1986), dependence on external consulting engineers undermined the development of technical expertise amongst the firm's own staff, entrenching technological conservatism. The contracting-in of consultants on an occasional basis also condemned British manufacturers to series of customized, one-off designs that prevented standardization and militated against economies of scale. However, the Second World War provided an important stimulus to British industrial research, too, expenditure rising from £4.5m in 1938 to £21.8m in 1945, and then to £40m in 1950. ICI, which had long had divisional laboratories, opened its first central laboratory in 1946 and AEI followed suit in 1947 (Edgerton, 1987).

Thus, in the years following the war, the large, central laboratory emerged as the leading institution for industrial R&D. Twiss (1986) summarized some of the main advantages of such centralization. Concentration of R&D activity within a single organization on the same site facilitates the coordination of programmes and reduces the risk of various sites all busily devoting themselves to 'inventing the same wheel'. Likewise, expensive equipment need not be duplicated at every site, and can be used more intensively at the one laboratory. A single laboratory is more able to achieve necessary 'critical mass' in activity, bringing together teams of scientists from diverse disciplines of sufficient strength to address the most complex problems. By locating scientists in central laboratories distant from the factories, they are removed from the everyday distractions of manufacturing and free to concentrate on longer-term research. Finally, a central laboratory can explore technological opportunities that may not fit in well with the narrow concerns of any of the company's existing divisions, but which may be vital to its long-term future as a whole.

Nonetheless, central laboratories come very expensive, and can only be afforded by the wealthiest companies. Department of Trade and Industry figures suggest that the cost of maintaining 100 scientific and support staff, equivalent to a fairly small laboratory, amounted to about £3.3m per annum in 1986 (*British Business*, 5 February 1988).

The benefits of centralized R&D reinforce the traditional Schumpeterian arguments in favour of large scale innovation and for the existence of an effective cut-off point below which R&D is too small to be worthwhile (Kamien and Schwartz, 1975). Examining *Business Week* data on US firms in 1975 and 1976, Soete (1979) did indeed find that the intensity of R&D expenditure increased with size of firm. However, two kinds of scale effect can be distinguished: those of the firm as a whole and those of the R&D facility itself. According to the Schumpeterians, the gains from scale in R&D itself relate to the ability of larger laboratories to attract better researchers, to develop specialists and to benefit from the cross-fertilization of different projects within a diversified overall research programme. The significance of firm size, on the other hand, derives from larger firms' access to cheaper finance for R&D and, within multi-product firms at least, wider opportunities to apply any particular scientific discovery (Fisher and Temin, 1973; Kamien and Schwartz, 1975). Of course, it is not unreasonable to expect that firm size and R&D facility size will be pretty much correlated; nevertheless, it remains theoretically possible for large firms to obtain scale advantages even in small, dispersed R&D units.

This distinction may be becoming more important. The post-war pattern of growing R&D efforts concentrated within the central laboratories of the largest manufacturing firms appears to be breaking up. The change has not been so much in the quantity of activity as in its organization. Despite an easing of growth in the 1960s, and even a slight decline in the mid-1970s, the secular trend for British industrial R&D still seems upwards, with total in-house expenditure reaching £5.146 bn. in 1985, in real terms 24% above the level of 1964 (*British Business*, 9 December 1983; 24 July 1987). What has changed is that this industrial R&D activity appears to be diffusing away from large firms and from central laboratories.

Perhaps a reflection of this diffusion is the increasing doubt being thrown on the Schumpeterian thesis. Using a larger sample than Soete's (1979), but one also based on U.S. firms in the mid-1970s, Bound *et al.* (1984) contradict the Schumpeterians by finding a u-shaped relationship between R&D intensity, as measured by expenditure, and firms' size. Here small firms appear to overcome the threshold effect by being proportionately more active than medium sized ones. Cohen *et al.* (1987), again relying on U.S. data from the mid-1970s, found that once industry effects and outliers had been removed there was no significant influence (amongst actual R&D performers) of either firm or unit size on proportionate R&D expenditure. Bound *et al.* (1984, p.50) speculated that the size–R&D intensity relationship was not fixed and might now be shifting against large scale.

These two studies question the Schumpeterian thesis, but confined to a single period in the 1970s, do not suffice to demonstrate some reversal of the historical trend. However, there are other signs that the post-war pattern of concentration and centralization is suffering increasing disruption. Firms appear to be reverting to the external supply of technology lamented by Mowery (1986). This time the reliance upon outsiders appears not to be just confined to the United Kingdom. Rubinstein (1985, pp.140–1) reports that in the United States:

> 'Although it was not clear until the past seven or eight years, there is now a definite trend towards the elimination and reduction of the size and scope of corporate research laboratories in large companies . . .. There is a clear trend towards considering or actually doing some "buying" of technology outside the firm rather than depending on internal R&D groups to produce it.'

In the UK Twiss (1986, p.196), too, notes a pattern of closures and cutbacks at central R&D labs during the early 1980s. A parallel trend had occurred for British extramural (sub-contracted) R&D. After an apparent decline during the 1960s and most of the 1970s (cf. Lowe and Silver, 1986), the level of extramural R&D as a proportion of intramural (in-house) expenditure rose from 2.9% in 1978 to 8.2% in 1986 (see Table 1). This trend is reflected also in the increasing numbers of organizations offering technical and research services on the open market. Between 1976 and 1985 (after which the classifications were changed), the UK Kompass Register of Industry and Commerce recorded a 59% rise in the number of separate entries for consulting mechanical engineers, a 62% rise in the number of consulting electrical and a 43% rise in the number of product research organizations. The Kompass Register indicates that many of these new entrants were in-house laboratories selling their services externally for the first time. This growth in external R&D should not be dismissed as a recession phenomenon, for it continued after the economic recovery beginning in mid-1981. More recently, the 44 members of the Association of Independent Research and Technology Organizations (AIRTO) reported a 24% increase

Table 1.   R&D expenditure (£m)

	1967	1972	1978	1981	1985	1986
Intramural	605	831	2324	3793	5146	5673
Extramural	20	27	68	na	332	463
Extramural/Intramural (%)	3.3	3.2	2.9	na	6.5	8.2

Note: data collection on extramural R&D was curtailed in the 1981 and 1983 surveys: however, for 1981 the amount of R&D carried out by research associations increased to £88, from £51m in 1978.
Sources: British Business, 9 December 1983; 24 July 1986; 5 February 1988; and 1986 extramural figures, J. Bowles, DTI.

in aggregate turnover between 1986/7 and 1987/8, and claimed to have served 93 out of the United Kingdom's top 100 industrial enterprises (data provided by G. Adler, Hon. Secretary, AIRTO).

This shift in R&D activity has been paralleled by a change in sources of actual R&D output. The Science Policy Research Unit's (SPRU) database of 4378 UK innovations between the late 1940s and the early 1980s reveals another u-shaped relationship between firm size and innovative success, with both smaller and larger firms out-performing medium sized firms over the whole period (Pavitt, Robinson and Townshend, 1987). Significantly, the share of innovations attributed to small firms has been increasing over the last decade or so: in the period 1960–64, firms with UK employments of less than 200 accounted for only 13.6% of innovations; by 1980–83, this proportion was 26.3%. By 1980–83, firms with employments between 100 and 199 were more than twice as innovative as their share of total UK employment would imply. This trend extended also to small units, which includes small establishments within very large firms: units with less than 200 employees increased their share of innovations from 17.4% to 32.1% over the same period. Though most of this growth in small unit innovativeness is accounted for by the increased performance of the small firms coming under this category, some is attributable to a shift of innovation by larger firms to smaller internal units. However, Pavitt and his colleagues (1987, p.202) concentrated on small firms when they concluded that 'the relationship between innovation intensity and firm size is not r-shaped, but if anything u-shaped, and is becoming increasingly so over time'.

The SPRU researchers suggested a number of advantages that small firms might have in innovation. Both Freeman (1982) and Rothwell (1985) cited evidence for the greater innovative efficiency of small firms in terms of innovations per £ expenditure: small firms are quicker and more effective in reacting to changes in markets; they enjoy the leadership of dynamic, entrepreneurial managers, keen to take advantage of new opportunities; and their compactness precludes communications barriers, especially between R&D and marketing. These alleged qualities of small firms may sound suspiciously like the conventional wisdom of today's 'enterprise culture' and they do not explain why small firms have suddenly become *more* innovative than before.

Before going on to examine various possible explanations for the resurgence of small firm and small unit innovativeness, it is worth noting a common theme of Freeman and Rothwell's catalogue: whether in terms of flexible reactions, or on account of opportunity-seeking entrepreneurs, or because of the greater influence of marketing, small firms are always characterized as more responsive to markets. Though interpreted in a slightly different way, the argument here is that this sensitivity to markets is particularly significant for the recent shift in R&D activity.

THEORETICAL EXPLANATIONS FOR THE RESTRUCTURING OF R&D

The evidence, then, is that since the 1970s the post-war pattern of industrial R&D activity has undergone significant changes. These shifts in the balance between large and small firms in innovativeness and research intensity, and between in-house and extramural R&D, need to be explained.

Table 2. Capital expenditure as a proportion of total intramural R&D expenditure

	1978	1981	1983	1985	1986
Capital (%)	9.5	8.5	7.4	10.0	9.3

Sources: British Business, 8 August 1980; 5 February 1988.

One possible explanation is that technologies have advanced to a level of complexity demanding a range of specialisms beyond the capacity of even large firms to maintain in-house (Stingelin, 1984; Kennedy, Read and Crossley, 1985; Stankiewicz, 1986). Accordingly, larger firms have simply ceased trying to maintain adequate facilities of their own, and instead have taken to shopping around wherever they can to get the requisite skills. Whilst this might help explain the rise of R&D consultancy, it fails to account for the increased innovativeness of small firms: because of their limited resources, these might be expected to suffer most from an overall rise in complexity. An alternative explanation almost reverses the argument by suggesting that Graham's (1985) 'Age of Big Science' is over, and that the efficiency rationale for the centralization of large-scale, expensive equipment (Twiss, 1986) no longer applies. Accordingly, small firms and small units have re-emerged because crucial technologies such as computing have become cheaper. However, Table 2's aggregate statistics on R&D capital investment do not suggest that equipment costs are declining (though it is possible that companies are spending the same on more): the dip in capital investment's share of all R&D costs during the early 1980s is probably attributable to recession, and appears to have been compensated by higher spending in the mid-1980s. A related explanation might be that technological and organizational improvements have now increased the ease with which dispersed small units can coordinate and communicate (Murray, 1983; cf. Child, 1985). Thus better organization and communications have tipped the balance away from large centralized laboratories to dispersed networks of smaller ones, in-house or extramural. Though there is some evidence for this effect in manufacturing generally, for R&D such organizational and technological improvements might equally be expected to lift the communications constraint that has traditionally depressed optimum laboratory size (Noltingk, 1985). But far from increasing laboratory sizes, the reports of Twiss (1986) and Rubinstein (1985) suggest that large firms are actually cutting back their laboratories.

The SPRU researchers have often suggested that the role of small firms changes according to the stages of Kondratieff long-waves or technology life-cycles (e.g. Freeman, 1982). The relationship between large and small firms over these cycles is said to be one of 'dynamic complementarity' (Rothwell, 1983), with the lead in innovation passing to and fro, each needing the other to complete the development. Thus, in the classic case of CAD/CAM, large firm laboratories initially performed the basic research; they then spun-off a mass of independent entrepreneurs who carried on the pioneering development work in their own small firms; finally, as the industry matured, these independent firms either became big themselves or were re-absorbed by large firms (Kaplinksy, 1983). The increasing innovativeness of small firms, then, would simply reflect a current upswing in the long wave or a bunching of new technologies in their pioneering stages. However, such an account confines itself to small firms and does not directly address the increased emphasis on small units in general. Moreover, Pavitt, Robinson and Townshend (1987) tested this argument in the technologically dynamic electronics sectors and failed to find that small firms were significantly more innovative in these areas than in industry overall.

The above explanations rely heavily on posited changes in technology. Technology has become 'smaller' or more complex; it has become more dispersive or more pioneering. Here, by contrast, R&D's shifting balance between large and small, in-house and extramural, will be explained in social and economic terms, situating it within a general crisis of 'organized capitalism' (Lash and Urry, 1987). The restructuring of R&D is not driven by its special dependence on technology, but is just part of a wider 'disorganization' of capitalist institutions. Crucial is the increasingly dynamic and competitive nature of contemporary capitalism.

Since the 1970s, the old national and regional oligopolies have been stripped of their protection from competitive market pressures. Product markets have become global, money markets have internationalized, the world has been linked by rapid, even instantaneous communications and equity markets have been deregulated. Comfortable niches have disappeared and ruthless predators stalk under-performing assets. As the U.K.'s growing trade deficit amply demonstrates for manufacturing as a whole, British companies are struggling to compete in these freer markets. The predicament is perhaps most severe for those large companies that hitherto carried out most industrial R&D. Reversing the previous trend towards concentration, the top 100 UK manufacturing firms' share of net output has fallen from 42.2% in 1974 to 37.2% in 1985 (*Business*

Monitor, 1985; cf. Hughes and Kumar, 1984). Some of this fall may be accounted for by disproportionate resort either to internationalization of production or to subcontracting of work formerly done in-house, but these large firms' exposure to world markets (Pratten, 1986) suggests it is *they* who are losing most market share in this period of increasing global competition. Added to growing pressure in product markets is the threatening rise in activity in the market for corporate control. In a highly active takeover market—expenditure rising from £2.3 bn in 1983 to £27.7 bn in 1987 (*Acquisitions Monthly*, various)—managements are trying to secure shareholder loyalty by concentrating on immediate profit maximization at the expense of longer-term investments whose returns are uncertain. Even the largest companies have become vulnerable to new techniques of leveraged takeover bids (Chiplin and Wright, 1987).

Under intensified market pressures, bureaucratic forms of organization come under increasing strain. Changing and competitive markets demand more flexible forms of organization and strategy (Piore, 1986). Direct control through organizational hierarchy is being increasingly replaced by indirect control through market regulation. Rather than carrying out operations for themselves, large companies shift responsibility to outside subcontractors whose costs and performance can be regulated by competitive tendering (Fevre, 1986; Shutt and Whittington, 1987). Retained operations are decentralized—financially through the introduction of divisional profit centres, geographically by the proliferation of branch plants—so that profit responsibility can be isolated and loss-making operations surgically removed (Marginson, 1985; Shutt and Whittington, 1987). Where disintegration or decentralization is infeasible or inadequate, remaining bureaucratic controls are reinforced by 'cultural controls', processes of indoctrination deliberately designed to change employee attitudes and create loyalty to company values (Ray, 1986).

Not only has the shock of competition been greatest for those large firms traditionally most active in industrial research, but R&D presents particularly acute problems for bureaucratic control. The chief input to the innovative process is the expertise of the scientists and engineers themselves. Labour accounts for 41.8% of all UK industrial research establishments costs, well above the 10–15% typical of manufacturing as a whole (*British Business*, 5 February 1988; Rubery *et al.*, 1987). However, the essentially creative and unroutine nature of much R&D work makes management control over this expensive and vital labour particularly elusive.

Originality cannot be produced to order. Scientific and engineering staff must be allowed discretion to find creative solutions to complex problems (Armstrong, 1987). This discretion can be abused. The value of R&D workers to their firms derives from technical excellence, but in developing this excellence workers may come to identify with their technologies rather than with their businesses. Their education celebrates specialization, generates expectations of autonomy and inculcates a desire to contribute to science as a whole, not just to the narrow commercial interests of a particular firm (Raelin, 1985; Keenan and Newton, 1986). Scientists are likely to exploit their autonomy by applying themselves to technical priorities rather than commercial business concerns. Management control became still more difficult during the 1970s as scientific staff increasingly participated in trade unionization (Gunz, 1979; Smith, 1987).

Bureaucratic control, therefore, had always been problematic with R&D workers and, with spreading trade unionization, was becoming more so precisely when, from the mid-1970s, their large firm employers were coming under tightening market pressures. As elsewhere in their operations, large firms responded by disintegration and decentralization. Outside market pressures are being translated into the firms themselves. Thus, as Smith (1989) argues, technical workers are experiencing increasing 'externalization' of their positions, being transferred from relatively stable internal labour markets into competitive markets as subcontractors and consultants. Where the scope for externalization is constrained, as Walley (1986) described in his Metalco case study, management attempts instead to expose their scientific and engineering staffs to market pressures internally. This 'exposure' strategy involves providing employees with more financial information on how the business is doing; increasing use of commercial criteria for promotions; and the conversion of technical centres into profit centres, trading with operating companies within divisionalized corporate structures at market prices. Whalley (1986, p.239) summarized the effect of this exposure strategy on Metalco's engineering staff thus:

> 'In the highly decentralized corporate structure of Metalco, the pressures on engineers' technical concerns came from customers rather than higher management... [The company] substituted market disciplines for direct bureaucratic control.'

The shifting balance between large and small units in innovation and the substitution of subcontracting for in-house R&D can be seen as part of a wider process of disintegration and decentralization following a crisis in 'organized capitalism'. Here, recalling the qualities attributed by Freeman (1982) and Rothwell (1985), the sensitivity of small firms to markets is crucial. However, this sensitivity is not some innate virtue, but born of vulnerability and powerlessness. Large firms can demand servile responsiveness from their small firm suppliers because they enjoy superior market power. They can exert their superior market power on recalcitrant R&D staff in similar ways. By transferring them away from the insulated environments of the old central laboratories into more vulnerable units, whether independent operations or autonomous divisions, management can extract the same 'market sensitivity' as that required of small firms. Isolated and exposed in these smaller units, R&D must live or die by the market. Thus the relative rise in small firm innovativeness since the 1970s can be interpreted as a shift by large firms to greater reliance on subcontracted R&D and, in terms of technology life cycles (Kaplinsky, 1983), a readiness to delay reincorporation of pioneering spin-off companies back into corporate bureaucracies. Likewise, the down-sizing of central R&D laboratories and the trend within large firms toward greater innovativeness in small internal units can similarly be understood as part of a general process of decentralization intended to increase exposure to market forces. In itself, size is not the crucial variable; what matters is exposure to markets.

THE EMPIRICAL RESEARCH STUDY

This research involved six case studies of in-house R&D organizations within large manufacturing firms and eight studies of independent contract R&D organizations. Table 3 describes their areas of activity and the numbers employed in 1987 (in the central laboratories of the in-house facilities, and in total for the independents). The company names are of course pseudonyms.

The data were obtained from 41 interviews with research managers, together with published and internal documentary data. In order to focus on situations where the processes of disintegration and decentralization

Table 3. Case studies

Company	Activities	R&D employees (1987)
In-house		
Electron	Electronics/Communications	900
Autoco	Vehicle industry supplier	22+6
Commodco	Construction Commodity	55
Glassco	Glass products	68
Engco	Various engineering activities	160
Lectric	Electrical/Communications	280 (closing)
Independents		
NORA	Industry research association	226
DORA	Industry research association	190
URA	Industry research association	85
Contech	Technical consultants' R&D wing	120
Gencon	General consultants' R&D wing	400
Testco	Test house	80
Indsci	Research organization	200
Prodco	Product development	250

were likely to be most advanced, the in-house laboratories were selected from those that, according to *Longmans' Directory of Research Organizations*, had suffered particularly sharp falls in employment in the first half of the 1980s. The term R&D is used broadly, covering activities ranging from fundamental research to rather routine testing and analysis.

The in-house laboratories were all products of the centralization of R&D beginning before the Second World War but accelerating thereafter. Lectric's laboratory was the oldest, established in 1934; Electron came next, founding its first central laboratory in 1946 and moving to its present site in 1959; Commodco began systematic research during the war, but did not establish a central laboratory until 1951; Engco set up its central laboratory in 1954; Glassco began research activities at one of its factory sites and finally set up a separate large laboratory in 1964. Autoco was exceptional in being the only central R&D laboratory to have always been adjacent to the main factory: developing organically from technical services, it gradually achieved status as a discrete laboratory with its own R&D programmes during the 1950s.

The three research associations were founded between the wars under government initiatives (cf. Kennedy *et al.*, 1985). Of the five true independents, all but Indsci (founded just after the Second World War) were established during the 1960s and 1970s. Several independents quickly developed substantial R&D staffs; but, unlike the in-house and research association laboratories, they had never enjoyed any special subsidies or client relationships. Instead they had always been dependent for survival on competitive success in the market-place.

The manufacturing companies' had little in common in terms of technologies and did not share equivalent positions on technology life-cycles. Technological sophistication ranged from Electron's commitment to high-tech communications, computing and defence to Autoco's involvement in traditional mechanical engineering. Engco's laboratory served a wide range of electrical and mechanical engineering businesses, few of which were based fundamentally on new technologies. Commodco and Glassco were both involved in materials raising considerable problems in process plant engineering, but whose basic scientific properties were by now well-understood.

What these companies had in common, however, were tightening economic environments. The focus here is on how increased exposure to market forces has been forcing the in-house laboratories to model themselves more and more upon the market-driven style of the independents. The research associations have a special position which will become clear later; in a sense they provide a half-way house between the once-insulated, in-house laboratories and the dedicated commercialism exemplified by 'Gencon' and 'Prodco'.

Old habits and new pressures

Until recently, the six in-house laboratories manifested all the problems traditionally associated with managing R&D personnel. Mostly situated in country-house locations distant from manufacturing units, they were widely accused of being 'ivory towers' or 'private universities without teaching'. Technical rather than business problems were paramount. One Electron research manager remembered the old attitudes thus:

'If there was an interesting technical problem, then that was sufficient justification. If they could come up with a way that it could be put into practice, that was a bonus. And if it just so happened that *Electron* could use it, then that was even better. The name of the game was technology, almost for its own sake.'

Scientific staff particularly disdained marketing their capabilities: one director described the attempts of his staff to sell themselves to clients as 'frankly embarrassing'. Co-operation with sister companies was often poor. At Commodco:

'The works would have a problem they couldn't solve themselves. A bunch of boffins would arrive; sit in the plant so it was overmanned with PhDs and BScs rather than hourly-paid blokes who left school at fourteen; run it for about three days; disappear back to (central R&D); and about two months later a report would arrive saying there wasn't much of a problem and the works should have been able to solve it themselves. The works manager would read the report and decide it was the last time he would use that lot.'

Isolated and able to hide behind a screen of impenetrable science, R&D escaped detailed management scrutiny. One non-technical personnel manager likened his R&D outfit to a 'Masonic lodge'. Comfortably subsidized by corporate levies, there was usually little pressure to perform to business criteria. A research manager at Autoco observed:

'One of the problems of a central research and development laboratory is that, not having to earn its *keep*, you get "cost centre creep". Obviously there can be a lot of expenses associated with development, things like travel, silly things like that, but when you've got products all over the world, it doesn't come cheap. A lot of research and development gets out of touch with the nuts and bolts, and there's a certain glamour associated with it, and the costs tend to go up. And that's when the axe falls. That's exactly what happened here.'

As market pressures intensified during the 1970s and 1980s, Autoco's laboratory was not the only one to feel the edge of the axe. Product markets in general were getting increasingly competitive. Heavily committed to the vehicle industry, Autoco was devastated by the decline of the UK assemblers. Cheap imports from the recently extended EEC had contributed to forcing down Commodco's U.K. market share from 60% to 40%. Glassco's market was steadily falling by 2% per annum as plastics substituted for glass. These product-market pressures were not confined to the manufacturing companies. NORA and DORA both served industries in chronic decline in the face of Third World imports. An equivalent challenge for the research associations was the change in their funding arrangements. Originally financed by statutory industry levies and matching Department of Trade and Industry (DTI) funds, since the Rothschild Report of 1971 they had been increasingly forced on to the 'customer-contractor' principle: by 1987, membership subscriptions had fallen to 0.5% of total turnover for NORA, 5.6% for DORA and 15.4% for URA. DTI funding switched to an increasingly rigorous project basis and was also declining: by 1987, to 15% of turnover for NORA, 18% for DORA and 25% for URA. Government funding was important to the other independents and was equally unreliable: Prodco's defence-related work fell from 50% of its business in 1986 to only 20% in 1987.

The parent manufacturing companies were not only under pressure in their product-markets but also under increasing threat of takeover. Electron, Commodco, Engco and Lectric had all been the subjects of bids in the mid-1980s, and all had introduced major changes in top management in response. Glassco was taken over in 1986. The main board technical director at Commodco described the pressure thus:

> 'We were fighting more or less for our very existence. We were expecting a takeover any time. And so we had to concentrate on the short-term—and get out there and *apply*. Put the resources where they would most benefit the Group, to get the share price up .... They (research) have got to realize that it's profits that count. They've got to understand that the company's share price is all important. The shareholders pay for us and pay for them .... They've got to be commercial. They're not owed a living; they owe others a living.'

Bringing R&D to the market

But conventional bureaucratic controls alone were unlikely to be sufficient in forcing R&D 'to be more commercial'. Under acute market pressures, the manufacturing companies responded by introducing market forces directly into R&D. In-house laboratories were established as discrete business units, pressed towards competitive contracting, and driven to an active marketing commitment.

As a senior manager at Electron put it with sardonic economy, 'The big stick of course is the funding'. The first stage in reforming central R&D was to change its status: five of the six in-house laboratories were established as separate divisions (Glassco was the exception). This enabled the transfer of

R&D funding from centrally determined levies on operating companies to transactions according to market-based criteria. As early as 1972, Engco's central laboratories were switched to charging subsidiaries 50% of cost, the difference being made up by central funds; in 1987, the laboratories were put on full-cost recovery, and required to meet targets for Return on Capital Employed. Despite being reorganized by McKinsey's in 1977, Commodco's subsidy to central R&D was still £6m p.a. in 1980; by 1987, however, it had been slashed to £0.5m. At Lectric, where the centre had provided two thirds of funds, the manufacturing divisions were provoked by tough performance targets in 1986 to demand direct control over their R&D spend: the result was closure of the central laboratory the following year. At Electron, where more than 50% of R&D spending had been provided centrally in the 1970s, by 1987 this had been cut to one sixth, with the bulk recovered by contracts at full cost. Everywhere these transformed funding arrangements stripped central R&D of its former autonomy: now, as a senior Commodco manager put it, 'the *will*, the *drive* and the *cost* control should come from manufacturing'.

Introduction of the customer-contractor principle increased readiness of operating companies to use external contractors for their research and a corresponding pressure for the in-house laboratories to compete on the open-market. These forces contributed to the recent resurgence in extramural R&D already remarked upon (see Table 1). At Glassco now:

> 'The first thing we always ask is: What's available outside? Don't just do it in-house: the horizon is as broad as you want. And if there is something and it does the job, we'll *buy* it. We're not going to reinvent the wheel here just because we think we can do it better.'

At Commodco, senior management welcomed the scope for wider technological inputs, but also well understood that subcontractors were under competitive pressures: 'They are all keen to deliver and there is an obvious incentive'. Central laboratories at Electron, Autoco and Engco were themselves getting involved in chasing external business—by 1987, 88% of Autoco's sales were outside the Group. At Commodco, central R&D's rather unsuccessful experiments on the open-market rebounded: seeing its pursuit of price-competitive external contracts, Group operating companies began to demand the same economies for their own internal contracts.

Reductions in subsidies and sudden entry into the market place were accompanied by severe cutbacks in personnel. All six in-house laboratories suffered redundancies, both amongst professional and support staff (see Table 4). These cutbacks achieved more than just economies; the uncertainty they created provided an important lever for forcing change upon surviving staff. As an Electron manager commented:

Table 4. In-house employment and redundancies

Case	Peak nos (date)	Redundancies (dates)	1987 Nos
Electron	1100 (1982)	1986	900
Autoco	180 (1979)	1981	22+6
Commodco	200 (1979)	1981,1983,1986	55
Glassco	240 (1975)	1982, 1986	68
Engco	210 (1980)	1982, 1984, 1986	160
Lectric	310 (1981)	1986, 1987	Closing

'When you've got a lot of question marks over the place, its much easier to change the style and the culture than if things are just putting along as they have been for the last five, ten, fifteen, twenty years.'

Senior research managers well recognized that the changes in external status and funding imposed an urgent need to achieve radical and pervasive internal reforms. As an Engco director put it:

'We've got to get ourselves from being a cost centre to being a profit centre; not just in the way in which you make a positive or negative profit each year, but in the whole sort of attitude and method of working.'

Often the response was to apply the logic of decentralization not only *to* R&D organizations, but *within* them. At Electron, the old internal departments were restructured into four divisional profit centres, organized to face Group businesses (e.g. defence-related) rather than technology-based as before. Lectric's remaining fragments, too, were split into three profit centres, each closer to specific businesses.

This sort of radical, financial decentralization was already well-established amongst the independents. NORA abolished its discipline-based departments in 1980, establishing four divisions as profit centres in themselves, and spinning-off small manufacturing activities into separate companies. At Testco, financial figures had been 'kept secret' from middle managers until 1983, when it transformed its seven departments into profit centres with financial results broadcast weekly. Prodco took this logic furthest: all six internal divisions were profit centres; the twenty sub-groups each had their own financial targets; all 150 projects were monitored weekly; and each consultant's 'billibility' (charged-out hours) and 'performance factor' (earned income divided by costs) were calculated every week.

Decentralization was spatial too, sometimes with a deliberate anti-union motive. Autoco's central laboratories concentrated largely on testing; in compensation, facilities had gradually grown up at subsidiaries' sites. The largest of these subsidiary facilities, 100 strong by 1987, had become a profit centre separate from its manufacturing parent and planned to move to a different site in order to introduce experimental machine tools free from AUEW restrictions. In the debate over relocating Lectric's surviving researchers, an internal memorandum emphasized the opportunity to escape union-bound 'custom and practice'. A similar motive led both Testco and Glassco to cease recognizing their research and technical unions in 1983.

Exposure to market pressures through decentralization and subcontracting entailed different attitudes to clients on the part research staff. Marketing became crucial. At the highly commercial Gencon, marketing expenditure represented 30% of all costs; at Prodco, 20%. URA, NORA, Electron, Commodco and Engco all experimented with special R&D marketing departments—with more or less success. But the main emphasis was on cultivating market sensitivity amongst the professional staff themselves—it was *they* who understood their product best, and *they* whom the clients would have to trust. Extensive training was employed in order to wean professional staff from their technical exclusiveness. At Gencon, all professional recruits began with a two-week consultancy skills course, after which they were expected to spend one week annually on further commercial courses. Both Electron and Engco recently instituted systematic commercial training for all their professional staff, with the emphasis on communications, sales and basic marketing principles. The message at NORA was that: 'Technical guys have got to stop being shrinking violets who don't like talking pound notes'. At Engco, 'Everybody's a salesman and a marketeer, as well as an engineer. That's the spirit we want to generate'.

The marketing ethos supported a very radical change in working practices. The professional autonomy of scientific discipline was replaced by responsiveness to customer requirements. At Autoco, they were:

'Making them (staff) customer accountable.... Our philosophy is that the customers deal direct with the sections .... The main motivation has got to be the involvement with the customer, with the guy who wants the job.'

Instead of being hierarchically supervised by internal management, work was to be regulated directly by the customers themselves. Customer dissatisfaction could be registered very immediately by a fall in personal 'billibility' or profit-centre profitability. Thus the marketing ethos achieved management control not by insisting on obedience to internal authority, but by emphasizing service to the customer and exposing failure to market disciplines.

Customers also demanded a different sort of service. Working practices and work content changed together. A management memo at Lectric warned:

'Working practices will change. It was not acceptable at [the old centre laboratory] and it will not be at [the new decentralized laboratory] that products are developed independently of processes at pilot stage or factory stage. No more kitchen hatch development—many more feedback loops, all the way through to the factory.'

Whereas before it was an unrealized desire, now management insisted on greater involvement in the immediate problems of production. Managers at all the in-house laboratories observed how the new subordination to operating companies altered the balance of work from longer-term, fundamental research towards short-term applied projects. At Lectric, head-office professed to encourage research but at the same time constantly enjoined its recently decentralized operated divisions to improve short-term profitability. One research manager noted with some exasperation:

'There is a strong determination that the [operating] units will do more R&D. But in the long run, the bottom-line is a strong influence upon people. One is always worried that decentralization will cut people's time horizons . . . . Which guy focussed on next December is going to do long-term research—or even medium-term development?'

CONCLUSIONS

This chapter has examined how the old, concentrated and centralized pattern of industrial R&D has begun to break up since the mid-1970s. The contemporary fragmentation of R&D is manifesting itself in cutbacks at central laboratories, a rise in subcontracting and an increasing share of innovations accounted for by small firms and small units. Though a number of possible explanations exist—mostly relying on posited changes in technology—they have been interpreted here as changes consistent with a general crisis of 'organized capitalism' (Lash and Urry, 1987). The transformation of industrial R&D is thus part of a widespread tendency for large corporations to respond to increased market pressures by imposing market disciplines on their labour.

Externalization (Smith, 1989) and exposure (Walley, 1986) are central strategies in this transformation. Externalization brings about the direct transfer of R&D work from sheltered in-house laboratories to external sub-contractors dependent on the market. The threefold growth between 1978 and 1986 in extramural R&D's share of total R&D expenditure bears witness to the importance of this strategy. Where complete disintegration is infeasible, market disciplines are being imported through increased decentralization. Isolated in profit centres, fragmented into small units, and financially accountable at ever lower levels, even in-house R&D workers are exposed to market pressures (Walley, 1986). By shifting R&D out of the large central laboratories in these ways, managements are able to reclaim control an accountability. One Commodco manager summarized the rationale behind these fragmentations thus:

'People have always been concerned about getting value for money from R&D .... The more you can break it down and make it accountable all relates to getting value for money out of this peculiar beast, R&D .... They (head office) feel that the bigger you make it, the more difficult it is to extract anything worthwhile out of it. The smaller you make it, the more you can apply your technology'.

But this pursuit of accountability through fragmentation and market disciplines involves a reversion to the same reliance on external sources of technology as characteristic of British industry in the inter-war period (Mowery, 1986). Again, companies risk an atrophying of their internal technological capabilities and a dependence on *ad hoc*, idiosyncratic designs. And market-driven demands for improved financial performance concentrate attention on short-term, incremental and applied development, at the expense of strategic research. As a manager at Electron put it bluntly: 'The easiest way to make a profit is to stop doing long-term R&D'.

ACKNOWLEDGEMENTS

The research on which this chapter is based was funded by the Nuffield Foundation. The author would like to thank David Edgerton, Chris Smith and the editors for their helpful comments on earlier drafts of the chapter.

REFERENCES

Armstrong, P. (1987). 'Engineers, management and trust', *Work, Employment and Society*, 1(4), 421–440.
Bound, J., Cummins C., Griliches Z., Hall B., and Jaffe A. (1984). 'Who does R&D and who patents?', in Z. Griliches (ed.), *R&D, Patents and Productivity*, University of Chicago Press, Chicago.
Chandler, A.D. (1985). 'From industrial laboratories to departments of R&D', in K.B. Clark, R.H. Hayes and C. Lorenz (eds), *The Uneasy Alliance*, Harvard Business School Press, Boston.
Child, J. (1985). 'Managerial strategies, new technology and the labour process', in D. Knights *et al.* (eds), *Job Redesign: Critical Perspectives on the Labour Process*, Gower, Aldershot.
Chiplin, B., and Wright, M. (1987). *The Logic of Mergers*, Hobart Paper 107, Institute for Economic Affairs, London.
Cohen, W.M., and Mowery D.C. (1984). 'Firm heterogeneity and R&D; an agenda for research', in B. Bozeman, M. Crow, and A. Link (eds), *The Strategic Management of Industrial R&D*, D.C. Heath, Mass.

Cohen, W.M., Levin, R.C. and Mowery, D.C. (1987). 'Firm size and R&D intensity: a re-examination', *Journal of Industrial Economics*, **35**(4), 543–565.

Edgerton, D. (1987). 'Science and technology in British business history', *Business History*, **29**(4), 84–103.

Fevre, R. (1986). 'Contract work in recession', in K. Purcell (ed.), *The Changing Experience of Employment*, Macmillan, London.

Fisher, F.M., and Temin, P. (1973). 'Returns to scale in research and development: what does the Schumpeterian hypothesis imply?', *Journal of Political Economy*, **81**(1), 56–70.

Freeman, L. (1982). *The Economics of Industrial Innovation* (2nd edition), Frances Pinter, London.

Graham, M.B.W. (1985). 'Industrial research in the age of big science', R.Rosenbloom (ed.), *Research on Technological Innovation, Management and Policy*, JAI Press, Connecticut.

Gunz, H. (1979). 'White collar unions in R&D: some U.K. experience', *R&D Management*, **9**(1), 29–32.

Hughes, T.P. (1985). 'Edison and electric light', in D. Mackenzie and J. Wajcman, *The Social Shaping of Technology*, Open University Press, Milton Keynes.

Hughes, A., and Kumar, M.S. (1984). 'Recent trends in aggregate concentration in the United Kingdom economy', *Cambridge Journal of Economics*, **8**, 235–250.

Kamien, M.I., and Schwartz, N.L. (1975). 'Market structure and innovation', *Journal of Economic Literature*, **13**, 1–37.

Kaplinsky, R. (1983). 'Firm size and technical change in a dynamic context', *Journal of Industrial Economics*, **32**, 1, 39–59.

Keenan, A., and Newton, T. J. (1986). 'Work aspirations and experiences of young graduate engineers, *Journal of Management Studies*, **23**, 2, 224–237.

Kennedy, A. J., Read, N.J., and Crossley, C.L. (1985). *Changes in the Research Associations Over the Decade 1972–82*, Technical Change Centre, London.

Lash, S., and Urry, J. (1987). *The End of Organized Capitalism*, Polity, Oxford.

Lowe, J., and Silver, M. (1986). 'R&D strategies and variable demand', *R&D Management*, **16**, 4, 325–333.

Marginson, P. (1985). 'The multidivisional firm and control over the work process', *International Journal of Industrial Organization*, **3**, 37–56.

Mowery, D.C. (1986). 'Industrial research, 1900–1950', in B. Elbaum and W. Lazonick (eds), *The Decline of the British Economy*, Clarendon Press, Oxford.

Murray, F. (1983). 'Production decentralization and the decline of the mass worker', *Capital and Class*, **20**, 12–31.

Noble, D.F. (1977). *America By Design*, Oxford University Press, Oxford.

Noltingk, B.E. (1985). 'A note on effective laboratory size', *R&D Management*, **15**(1), 65–9.

Pavitt, K., Robinson, M., and Townshend, J. (1987). 'Size distribution of innovating firms in the UK: 1945–1983', *Journal of Industrial Economics*, **35**,3, 185–204.

Piore, M.J. (1986). 'Perspectives on labour market flexibility', *Industrial Relations*, **25**(2), 146–66.

Piore, M.J., and Sabel, C.F. (1984). *The Second Industrial Divide*, New York, Basic Books.

Pratten, C. (1986). 'The importance of giant companies', *Lloyds Bank Review*, **159**, 33–48.

Raelin, J.A. (1985). 'The basis for the professional's resistance to managerial control', *Human Resource Management*, **24**(2), 147–75.

Ray, C.A. (1986). 'Corporate culture: the last frontier of control?' *Journal of Management Studies*, **23**(3), 287–95.

Rosenberg, N. (1985). 'The commercial exploitation of science by American industry', in K.B. Clark, R.H. Hayes, and C. Lorenz (eds), *The Uneasy Alliance*, Harvard Business School Press, Boston.

Rothwell, R. (1983). 'Innovation and firm size: a case for dynamic complementarity; or is small really so beautiful?' *Journal of General Management*, **8**, 3, 5–25.

Rothwell, R. (1985). 'Venture finance, small firms and public policy in the UK', *Research Policy*, **14**, 253–63.

Rubery, J., Tarling, R., and Wilkinson, F. (1987). 'Flexibility, marketing and the organization of production', *Labour and Society*, **12**(1), 131–51.

Rubinstein, A.H. (1985). 'Trends in technology management', *IEEE Transactions on Engineering Managment*, November, 141–3.

Shutt, J., and Whittington, R. (1987). 'Fragmentation strategies and the rise of small units: cases from the north west, *Regional Studies*, **21**(1), 13–23.

Smith, C. (1987). *Technical Workers*, Macmillan, London.

Smith, C. (1989). 'Technical workers: a class and organizational analysis', in S. Clegg (ed.), *Organization Theory and Class Analysis*, de Gruyter, New York.

Soete, L.L.G. (1979). 'Firm size and R&D investment activities', *European Economic Review*, **12**, 319–40.

Stankiewicz, R. (1986). *Academics and Entrepreneurs*, Frances Pinter, London.

Stingelin, D.V. (1984). 'The role and importance of contract research, in *The Survival of Industrial Research and Development*, The Research and Development Society, London.

Twiss, B. (1986). *Managing Technological Innovation* (3rd Edition), Longman, London.

Whalley, P. (1986). 'Markets, managers and technical autonomy', *Theory and Society*, **15**, 223–47.

9

Determinants, Processes and Strategies of Technological Innovation: Towards an Interactive Paradigm

Michael Saren

University of Bath

Research into the managerial and economic aspects of technological innovation at the micro level of the industry and the firm has undergone some substantial changes in direction and emphasis since the early work of Schumpeter (1942). The modern literature on managerial economics and strategic management now focuses more on the role of the firm's *strategy*, rather than on the relative effects of, for example, demand-pull and technology-push explanations of its technological innovation behaviour.

Despite the competitive need for secrecy about such activities, the technological innovation process does not necessarily occur entirely within the single firm. (This is illustrated in an example later in this chapter.) This phenomenon does not appear as yet to be fully reflected in the empirical and theoretical literature on the strategic aspects of technological and innovative

The Strategic Management of Technological Innovation. Edited by R. Loveridge and M. Pitt
©1990 by John Wiley & Sons Ltd

behaviour by companies. For this reason, this chapter examines the role of technological innovation and its affects on companies from an interactive, process-oriented perspective.

EXPLANATIONS OF TECHNOLOGICAL INNOVATIVENESS

A lot has been written about firms' technological innovativeness in terms of the causal determinants. Until fairly recently much of the literature on market structure and technological innovation was concerned with Schumpeter's (1942) twin hypotheses regarding the relative effects on firms' propensity to innovate of perfect versus monopolistic competition and large versus small firm size. Kamien and Schwartz (1982) restated his arguments as follows:

1. Innovation is greater in monopolistic industries than in competitive ones because:
   (a) a firm with monopoly power can prevent imitation and thereby can capture more profit from an innovation;
   (b) a firm with monopoly profits is better able to finance research and development.
2. Large firms are more innovative than small firms because:
   (a) a large firm can finance a larger research and development staff;
   (b) a large diversified firm is better able to exploit unforeseen innovations;
   (c) indivisibility in cost-reducing innovations makes them more profitable for large firms.
3. Innovation is spurred by technological opportunity.
4. Innovation is spurred by market opportunity (demand-pull).

Following Schumpeter there have been a considerable number of empirical studies aimed particularly at testing the effects of the related factors of market concentration and firm size on innovation (i.e. elements 1 and 2 above). Studies include Anglemar (1985), Comaner (1967), Eisner and Strutz (1963), Freeman (1971), Grabowski (1968), Hamberg (1964), Kamien and Schwartz (1982), Kaplinsky (1983), Mansfield (1963), Rothwell and Zegveld (1981), Scherer (1980). (For a partial review see Saren 1987.) Taken together this body of literature has produced no clear conclusions regarding either factor.

The corresponding contrast between technology-push and demand-pull explanations of the stimulus to innovate (i.e. elements 3 and 4 above) has until more recently perhaps had a clearer outcome in favour of the latter.

Obviously both demand and technological conditions must be present (or perceived) for firms to innovate (as 'two blades of a pair of scissors', according to Schmookler, 1966). The famous empirical studies of Carter and Williams (1957), Langrish et al. (1972), Myres and Marquis (1969) and Project SAPPHO (1974) have generally been interpreted as favouring market-need or demand-pull as the more dominant explanation. Mowary and Rosenberg (1979) provided a review and critique of such studies, concluding that this interpretation (i.e. as supporting demand-pull) is not justified, primarily on account of the nebulous and varied semantics of the terms 'need', 'use' and 'demand'. Most studies' methodologies are biased against technology-push factors, they argued.

Recent empirical work—particularly by SPRU-has gone a considerable way to redress this balance in support of a technology-push explanation of innovation. Townsend et al. (1981) collected data on 2000 innovations commercialized in Britain since 1945. Pavitt (1983) has shown on the basis of these that 'broad sectoral regularities in the sources and directions of technological progress do emerge as a function of innovating firms' principal activities'. This work takes the basic demand-pull/technology-push debate to a new frontier. As Pavitt (1984) put it:

'A description and explanation of sectoral differences in patterns of innovation and technological accumulation has certain uses. In innovation studies it will at least help prevent general and sterile discussions about the relative importance of large and small firms in making innovations and of technology-push and demand-pull as stimuli to innovative activities. In strategic management research, it will help identify more precisely the nature of the technological opportunities and constraints faced by firms with different types of principal activity.'

This research links up with concepts such as 'Natural Trajectories of Technological Change' (Nelson and Winter, 1977), 'New Technological Systems' (Freeman, 1982), 'Technological Regimes and Dominant Designs' (Abernathy and Utterback, 1975) and 'Technological Paradigms' (Dosi, 1982) which have turned attention towards the supply-side of technological opportunities as an explanation of firms' innovative activities and, as Pavitt suggests *a fortiori*, of their *strategic* behaviour generally.

TECHNOLOGY AND CORPORATE STRATEGY

Technology strategy

This had led to the idea of company technology strategy. Only ten years ago Kantrow (1980) argued that:

'The major unfinished business of the research literature is to provide managers with needed guidance in their formulation of a technological strategy for their companies.'

The role of technology in the achievement of competitive advantage has, of course, become widely recognized (e.g. Porter, 1985). However the conceptualization of a *technology strategy* suggests a view of the firm in which managers make strategic decisions—explicitly or tacitly, consciously or unconsciously—about technology. Whether technology is interpreted as 'means of production' or as a 'set of competences' this approach clearly conflicts with the neoclassical theory of the firm. However, there is strong support for the notion of the strategically proactive firm emerging from developments in managerial and behavioural theories of the firm and from new research on technological innovation.

Penrose (1959) proposed a theory of growth based on the firm's unique productive resources which provide a 'bundle' of potential services, limiting and directing company growth. Bounded rationality and satisficing—not maximizing—decisions characterize other behavioural theories of the firm (Cyert and March, 1963; Simon, 1976). This accords with Nelson and Winter's (1982) evolutionary theory of the firm operating in a 'selection environment' and with its distinctive, acquired decision-rules. The latter are affected by innovation, thus limiting or modifying strategic options.

According to these types of models, a picture can be constructed of strategic decisions about technology taking place within differentiated managerial hierarchies, responding to uncertainty by utilizing their acquired distinctive resources and skills in a boundedly-rational manner (Moss, 1981). This is a picture, which as Kay (1984) shows is relatively close to the business policy literature, of strategy as the *outcome* of distinctive competences resulting from firms' histories and relationships.

A firm's technology strategy can be posited similarly, wherein managers husband, utilize and transfer their technological competences and resources according to changing environmental and internal opportunities and competitors' actions (Kay 1984). Companies continually face strategic decisions on which technologies they should invest their resources, individually and as part of their *portfolio* of technologies. Rarely can they afford, or have the competence to achieve or maintain leadership in each. Further, they must choose the appropriate *means* of acquisition through their own internal R&D or through contracted-out research, licensing-in or a joint venture. Achievement of the optimum return on a company's technological 'assets' also involves decisions on the timing and means of exploitation—whether by incorporating a technology into the company's own products or processes, licensing it to others or through some form of joint-venture or contracted-out marketing or manufacturing (Ford, 1988).

These arguments provide a rationale for the view that choice of overall strategy has a crucial role in determining technological innovation in the business organization. The central concern of strategic management is making and implementing strategic choices which match the organization's capabilities with opportunities present in the environment to achieve long-term objectives. Empirical research by Johne (1982) based on sixteen firms in the U.K. test instrument industry showed that the nature and pace of technological innovation was closely related to the choice of a 'leader' and follower' innovation strategy. Firms pursuing the former placed more emphasis on product innovation, whilst the latter group emphasized process innovation.

On the other hand, many research studies of innovating firms indicate that a large number of factors *other* than the choice of strategy are related to the rate and success of their technological innovations. (For a review of the empirical findings see Saren, 1987.) These factors can be categorized into four groups:

(1) Economic factors, e.g. size of firm;
(2) Social and behavioural factors, e.g. values, education, attitudes;
(3) Information and communication factors, e.g. contacts with the scientists and technologists;
(4) Organizational and managerial factors, e.g. delegation of responsibility.

It seems that the formulation of strategy, including technological choices, may determine the initial direction and pace of the firm's innovation activity, but other factors such as those above can alter or hinder its successful outcome. However, managerial discretion for strategic *choice* in technological innovation does not appear to be wholly limited by and contingent on these other factors. This can be demonstrated by the fact that even within the same business/market environment firms do adopt substantially different innovation strategies.

Strategic options for technological innovation

In modern business organizations technological innovation is normally aimed at the development of new products and processes. A company's technology strategy for innovation, then, has its roots in overall corporate strategy—Mintzberg's (1988) 'mediating force' linking the organization and its environment. It is with reference to the firm's posture *vis-à-vis* its environment that most writers have expressed the various strategic options for innovation. Few, though, have explored in detail the role of the firm's relationship with *other firms* in its technological innovation activity.

Urban and Hauser (1980) contrasted two broad alternatives: (i) *reactive* strategies, where firms respond to customer demands and competitors activities and (ii) *proactive* strategies, where they seek to forecast and anticipate environmental changes. Depending upon the internal and external circumstances of a particular firm or Strategic Business Unit (SBU), either of these approaches may be appropriate. Within each of these categories they distinguish four sub-options.

Reactive strategies can be (i) responsive, where the firm reacts directly to customers' requests for innovation, (ii) imitative, in reacting to competitors' new product introductions by copying them, (iii) 'second-but-better', by developing and improving on competitors' innovations or (iv) defensive, where their reaction to competitors' innovative challenges is to modify their existing product in some way rather than (or perhaps as well as) developing an entirely new product.

Proactive approaches may be (i) R&D based, with innovations resulting from the technological initiative of the research and development function, (ii) entrepreneurial, where the innovation activity is high-risk and opportunistic, but not necessarily very technically novel, (iii) acquisitive, where innovation is achieved through the purchase of new products or companies, or (iv) marketing-based, from the initiative of the marketing function for an anticipatory (and usually competitively aggressive) product innovation.

Other sources have constructed similar taxonomies to that of Urban and Hauser. Freeman (1974) identified six strategies for technological innovation as Offensive, Defensive, Imitative, Dependent, Traditional and Opportunistic. He linked these to the firm's internal scientific and technical capabilities (see Table 1).

Freeman argued that only a small minority of firms will adopt an 'Offensive' strategy and even they are unlikely to follow this approach consistently over a long time period. He emphasized that the other categories are not entirely distinct and can 'shade' into one another.

Pessemier (1977) counterposed firms with a 'market-orientation' towards product innovation with those which exhibit an 'R&D-orientation'. The first type use relatively stable technologies and follow changing user *demands* to guide their innovation, and thus most new product ideas come from external sources. The second group of firms conduct innovation projects which are initiated and driven primarily by technological advances. This is similar to technology-push and market-pull explanations of innovation. Pessemier, however, contended that in each *industry* one of these strategies tends to predominate because firms face similar environments and competitive conditions. Thus industries tend to be mainly technology-led or market-led in their innovative activities.

Twiss (1980) viewed technological innovation as a 'conversion process' and contrasted two dichotomous approaches which appear *prima facie*

Table 1. Strategies of the firm

Strategy	In-house scientific and technical functions within the firm									
	Fundamental research	Applied research	Experimental development	Design engineering	Production engineering –Quality control	Technical services	Patents	Scientific and technical information	Education and training	Long-range forecasting and product planning
Offensive	4	5	5	5	4	5	5	4	5	5
Defensive	2	3	5	5	4	3	4	5	4	4
Imitative	1	2	3	4	5	2	2	5	3	3
Dependent	1	1	2	3	5	1	1	3	3	2
Traditional	1	1	1	1	5	1	1	1	1	1
Opportunist	1	1	1	1	1	1	1	5	1	5

Range 1–5 indicates weak (or non-existent) to very strong. (Freeman, 1982)

Product orientation

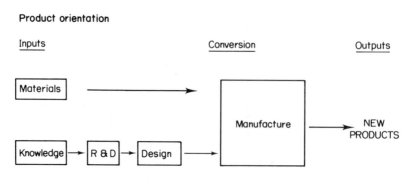

Figure 1 Innovation as conversion process

similar to Pessemier's. These he called a product-orientated approach and a technology/market orientation. The former, which is illustrated in Figure 1, was predominant, he argued, at a time when customers' power in relation to the manufacturer was weak. The change in their relative positions has led to the more widespread adoption of market research in order to direct the innovation process more accurately towards customer needs. The technologists located in the R&D and Design functions remain, though, largely isolated from the users. Thus there is a gap between the *initiator* of the new products and the *market* in this representation of the innovation process.

Twiss's alternative approach is shown in Figure 2. Here the process is

Marketing orientation

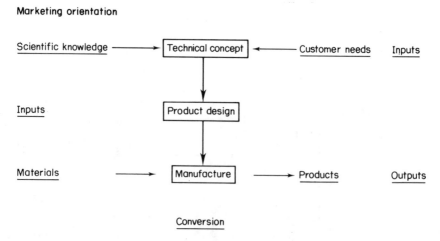

Figure 2 Innovation as conversion process (Twiss, 1980)

viewed as the conversion of scientific or technical knowledge *directly* into the satisfaction of customer need. He argued that this approach represents a considerable shift in technologists' attitudes to innovation, because it focuses their minds on customer needs. These alternative strategies are thus slightly different from Pessemier's. For Twiss the organizational *position* of the technologist, as the initiator of new products, in relation to the users, is the key difference.

There are many other taxonomies, paradigms and models of strategic options for technological innovation (e.g. Ansoff, 1979; Johne, 1982). Like the examples above, these essentially contrast two polarized strategies, variously described as 'leader/follower', 'entrepreneurial/positional', 'internal-led/external-led', 'proactive/reactive'. As we have seen, there are, of course, positions *between* the extremes.

Foxall (1984) posited four 'broad possibilities' of 'strategically relevant' sources of innovative opportunity. His table (Table 2) provides an excellent exposition of a point which is implicit in the approaches to innovation strategy which have already been discussed, namely that there is a relationship between a firm's (or SBU's) innovation strategy and the *source* of both opportunity and entrepreneurship. This ranges from extremely reactive firms who take their product development 'cues' from competitors to the most proactive firms who search through a multitude of sources for innovative ideas—internal and external—in an entrepreneurial manner. In the former case the initiative and entrepreneurial locus is external and in the latter it is entirely within the firm.

Table 2. Strategically-relevant sources of innovative opportunity

Relative strategic response	Nature of source of innovative opportunity	Locus and direction of entrepreneurship
Reactive	1. Producer innovates only when forced to do so in order to survive or protect strategic business unit.	Competitors. From external environment to producer.
	2. User needs identified for producer.	Buyers and/or users. From external environment to producer.
	3. User needs identified by producer.	Producer. From customer-oriented search.
Proactive	4. Relatively unhampered exploration.	Producer. From creative product-market search and entrepreneurial alertness to opportunity.

(Foxall, 1984)

This argument links to von Hippel's (1978) comparison of the 'customer active paradigm' (CAP) and 'manufacturer active paradigm' (MAP) of product and process innovation in industrial markets. In MAP manufacturers scan the market for new product ideas, whereas in CAP the customer initiates the new concept, develops it and *selects* a supplier to manufacture it. These paradigms can be described in terms of Foxall's categories as special cases of, respectively, proactive and reactive strategies.

Although there is some difficulty in comparing categories of external sources, empirical studies of the location of ideas for innovation have suggested that approximately half are generated internally and half emanate from a variety of external sources. Examples of studies by Langrish *et al*. (1972) and SPRU (1972) are shown in Table 3.

A further problem in relating empirical findings about idea sources to strategic issues of innovation is that, as Foxall's and von Hippel's work show, the source of the *idea* is not necessarily the source of the *initiative* or entrepreneurship. For example, in MAP the idea can often be sought outside the firm. Here, the *initiative* is internal but the *source* is external. For technological innovation strategy the key issue is whether the firm is proactive or reactive in *acquiring* new concepts, not where the ideas themselves come from.

If, as was argued previously, strategic choices are not wholly explained by external environmental factors, then firms' differing innovative behaviour must be influenced to some extent by *internal* resources and capabilities. This leads to a consideration of what organizational capabilities are required for technological innovation.

ORGANIZATIONAL CAPABILITIES FOR INNOVATION

Whether innovation strategy is proactive or reactive, certain organizational capabilities are required. The results of a questionnaire survey conducted by the author of a small sample of medium and large sized U.K. companies, suggested that managers, themselves, rate their firms' performance at tasks generally associated with the earlier stages of the innovation process as *lower* than performance at tasks associated with the later stages (Saren, 1986). Figure 3 shows *the average score* at specific product development activities in twelve companies for which managers rated their performance. The arrows indicate the stages of the process with which each task can be associated. Despite the fact that the *range* of scores is not great, it can be seen that their performance was rated lower for: (i) recognizing the need for change, (ii) spotting opportunities and (iii) encouraging staff to innovate. Higher ratings were given for tasks such as: (iv) facilitating project proposals and decisions, (v) controlling projects and (vi) imitating ideas.

Table 3

(a) 158 'key technical ideas'—in 51 innovations (Langrish *et al*., 1972)

Source of ideas		Number of ideas
Within the firm	56	
Outwith the firm	102	
External:		
of which		
Person joining the firm		20.05[a]
Industrial experience		15.00
Educational experience		9.00
Commercial agreement (e.g. takeover)		10.05[a]
Literature (technical, scientific and patent)		9.05[a]
Personal contact in the country (UK)		8.05[a]
Collaboration with supplier		7.00
Collaboration with customer		5.00
Visit overseas		6.05[a]
Government organizations		6.00
Conferences in the country (UK)		2.05[a]
Consultancy		2.00
		102.00

[a] In some cases the sources are not mutually exclusive.

(b) Project Sappho : 86 innovations (SPRU, 1972)

Source of ideas	Percentage of cases
Internal—40 innovations	
External—46 innovations:	
of which	
Universities	22.0
Government agencies	28.0
Other industrial companies	39.0
Outside individuals	11.0
Research associations	0.0
	100.0

These results might be explained by the conditions at the earlier stages of the innovation process where the managers are often operating with more uncertainty, less information, greater novelty, little or no previous experience, lack of organizational commitment and thus lower managerial *confidence* compared with the better-defined and 'manageable' tasks later in the process. This would also be consistent with the much higher drop-out rates which have been found to occur at the initial stages, despite more recent evidence that the failure rate has been falling (Booz, Allen and Hamilton, 1968; 1982).

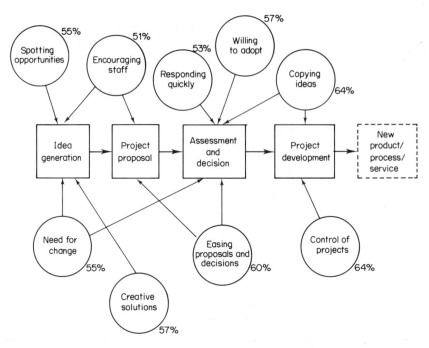

Figure 3 Performance at product development tasks during the innovation process (Saren, 1986)

The capabilities to encourage internal initiative and respond to external developments are clearly important for innovation strategy at *any* position in the proactive/reactive continuum. Even viewed from a formal 'strategy-as-plan' perspective these are key implementation functions. When innovation strategy is conceptualized in terms of the firm's posture or perspective, then internal performance capabilities are themselves an influence on the firm's strategic behaviour.

The innovation process has been represented in many different 'models' and these have tended to concentrate on the intra-firm aspects, usually as a series of stages (Saren, 1984). The simplest of these is the 'activity-stage' representation found in many textbooks (Figure 4), through which the embryonic innovation progresses from idea generation to commercialization. A number of problems with this conceptualization have been recognized, such as: the implication of clear decision and task points, the suggestion of order and structure and the assumption that the process is sequential rather than recursive. Also, it might be expected that recent developments in CAD/CAM and FMS technologies could make the process more elliptical.

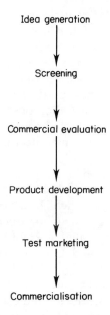

Figure 4   The innovation process

By its very nature innovation tends to be chaotic (Quinn, 1985) and no two innovations seem to follow the same path from conceptualization to commercialization—even within the same organization (Kanter, 1983). Therefore, it is unlikely that technological innovation can be entirely planned, nor entirely opportunistic, whichever strategy is followed. The firm's innovation performance capabilities are also likely to depend qualitatively on the extent to which innovation in the organization is 'top-down' or 'bottom-up' and the extent to which innovation results from 'induced' and 'autonomous' behaviour (Burgelman, 1983). It might be postulated that the greater the extent of autonomous, bottom-up, entrepreneurial initiatives for innovation in the firm, the more strategic 'forcing' is likely to occur, which—in Burgelman's terms—results in innovations that lie outside the current strategic thinking.

A CASE STUDY OF TECHNOLOGICAL INNOVATION AND NEW PRODUCT DEVELOPMENT

The firm in this case is a traditional U.K. company which had achieved market leadership in high quality inspection equipment. It had begun to decline over the last ten years and was acquired by a larger manufacturer in 1984. At that time the firm's product range was technologically outdated and the market segments in which it had concentrated were declining in value terms at 18% p.a. The new managing director, after consultation with Group HQ, decided to undertake a modernization programme for the entire product range. Staff were assigned to study new technologies with the potential to form the basis of a new product range and to research possible new (to the firm) markets which the company could enter. Thus, the innovation strategy which was adopted could be described in hindsight as induced (by top management) and manufacturer, as opposed to customer, active.

The external search produced a 'map' of the metrology market and a number of potential technological developments in various stages of completion. The decision as to which to pursue was predominantly *technology driven* in that the new products—a range of advanced measurement devices—incorporated a combination of developments in computing, electronics, optics and laser technologies (*all* of which were new to the firm) but they were targeted at the same metrology market as the existing mechanical products, and with the eventual aim of replacing them. The invention of the basic device and the technological development of the prototype had been conducted by a university engineer in his own spin-off company. He had been 'discovered' during the technology search initiated by the MD.

Because the firm had very limited capabilities in these technologies— no skilled staff, no manufacturing equipment, no distribution or sales expertise—critical consideration and some considerable management time were devoted to the selection of the *means* of acquiring the required technologies, and—linked to this—the method and source of manufacturing. The implementation of the chosen innovation was complex for the management of the firm, involving relations with Group management; negotiations with the inventor, suppliers and legal representatives; technical development and, more critically, *control* of the manufacturing process; and managing the organizational and staffing implications of the adoption of the innovation.

Although in this case the innovation was 'MAP' and the company initiated the search for new technologies, *at the same time* the inventor-supplier of the prototype product was searching for a company with suitable distribution, sales development and (perhaps) manufacturing capabilities. The eventual *realization* of this innovation was the result of the meeting and combination of two separate organizations each of which was—in a sense—looking for each other. It would be hard to determine *objectively* where the key initiative was located—with the inventor or the metrology company—although *subjectively* each initiated both a 'search' and key tasks in the innovation process. The initiation and development of this new product range appears to provide a case of what is termed here the *interactive paradigm* of technological innovation.

CONCLUSION: TOWARDS AN INTERACTIVE PARADIGM

Many empirical studies have concluded that external market and technology links are important for successful innovation (e.g. SAPPHO), but what is postulated in the interactive paradigm is that the technological innovation process involves activities and *initiatives* on the part of a number of *organizations*. Firms are operating in complex technology 'networks' (Johanson and Mattson, 1987), 'chains' (Clarke, Ford and Saren, 1988), 'filière' (Malsot, 1980) or 'Gesellschaften' (Tonnies, 1957) and it is within such a context that the process of technological innovation occurs. This is separate from the process of *diffusion* of innovation, which takes place in and across networks of adopters (Rogers, 1962). Here the argument is that the earlier process of *initiation* and *development* (i.e. up to the time of market launch) of a technological innovation can take place in conjunction with, and are dependent on, other organizations at various stages. The emergent models of 'high-tech' innovation through the mechanisms of science and technology parks, joint-ventures and technological 'alliances', and the proliferation of independent R&D companies in industries such as computer software and biotechnology, provide some anecdotal evidence to suggest that the interactive paradigm is becoming more common.

If this paradigm has validity, then it requires at the level of the firm that the management of technological innovation takes much more account of the potential of inter-firm relationships and innovation networks in the development of strategy. It means that the firm's strategic behaviour regarding technology should be viewed not only in terms of its position along a proactive/reactive or leader/follower continuum or in respect of Schumpeterian technology-push/demand-pull forces, but also in terms of its inter-firm relationships and networks. This suggests that decisions about future technologies need to be taken within a broad contextual appreciation of the firm's competitive *and* collaborative prospects.

Thus, when reviewing technological competences, managers are advised to consider explicitly how these may be enhanced or, if necessary, substituted through the medium of pro-active networking. As the case study illustrates, such possibilities may exist even in seemingly unpromising situations. However, it is equally clear that 'importing' new technologies to the firm is inherently risky, particularly when it is undertaken as a response to a declining market position under competitive threat.

REFERENCES

Abernathy, W.J., and Utterback, J.M. (1975). 'A dynamic model of process and product innovation', *Omega*, III(6), 634–56.

Anglemar, R. (1985). 'Market structure and research intensity in high-technological-opportunity industries', *Journal of Industrial Economics*, 36, 69–79.

Ansoff, H.I. (1979). *Strategic Management*, Macmillan, London.

Booz, Allen, and Hamilton (1968). *Management of New Products*, Booz, Allen and Hamilton, New York.

Booz, Allen, and Hamilton (1982). *New Products Management for the 1980s*, Booz, Allen and Hamilton, New York.

Burgelman, R.A. (1983). 'A process model of internal corporate venturing in the diversified major firm', *Administrative Science Quarterly*, vol.28.

Carter, C.F., and Williams, B.R. (1957). *Industry and Technical Progress*, Oxford University Press.

Clarke, K., Ford, I.D. and Saren, M.A.J. (1988). Strategic Management and Technology Strategy, Eighth Annual Strategic Management Society Conference, Amsterdam.

Comaner, W.S. (1967). 'Research and technical change in the pharmaceutical industry', *The Review of Economics and Statistics*, 182–190.

Cyert, R.M., and March, J.G. (1963). *A Behavioural Theory of the Firm*, Prentice-Hall, Englewood Cliffs, New Jersey.

Dosi, G. (1982). Technological paradigms and technological trajectories, *Research Policy*, 11, 147–162.

Eisner, R., and Strutz, R.H. (1963). 'Determinants of business investment', in *Commission on Money and Credit—Impacts of Monetary Policy*, Prentice-Hall, Englewood Cliffs, New Jersey.

Ford, I.D. (1988). 'Develop your technology strategy', *Long Range Planning*, 21(5), 85–95.

Foxall, G.R. (1984). *Corporate Innovation: Marketing and Strategy*, Croom Helm, London.

Freeman, C. (1971). 'The role of small firms in innovation in the UK since 1945', Report to the Bolton Committee of Inquiry on Small Firms, Research Report No.6, HMSO.

Freeman, C. (1974). *The Economics of Industrial Innovation*, Penguin, Harmondsworth.

Freeman, C. (1982). *The Economics of Industrial Innovation* (2nd edition), Frances Pinter.

Grabowski, G. H. (1968). 'The determinants of industrial research and development: a study of the chemical, drug and petroleum industries', *Journal of Political Economy*, 76, 292.

Hamberg, D. (1964). 'Size of firm, oligopoly and research: the evidence', *Canadian Journal of Economics and Political Science*, 30(1), 62–75.

Johanson, J., and Mattson, L.G. (1987). Interorganizational relations in industrial systems: a network approach compared with the transaction-cost approach', *International Studies of Management and Organization*, XVII(1), 34–48.

Johne, F.A. (1982). Innovation, organization and the marketing of high technology products, PhD thesis, University of Strathclyde.

Kamien, M.I., and Schwartz, N.L. (1982). *Market Structure and Innovation*, Cambridge University Press.

Kanter, R. M. (1983). *The Change Masters*, Simon & Schuster, New York.

Kantrow, A. (1980). 'The strategy-technology connection', *Harvard Business Review*, July–August.

Kaplinsky, R. (1983). 'Firm size and technical change in a dynamic context', *Journal of Industrial Economics*, **32**, 39–59.

Kay, N.M. (1984). *The Emergent Firm: Knowledge, Ignorance and Surprise in Economic Organization*, Macmillan, London.

Langrish, J.M., Gibbons, M., Evans, W. and Jevons, F. (1972). *Wealth from Knowledge*, Macmillan, London.

Malsot, J. (1980). 'Filières and pouvoirs de domination dans le système productif', in *Les Filières industrielles*, Annales des Mines, January.

Mansfield, E. (1963). 'Size of firm, market structure and innovation', *Journal of Political Economy*, **LXXI**, 556–76.

Mintzberg, H. (1988). 'Opening up the definition of strategy in J.B. Quinn, H. Mintzberg, and R.M. James, *The Strategy Process*, Prentice-Hall.

Moss, S. (1981). *An Economic Theory of Business Strategy*, Martin Robertson, Oxford.

Mowery, D. and Rosenberg, N. (1979). 'The influence of market demand upon innovation: a critical review of some recent empirical studies', *Research Policy*, nineb 8, 102–53.

Myres, S., and Marquis, D.G. (1969). *Successful Industrial Innovation: a study of factors underlying innovation in selected firms*, National Science Foundation, NST 69–17, Washington.

Nelson, R., and Winter, S. (1977). 'In search of useful theory of innovation', *Research Policy*, **6**, 36–76.

Nelson, R., and Winter, S. (1982). *An Evolutionary Theory of Economic Change*, Harvard University Press, Cambridge, Mass.

Pavitt, K. (1983). 'Some characteristics of innovative activities in British industry', *Omega*, **11**, 113–20.

Pavitt, K. (1984). 'Sectoral patterns of technical change: towards a taxonomy and a theory', *Research Policy*.

Penrose, E. (1959). *The Theory of the Growth of the Firm*, Blackwell, Oxford.

Pessemier, E.A. (1977). *Product Management: Strategy and Organization*, Wiley, New York.

Porter, M. (1985). *Competitive Advantage: Creating and Sustaining Superior Performance*, Free Press, New York.

Quinn, J.B. (1985). 'Managing innovation: controlled chaos', *Harvard Business Review*.

Rogers, E. (1962). *Diffusion of Innovations*, Collier-Macmillan, New York.

Rothwell, R., and Zegveld, W. (1981). *Industrial Innovation and Public Policy*, Frances Pinter, London.

Saren, M.A.J. (1984). 'A classification and review of models of the intra-firm innovation process', *R&D Management*, **14**(1), 11–24.

Saren, M.A.J. (1986). Intrapreneurship and Innovation: Implications for Management, Paper presented at the 4th Plenary Conference of the Business Strategy Network, Oxford.

Saren, M.A.J. (1987). 'The role of strategy in technology innovation', in I.L. Mangham (ed.), *Organization Analysis and Development*, John Wiley, London.

Scherer, F.M. (1980). *Industrial Market Structure and Economic Performance*, 2nd edition, Rand McNally, Chicago.

Schmookler, J. (1966). *Invention and Economic Growth*, Harvard University Press, Cambridge, Mass.

Schumpeter, J.A. (1942). *Capitalism, Socialism and Democracy*, Allen & Unwin, London 1976: first edition, Harper & Row, New York, 1942.

Science Policy Research Unit (1972). *Success and Failure in Industrial Innovation*, Centre for the Study of Industrial Innovation, London.

Simon, H.A. (1976). *Administrative Behaviour*, 3rd edition, Macmillan, New York.

Tonnies, F. (1957). *Gemeinschaft und Gesellschaft*, first published in 1887, translated by C.P.; Loomis, published as Community and Society, 1957, Michigan State University Press, Michigan.

Townsend, J. *et al*. (1981). *Innovations in Britain Since 1945*, Occasional Paper, No.16, Science Policy Research Unit, University of Sussex.

Twiss, B. (1980). *The Management of Technological Innovation*, Longman, London.

Urban, G.L., and Hauser, J.R. (1980). *Design and Marketing of New Products*, Prentice-Hall, Englewood Cliffs, New Jersey.

Von Hippel, E. (1978). 'Successful industrial products from Customer Ideas', *Journal of Marketing*, **42**(1).

10

Innovation-design and Innovation Poles

Peter Clark

Aston University, Birmingham

and

Chris DeBresson

University of Quebec in Montreal, Canada

This chapter explains the analytic requirement for the concepts of innovation poles and innovation-design and shows how they are part of the line of analysis originally developed by Utterback and Abernathy (1975). The perspective on strategy and technology now associated with Utterback and Abernathy emerged through several complementary investigations spread over some fifteen years and is still evolving. The perspective is not tightly bonded though it aims to treat technology and strategy as a total process (Kantrow, 1980).

Abernathy, in various publications until his death in 1983, sought to extend the original insights through organization theory and tests of the Utterback and Abernathy (U–A) model. In addition, the analysis of design and development (D and D) in Japanese car firms raised the spectre of a

The Strategic Management of Technological Innovation. Edited by R. Loveridge and M. Pitt
©1990 by John Wiley & Sons Ltd

new best practice which would require the 're-paradigming' of corporate America and its business schools. Abernathy gradually shifted his attention from all-embracing meta prescriptions to the requirement for detailed descriptions of technologies, to ascertain the degree to which they were *entrenching* or *altering* in their consequences for the existing organization structures.

The contributions of this chapter are twofold: first, to direct attention to the role of specialist professionals situated in the technostructure of large corporations and to show how their expertise (or its absence) provides an internal design capability; second, to indicate the importance of innovation poles in corporate survival.

The central points are illustrated in an analytically structured narrative over eight decades for a British automobile firm: Rover (Whipp and Clark, 1986). The situation of Rover, its moments of triumph and of crisis, are particularly relevant to understanding the strategic problems of:

(i) resolving the dilemma between an orientation to efficiency versus an orientation to innovation (Abernathy, 1978);
(ii) developing strategic policies for competing through technology as part of design (Clark, 1987).

The relationship between Rover and Honda during the 1980s and the incorporation of Rover into British Aerospace in 1988 exemplify the continuing need for a useful theory of the relationships between strategy, innovation and technology.

UNRAVELLING INNOVATION-DESIGN

The problem

There are sharply different interpretations of the *patterning* of radical and incremental innovation over time for any sector. Two figures may be used to illustrate the differences.

At one time it seemed that the problem of formulating strategic policies for technological innovation had been resolved. The solution seemed to be embodied in the stylized U–A 'model' illustrated in Figure 1 (Utterback and Abernathy, 1975). Their model compared profiles of the ideal-typical firm during the founding and emerging eras of a sector and its profile during the 'mature stage' many decades later. The logic of the model was that in the mature state technological innovation should be incremental in both the product and in the production processes. As applied to the automobile industry, the strategic logic was encapsulated in the notion of the world car produced cheaply through economies of scale.

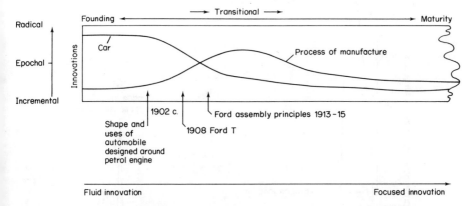

Figure 1 Radical-incremental patterns of innovations for products and for processes over the life course of the sector: case of automobiles. (Modified from J.M. Utterback, and W.J. Abernathy (1975). A dynamic model of process and product innovation, *Omega*, 3(6), 639–656. Copyright (1975) Pergamon Press plc)

But an alternative model has now appeared. In this model the modern age is depicted as being typified by dramatic, radical innovations in raw materials, in the performance of equipment and in the use of information to control processes. The new model is illustrated in Figure 2 which shows that there have been dramatic bursts of innovative activity in the modern period (Tushman and Anderson, 1986), as well as irregular bursts in the past. The interpretation which we place on Figure 2 is that any sector is impacted in an irregular manner by innovations which occur outside the sector in the wider economy. These macro level bursts of innovative activity lead to massive economic restructuring, during which past forms of corporate skill and technology become redundant.

Utterback and Abernathy: vision of a solution

Figure 1 displays the degree to which innovation in technology is radical or incremental in a longitudinal pattern covering many decades and it distinguishes between the degree of radical innovation in the product compared with radical innovation in the production process. Radical innovation occurs at the founding of sectors, initially in the product and then in the production process. The end of the founding period is signalled by a decline in radical innovation in product and by the onset of a short period of radical innovation in the production process. Thereafter, both are typified by incremental innovation.

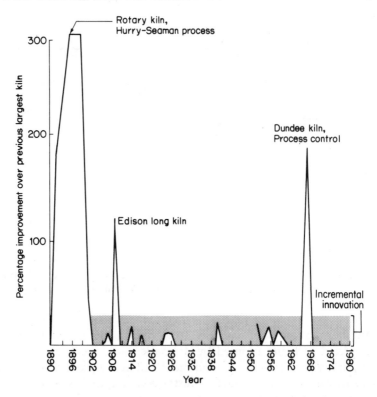

Figure 2 Irregular epochal bursts of innovation in the largest US cement kiln (1890–1980). (Derived from Tushman and Anderson, 1986.) (Barrels per day production capacity)

Figure 1 was derived from large data bases on innovation. It is an *icon*, representative of a whole genre of prescriptive models produced for policy makers in the 1960s and 1970s. All claimed impeccable empirical pedigree and all were equally sharp in their strategic implications. Most required considerable explication and some now require rejection.

The implications of this long-term pattern for strategic decisions about technology are clear: the modern firm in a long established sector like automobiles should produce a standard product in huge volumes, from components manufactured in routine technologies situated in the least expensive locations. These components are then channelled to assembly plants convenient to the market. A further implication is that large-scale firms will continue to prosper whilst medium scale plants will be forced out of business because of their excessive costs.

Any projections for the automobile sector made in 1970 based on this model would have forecast the disappearance of small-scale manufacturers (e.g. Saab, BMW, Jaguar, Mercedes and Volvo) and the prob. concentration of the Japanese industry into a single firm. Since 1970 the large-scale American firms have found their international position and their position in the North American market *challenged* by Japanese firms (e.g. Toyota, Honda and Nissan) as well as by speciality models imported from Europe. Moreover, it is apparent that the approach to technology of the Japanese and the Europeans has differed from that of the North Americans (Clark, 1987). The U–A model therefore required both unravelling and refinement.

The dilemma of efficiency versus innovation

The U–A model suggested that sectors gradually acquire highly standardized product-markets. In other words, that an orientation towards efficiency should replace an earlier orientation towards innovation. However, by 1980 Abernathy, Hayes and their co-workers were offering a very different interpretation of the international future of the American automobile industry. The explanation for this shift was that testing of the U–A solution revealed very serious problems. Testing commenced by applying the U–A model to a longitudinal study of strategy and technology in one major American car firm (Abernathy, 1978) and continued with a detailed analysis of why the Japanese produced cars more cheaply than the Americans (Abernathy, Clark and Kantrow, 1983).

Abernathy reasoned that if the U–A model was a useful diagnostic approach then it should be possible to demonstrate that Ford—a successful firm—had moved along the curves shown in Figure 1. The test had two themes. First, the selection of twenty key innovations in the product, its assemblies and in the production process as *markers* of the transition from radical, 'fluid' innovation to incremental, 'specific' innovation (see Figure 3). Second, the recognition that the degree of movement from radical to incremental was likely to vary between different stages of the production process. Stages were usually undertaken in separate establishments (defined as productive units, e.g. engine plants and assembly lines). The productive unit was the focus of analysis and the U–A model predicted that the portfolio of productive units in a firm would shift from fluid to specific along various dimensions as in Figure 3. Abernathy concluded that the productive units in Ford had moved in the directions postulated by the U–A model.

However, Abernathy's interpretation of the U–A model and of the Ford study deserves and requires careful assessment (DeBresson and Lampel, 1985a; Clark and Starkey, 1988, pp.22–36). It is necessary to examine the use of organization theory, especially the perspectives of Woodward (1965) and Perrow (1967) because Abernathy did *not* use these perspectives to analyse

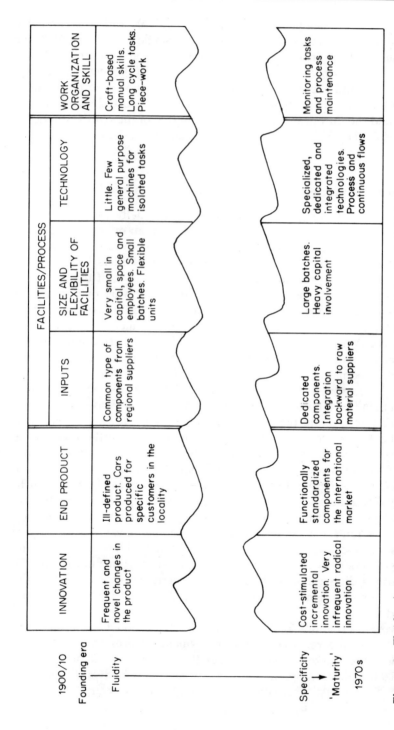

Figure 3 Fluidity/specificity in production units. (From R. Whipp and P.A. Clark (1986). *Innovation and the Auto Industry: Product and Work Organization*, Frances Pinter, London. *Original source*: derived from W.J. Abernathy, *The Productivity Dilemma*, 1978, 148–9.)

the implications of the changing market and technological environments. In many respects the programme for 're-paradigming' corporate America provided too little analysis on the medium-term organizational problems of re-adaptation.

Unravelling design with organization theory

Abernathy (1978) used organization theory, especially the perspectives of Woodward and Perrow, but the usage was confined to explicating the U–A model and its tendencies. Consequently the implications of organization theory for a redirection of innovation towards epochal and altering transitions were understated.

Abernathy highlighted the role of designers in providing specific solutions to the dilemma of combining *productivity* with *innovation*, yet the analysis of design was incomplete (Nelson and Winter, 1982: Chs. 4 and 11; Clark, 1987: Ch. 5: Clark and Starkey, 1988). The role of the technostructure in the design of new cars and in the structuring of activity at the operating units was central for Abernathy, who fully intended to develop the U–A model by unravelling the role played by the functionally specialized professionals within design in specifying events within the portfolio of productive units.

To continue this direction, one must examine the role of organization theory in the analysis of the market and the role of technostructure. Organization theory provides a diagnostic framework (Galbraith, 1977; Daft, 1986) which goes beyond the simple empirical profile of the U–A model and also provides a useful theory for assessing the longer-term viability of car firms like Ford.

By the late 1960s organization theory had reached the point where the causal linkages between the market, the equipment, corporate knowledge and structure could be expressed in a clear diagrammatic format as illustrated in Figure 4, based loosely on the approach of Woodward (1965).

Her contribution was to argue that technology strategy should reflect the degree of market variability as indicated by the degree of standardization in consumer preferences. Figure 4 depicts this analysis and takes the case of consumers with expectations of highly standardized products which are relatively cheap (e.g. American roadsters in the 1950s). The model prescribes the development of a technostructure with expertise in long-range planning and in the standardizing of shopfloor operations. Mechanized and automated equipment should be substituted for direct labour. There should be high formalization of structure coupled with decentralized decision making within defined zones of authority. The model extends into the relationship between R and D and Manufacturing. It also

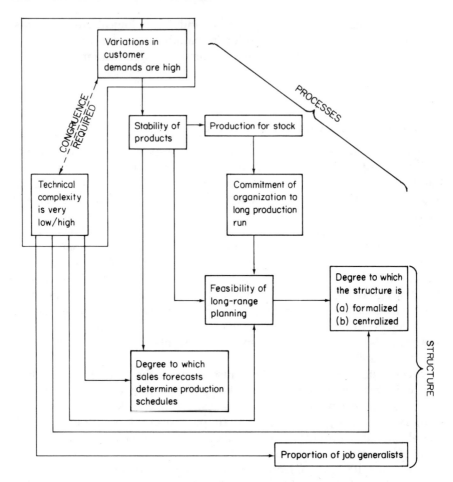

Figure 4 Prescriptive causal linkages: markets, technology and organization

covers the situation when consumer variability is high, in which case the signs at the front of the arrows are reversed.

Two conclusions may be drawn from Figure 4. First, Woodward was not a technology theorist. Rather, she was a multi-variable theorist and her notion of technology included equipment and knowledge (as Perrow, 1967). Woodward defined technology broadly, consistent with organizational analysis, to include the operations technology, the knowledge technology, the tasks required by operatives and the 'control' systems to be used by management. Second, organization theory should be used in a discriminating fashion to analyse the hidden and future tendencies in consumer preferences. Consumer preferences can and do change.

Woodward's contribution requires an analysis of future market structures in order to derive prescriptions for strategy and technology (Clark, 1972, pp.79–84).

Abernathy made partial usage of the knowledge technology perspective of Perrow (1967) to demonstrate that the growing routinization of problem solving at the shopfloor level had been fashioned by the growing technostructure. Perrow's theory of corporate knowledge was also used by Abernathy and this too requires clarification. Perrow focused upon the linkages between knowledge, tasks and control and distinguished between three major decision levels: the strategic, the technostructure and the operating systems levels. He proposed a fourfold typology of problem solving in which there should be a congruence between these three levels for each type.

These principles are illustrated in Figure 5, derived from an investigation of long-term changes in organizational knowledge in the British construction sector (Clark, 1986, 1988). However, the crucial issue of the problem solving regimes of the designers in the technostructure was not fully analysed (cf. Whipp and Clark, 1986).

The logic of organization theory should have led Abernathy to investigate the distinctive features of the American market as a context which entrains

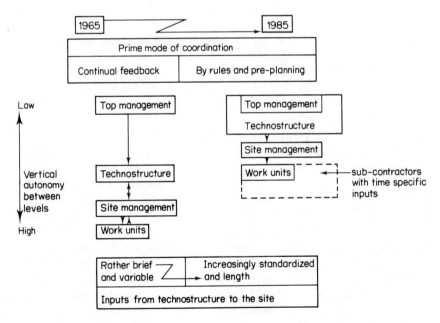

Figure 5 Organization knowledge and the role of the technostructure in construction—1965 and 1985.

its sectors in particular ways (Clark, 1987). Organization theorists reading the analytically structured narrative of Ford would conclude that the organization structure was suited to a stable market, but that the market was changing and that Ford's organization seemed very rigid: a machine bureaucracy (Mintzberg, 1983). Also, they would note the small references to changes in consumer preferences for models which differed from those in which Ford had invested its knowledge and capital. The warning signs of increased uncertainty in the North American market and in the technology were there: indeed Abernathy noted that some productive units seemed to be reversing their movement towards specificity.

The comparative analysis of longer-term dynamics in macro markets is still in its infancy (Clark, 1989), but American approaches to strategy, technology and organization tend to assume that the North American market does not require analysis in a comparative fashion (e.g. Kantrow, 1980; Lawrence and Dyer, 1983; cf. Clark and Starkey, 1988). Likewise, Abernathy should have examined the degree to which speciality markets had emerged (see DeBresson and Lampel, 1985b) in the late 1920s (e.g. Volvo and Rover) and had burgeoned in the 1960s (e.g. Mercedes, BMW, Volvo). Moreover, the features of the Japanese market provided a distinctive competitive context for the emerging industry (Cusumano, 1985).

Technology ferment

Abernathy recognized the need to substantially revise the original U–A model and to re-paradigm American management with an alternative model (Abernathy, Clark and Kantrow, 1983). A new framework was proposed in which the degree of ferment in the technology was related specifically to two contextual aspects of the firm: its market environment and its production systems.

The technology ferment framework is depicted in Figure 6. The figure represents an important shift away from the earlier U–A model of Figure 1. The dichotomizing of innovation into radical versus incremental is replaced by a more or less continuous scale. The significance of this is clearly revealed in its application to the competition between Chrysler and Ford during the 1930s and 1940s, because this reveals considerable innovation during a period which the U-A model (Figure 1) had shown to be largely incremental (Abernathy and Clark, 1985; Clark, 1985). Also, in the new framework users are meant to concentrate upon *description* because it is the mapping of specific contemporary innovations which provides the key input to managerial decision making. The significance of this shift for managerial 'work' at the strategic level should not be underestimated (Abernathy, Clark and Kantrow, 1983). However, the detailed relationship of the framework to strategic policy making still needs further development.

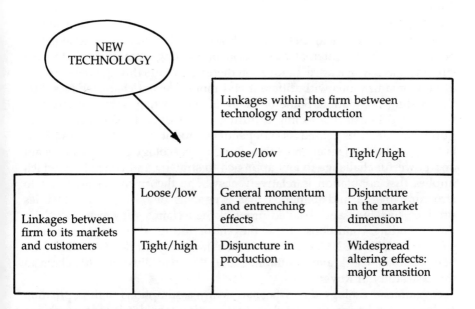

Figure 6   Entrenching and altering dimensions of new technologies. (Derived from W.J. Abernathy and K.B. Clark (1985) with elements from Miller and Friesen (1982).)

INNOVATION POLES AND ORGANIZATION ECOLOGY

Much of organization theory has neglected the cooperative and competitive environments of organizations and the position of a focal organization in relation to its suppliers, customers, financial institutions and regulative bodies: the organization ecology dimension. The quality of interorganizational relationships directly influences the capacity of any major organization to survive (Clark and Staunton, 1989, Ch. 8). The *density* of innovative transactions within the networks may be examined together with the *linkages* between sectors and chains (DeBresson and Murray, 1982, Ch. 5). The connections constitute an *innovation pole* and these poles vary in their capacities and in the position on the advanced frontier of technological development (DeBresson, 1989).

There are two perspectives on the interrelations between organizations having particular implications for strategy and technology: (i) the population ecology perspective; (ii) the organization ecology perspective. The latter perspective tackles the issue of innovation poles. However, the former perspective provides certain awkward challenges to the notion of innovation poles.

The *population* ecology perspective contends that when organizations are founded their cognitive apparatus and operating procedures are imprinted (Boeker, 1988) and become so crystallized that there is long-term inertia (Hannan and Freeman, 1984). It follows that innovation is unlikely to occur through the internal adaptation of existing organizations to new circumstances which arise. Innovation therefore occurs through the entrance of new organizations with different and more relevant characteristics than the existing population. Some or many of the existing organizations may be expected to exit. For example, during the 1920s and 1930s a number of European car firms ceased as independent producers whilst several firms entered, including Jaguar and Volvo. Population ecology provides a relevant and powerful challenge to any approach to strategy and technology which implies that adaptation is relatively easy. The theory draws attention to structural inertia and demands that exponents of 'fast change' theories produce sustained and detailed proof of their claims. However, the claim that founding conditions shape the entire future of an organization is simplistic (Dill, 1962; Clark, 1972, 1979; Whipp and Clark, 1986; Clark and Starkey, 1988; Clark and Staunton, 1989). We shall illustrate this claim in the case study of Rover.

Organization ecology theorists share many assumptions with population ecologists, whilst rejecting the claim that existing organizations are unable to adapt. Some organizations are enabled to adapt, some are not. Our interest in organizational ecology is on the division of labour between organizations over the various, iterative phases of innovation. Three features should be noted. First, there are extensive symbiotic relationships between sectors in the same economy in developing innovations (DeBresson and Murray, 1982, Ch. 5). The nature of linkages varies between nations and their formation shapes the channels along which the hierarchy and neighbourhood effects associated with 'A' type and 'B' type learning curves unfold (Clark and Staunton, 1989). Second, sectors vary in the extent to which they are the locus of innovation and facilitate the appropriation and commercial diffusion of new knowledge and equipment (Pavitt, 1984). Third, there is clear evidence that some organizations pursue strategies in which they create mutualities with other organizations at different positions in the supply chain. For example, Marks & Spencer have spent more than five decades developing a science and technology infrastructure with the science based developers of new materials, with equipment manufacturers and garment assemblers (Clark and Starkey, 1988, Ch. 6). Likewise, Volvo has developed international linkages covering many aspects of existing and new raw materials, of information technologies and of new equipment (Kinch, 1984).

The concept of an innovation pole can be used as a variable (DeBresson, 1989) to assess the degree to which a set of organizations evolves a division of labour to maintain and develop the whole array of innovations on which a focal organization is dependent. These cannot all be developed within the organization, although some large organizations will attempt to internalize many different technologies. For example, when major car firms buy electronic data firms, or when firms with specialist expertise in new materials in the aerospace sector become associated with car assemblers.

STRATEGY, TECHNOLOGY AND ORGANIZATION IN THE LONG TERM

Innovation as an activity cannot be adequately dealt with by constructing dramatic contrasts between radical and incremental innovation, because over the long term, sectors are subject to varying intensities of innovation— some arising from within a sector and many arising from uneven developments in basic and emerging technologies.

Freeman (1982) contended that the transformation of scientific discovery into new technologies occurs unevenly through time and through sectors. There appear to be somewhat regular occurrences of technological expansion when a new techno-economic paradigm surfaces (Perez, 1983). Freeman gave careful consideration to whether there are regular longwave patterns in technological innovation and concluded that economic restructuring does recur. In other words, existing sectors may be impacted by radical technological innovations which also create new sectors. This economic perspective on technology raised further doubts over the U–A model.

Much of business strategy has ignored long pulsations in technological activity and has confined attention to the management of R&D inside the firm. Here it is argued that the position of the firm rests on the innovation pole which it is able to create jointly with other, complementary firms. The achievement of the approach of Utterback, Abernathy and their colleagues was to show that the *balance* between innovation and efficiency is both dynamic and problematic.

THE CASE STUDY OF ROVER: 1896–1988

How, over a period of nine decades, did the British car firm known as Rover acquire and then lose access to an innovation pole and how did its management develop and then lose an internal design capability? This case study is based on extensive access to the archives of Rover and its suppliers and to documents from various other parties, including the trade unions in the West Midlands (Whipp and Clark, 1986). More than sixty key individuals were interviewed, including suppliers and trade union officials.

The analytically structured narrative is constructed around the problem of acquiring an innovation-design capability. The key episodes over time include:

1912   Rover claimed to be the world's largest assembler of bicycles, motorcycles and automobiles;

1928   It faced exit at the very moment when several new firms were entering the European industry (e.g. Volvo, Jaguar and Rootes);

1930s   Rover developed considerable design skills for the product along with adequate production engineering;

Post-1945 Rover introduced the assembly line for saloon cars and launched the highly successful Land Rover;

1964   Rover launched a best-selling British speciality car known as Rover 2000;

1968   It was incorporated into British Leyland which was received into state control in 1975;

1970s   It was the focus of the largest British investment in a new car, new plant and organization. The existing innovation and design capability was restructured and impaired;

1980s   Rover was the marque which became underpinned by the imported design expertise of Honda;

1988   Rover was sold to British Aerospace. Merger proponents claimed that Rover provided expertise in modern production whilst British Aerospace provided expertise in advanced materials relevant to automobile design and production.

This narrative about design falls into five partially overlapping phases: 1896–1932; 1928–34; 1932–64; 1960–71; 1968–88. Each phase was distinguished by a different design capability and by a distinctive position in relation to the relevant innovation poles.

In Europe the automobile industry developed initially in France where several successful firms had emerged by 1914. French car design was advanced and Louis Renault quickly recognized the significance of Ford's success and endeavoured to learn which lessons were most relevant to the European context (Clark and Windebank, 1985). A major difference between Europe and North America was that in the USA the combination of cheap food and high wages enabled many consumers to enjoy a disposable income sufficient to purchase a cheap car, as Ford discovered and realized from 1908. In Europe the market was the wealthy and the middle classes who sought distinctive cars. European firms tended to concentrate their technology strategies on the *car* until around 1920.

British car makers pursued market strategies in which the market inferface was mediated and controlled by independent distributors. This left the car makers in a vulnerable position for gaining market information. Also, the distributors whose concern was the car rather than the *production process* often had an indirect influence on the assemblers.

1896–1932: diversified generalist assemblers

The origins of Rover lay in the cycle industry and in the patent of J.K. Starley who is credited with key inventions which transformed the penny-farthing into the modern two-wheel cycle driven by a chain. He founded a company to assemble the cycle and registered it in 1896 as a cycle assembler and distributor. The cycle boom was already entering a more competitive phase based on larger scale production. Starley died in 1902 by which time his fellow directors were expressing their interests in a new product: the automobile.

After Starley's death the directors, who were generalists, dropped all reference to cycles from the company's name and despite the nervousness of shareholders, commissioned an independent designer, Mr Lewis, to design a six horse-power automobile and to itemize the components necessary for its assembly. Lewis remained as a contract designer to 1910. In 1906, their first selling season, Rover's 6 hp automobile sold as well as the board had hoped. Within two years Rover was selling five models with total sales of 1200 per year and was the third largest domestic supplier. Rover always distributed a high proportion of its profits to shareholders, a common pattern in the sector. 1912 was a turning point in the British market although the consequences did not emerge until after the First World War. It was a year of peak profits for Rover.

Meanwhile Ford opened its kit assembly plant for the Ford Model T at a special site in Manchester. Ford immediately captured one-quarter of the market, becoming the largest single supplier. The 'T' was sold at the cheap end of the market which could not be supplied by domestic assemblers, partly because their suppliers could not produce cheap, standardized components and subassemblies. British assemblers catered for the wealthier clients able to afford speciality cars. The impact of Ford was interrupted by the First World War. However, during the war, the Austin Company acquired inexpensive premises and equipment deriving from armaments contracts, so became well prepared to introduce a small British car, the Austin 7, in the 1920s. That car altered the structure of the British market.

Back in 1910 Rover hired a new contract designer, Mr Clegg, and his range of models covered the period to 1920, when the Board recognized the opportunity to enter the small car market in addition to their existing market for medium and more expensive automobiles. Mr Clegg assisted with the problems of designing the dies and similar basic tools for the new models and he instituted a standard operating procedure for updating. However, Rover's internal design capability was *ad hoc* and incomplete and its board lacked expertise in car-making and design. In 1920 Rover entered the small, cheap car market through the acquisition of a local firm whose owner, Sangster, had designed the car and a factory to produce it in small volumes.

The Sangster factory was designed as a tight workflow with systems of flashing lights for supervisors to signal faults. However, sales did not match expectations and the small car market was soon led by the Austin 7 built at Longbridge in Birmingham and by the 'Bull-nose' from Morris at Oxford. Soon Austin and Morris emerged as the leading British assemblers: a position they retained until their merger in the early 1950s. So the Rover board had correctly anticipated the shift in the market previously indicated by the Ford T, but failed to gain a sufficient market share for its own models.

By 1922 the board was increasingly conscious of the lack of expertise in car design and car production within the firm at senior levels. A twofold strategy was developed: they recruited an ageing, though experienced director and attempted to establish an internal design capacity by recruiting experienced engineers from rival firms to develop the mechanical aspects (e.g. engine systems). For electrics they increasingly relied upon the Lucas Company which was developing considerable skills supplying the car assemblers, an early indication of a developing innovation pole situated in the West Midlands. Whilst Lucas succeeded in providing electrical expertise from outside the firm, the position of the mechanical engineers within Rover was more awkward. There were strains of accommodation to the changing division of expertise within management as the firm shifted from an assembler of kits to a designer of cars and their production systems.

These shifts in corporate knowledge took place during the restructuring of the British market concurrent with the rise of Austin and Morris. From 1922 onward Rover's profits deteriorated whilst dividends were maintained. The board was aware of the overall situation, yet seemed powerless to reduce dividends in case the shareholders lost confidence and ceased investment. Within the firm there was poor cost control and little understanding of the economics of competition available through increasing scale and decreasing scope. With decreasing investment funds, Rover could not develop new models to match rivals. Still, annual reports maintained a hopeful vision of the next 'selling season' until 1928 when the capital had to be written down by 75 per cent.

Photographs of the plant and of the work-force at this period reveal a collection of workshops, each with very little equipment and negligible amounts of powered equipment. The workflow was hand-driven. The cars were largely assembled by skilled craftsmen who shaped parts and crafted the vehicle. Painting was a skilled craft. Control over the pace of work was embedded in piecework systems and there was considerable shopfloor influence on the car. There were no assembly lines.

1928–1934: facing exit

During the 1920s the automobile and its markets evolved in both North America an in Europe with much more *ferment* than the Utterback and Abernathy framework would suggest (see Clark, 1985). In Europe new firms entered the sector: Volvo in Sweden; Rootes and Jaguar in Britain. Yet existing firms were disappearing through bankruptcy and acquisition. The ecology of car firms was changing.

The board of Rover sought to decode and understand the external transitions. For example, it was surprised to learn from its South African distributor that its selling price was above the production cost of rival models and further, dismayed to discover that the accounting section could merely record events after they had occurred. So, the board sought outside assistance and recruited Mr Searle in 1928 as managing director. He was experienced in production and introduced the American consultants Bedeaux to reduce shopfloor influence by systematic time measurement and production control. A considerable struggle ensued.

Fortuitous changes in the market coupled with the removal of certain areas of wastage provided a temporary period of profitability, but losses resumed in 1932 and the firm was bankrupt with heavy debts to its bankers and to its principal supplier, Lucas. At that moment Lloyds Bank and Lucas intervened to restructure the board, the management and the role of the technostructure. Two key figures were appointed as managing directors and Mr Searle was dismissed. Mr Graham became responsible for financial affairs with the task of introducing and establishing standard costing systems and ensuring the repayment of debts. He succeeded in both tasks. Mr S.B. Wilkes, a brilliant engineer, was appointed as managing director responsible for the car and its manufacture. He stayed until his death in the early 1960s. These appointments symbolized the adaptation of Rover to produce cost effective cars and signalled the 'unlearning' of the obsolete skills on which the firm had been founded. The founding recipes were wholly transformed at board and managerial levels, yet on the shopfloor the existing practices were to remain largely intact until the late 1950s.

1932–64: focused marketing and robust design capability

Turnaround: 1930s

It is important to keep the new approach in accounting in mind whilst focusing on the actions implemented by S.B. Wilkes. The strategy of Rover was to target a segment of the market which it believed would establish a secure market niche as a firm base for recovery and future expansion: doctors and professionals. Having selected a target consumer, Wilkes established a procedure for designing what became known as the 'P series' of cars. The aim was to achieve volume sales by narrowing the range to cater for this stable niche and thus to achieve volume by selling the same model for several years with only incremental modifications. So, although the market was highly seasonal and difficult to predict, Rover sought dominance in a niche which was highly secure, even during economic recession.

This choice of market segment permitted Wilkes to retain features of the pre-existing work organization for the manufacture of cars and therefore to avoid—for the moment anyway—the conflicts which surrounded Searle's attempt to achieve high volume production. During the 1930s Wilkes made improvements in the flow of work, benefiting from the focusing of models and from using simple gravity conveyors. The novel feature of Wilkes's strategy was to blend traditional work organization on the shopfloor with a much developed technostructure and managerial capability as an over-layer. Many appointments to this over-layer were brothers and relatives.

With Wilkes as managing director Rover's internal design capability was created and institutionalized in the firm, with the highest level of access to the key decision makers concerned with car design and with it productionizing. Yet the scope of design capability was narrow and its basis more precarious than the successes of the next three decades might suggest. Wilkes established several principles of car design which were essential and successful. First, each new 'P' was designed to embody sufficient of existing best practice to satisfy the target consumer, yet with simple annual refinements. The car combined leather and wood interiors with a sedate shape. Existing models could be stretched into new models.

Second, verbal and written protocols were established for design including the 'factory bible' in which key refinements were recorded. The protocols were partly formalized, yet their interpretation and activation depended upon the clan-like relationships within line management and the technostructure. Also, certain external dependencies on the expertise of suppliers of basic materials were maintained. Third, focus was mainly on mechanical engineering skills rather than on electrical engineering or science. Very few graduates were recruited—even in mechanical engineering. Fourth, Wilkes had a clear vision of the relationship of car design to manufacture, so that the former was shaped by the capabilities of the existing work organization. Fifth, the existing work organization was subjected to slow incremental change through small-scale mechanization, most notably in 1946 when a slow moving conveyor system was installed.

Design became an *integrated* activity, meshing product and production. Moreover, robust design was used to blend the discontinuity of the new management superstructure with the continuity of existing shopfloor practices. Relative to the pace of change in its consumers' preferences Rover was sufficiently flexible that there was little need for specialized equipment.

Take-off in domestic markets: 1940s and 1950s

Rover played an important part in the Second World Way by establishing a 'shadow' factory in nearby Solihull where aero-engines were assembled. After 1945 Rover re-entered the saloon market with a new plant at Solihull with a moving assembly line. At the time it was claimed that the new line would pace work and reduce inventory. In practice the line moved slowly and in retrospect is better interpreted as shifting heavy weights in an orderly fashion. Moreover, the honeycomb of small workshops continued for all sub-assembly operations.

Rover faced shortages of basic raw materials and the requirement of the Labour government to promote exports. It was an extremely testing period. Wilkes (in 1947) searched for alternative products and conceived the opportunity for a general farm vehicle based on the 4-wheel drive American Jeep. It was designed and named the Land Rover. It could be constructed from readily available aluminium. Like the saloon car it was assembled around a frame, with the traditional forms of work organization. Indeed assembly lines were not introduced for another three decades. The Land Rover quickly filled a market niche both in Britain and in Australia. Its potential as a utility vehicle for the police, the fire service and the armed forces soon became apparent. Later, it became a cult vehicle and was stretched into the Range Rover, a car much used by royalty. The Land Rover and the Range Rover became significant contributors to corporate sales and profits after 1950. We shall however concentrate upon the saloon cars.

Rover improved its motor-car market position during the 1950s and achieved some success with exports to the Commonwealth (e.g. India). In 1958 the design team commenced the P6 model (Rover 2000) with the intention of launching it in the early 1960s.

1960–71: international competition in speciality markets

The later 1950s and early 1960s saw evolution in the world car market. In Europe the rising standard of living in Germany and France provided a new market segment for speciality saloons (e.g. BMW, Mercedes, Volvo and Saab) and suppliers found demand leaping. Meanwhile in Japan the nascent small car industry was entering a transitional period of great expansion and upgrading (Cusumano, 1985). Rover saloon models benefited from these developments with the P6, which sold over 340,000 units between 1964 and 1976, becoming the best selling domestic executive car. However, in the European market and in the American market Rover fell behind other European speciality car firms.

During the lengthy gestation of the P6 certain very significant transitions were undertaken in the car, in the plant and in work organization. The death of S.B. Wilkes just prior to the launch of the P6 range signalled a deeper transition in managerial skills and the role of the technostructure of designers. A new set of designers was emerging.

The board recognized that it had to shift saloon production from small batch methods to reduce the average cost and recoup the rising costs of capital investment in the new equipment now available for all stages of production. The costs of tooling and of new, high quality paint plants were also increasing. These shifts had major consequences for long established shopfloor forms of governance, with the extensive devolvement of responsibility to skilled operatives and their union. The traditional shape of the Rover car was transformed for the P6. The construction of the car was simplified to facilitate the introduction of large batch assembly operations, to be undertaken on very much shorter cycles and with semi-skilled operators. During 1963–5 large numbers of these were recruited. The transition was accompanied by bitter struggles between the long-established skilled employees and the large wedge of new semi-skilled employees whose interests were organized through a general union. The paternalistic management team with its close personal control over the innovation-design processes was unable to resolve the disputes which its new requirements generated. Rover experienced bitter industrial relations.

Paradoxically Rover's greatest market success merely masked its underlying problems. First, there was the failure to penetrate the continental markets of Europe and of North America. Second, the industrial relations problems coincided with extensive changes in management arising from retirements. Third, the car itself was manufactured by more advanced and complex means (e.g. painting requirements) so that new materials were introduced. These made existing forms of managerial and engineering expertise redundant. Yet access to new expertise was uneven: islands of great strength in expertise were surrounded by many weaknesses.

In 1967 Rover merged with the highly successful Leyland Corporation, but within a year Rover was part of the government- led rationalization of the U.K. automobile industry which created British Leyland. At first little changed. The design team, now led by Bache, began designing the next 'P' models and within three years were ready to modify their existing production facilities to accommodate the new models.

1968–88: enter Honda

In 1971 the board of British Leyland (B.L.) concluded that the production methods even of *speciality* cars had become routinized in the high volume construction of basic assemblies and components and their final assembly on flow-line principles (see Figure 7). So B.L. rejected the new Rover 'P' designs and the market estimate of 30,000 units and set a target of 150,000 unit sales per year, mainly in Western Europe. So what was needed was a car designed for European consumers.

This period commenced, then, with the declaration that Rover would design and make 'a car for Europe' in a brand new, integrated, high volume assembly facility and paint plant on the existing site. The period ended with the plant closed and production transferred to Oxford where the Rover name fronted the input of innovation-design expertise from Honda of Japan. During this period British Leyland was received into state ownership in the mid-1970s, then 'sold' to British Aerospace in 1988. The collapse of the U.K. owned industry had profound effects both for close suppliers (e.g. Lucas) and for the standard of living of people in the West Midlands.

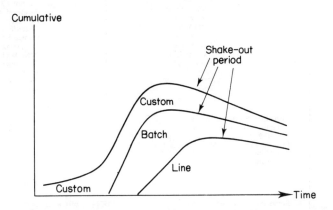

Figure 7 Automobile sector: segments and co-existence. (Reprinted with permission from C. DeBresson and J. Lampel (1985a). Beyond the life cycle: organizational and technological design. I. An alternative perspective, *Journal of Product Innovation Management*, **3**, 170–87. Copyright 1985 by Elsevier Science Publishing Co., Inc.)

The story of the period to 1982 has been compacted into Figure 8 which displays the innovation-design cycle as four sequential states: design vision, translation, commissioning, operating. These are states rather than stages and their unfolding is very uneven, especially across the different aspects of the car, its build environment and the organizational dimension. The framework of Figure 8 could be used to examine any similar set of

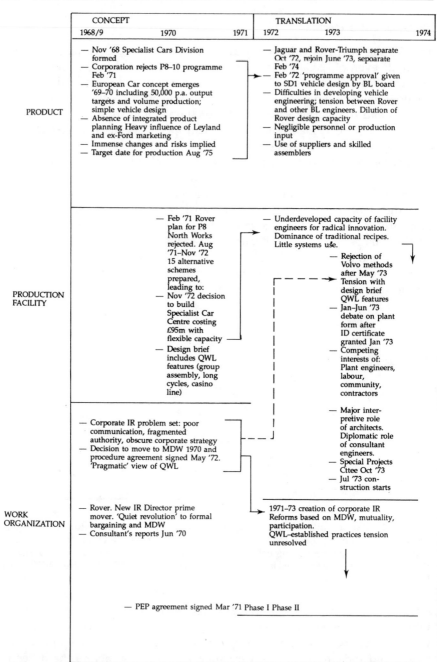

	CONCEPT			TRANSLATION		
	1968/9	1970	1971	1972	1973	1974
PRODUCT	— Nov '68 Specialist Cars Division formed — Corporation rejects P8–10 programme Feb '71 — European Car concept emerges '69–70 including 50,000 p.a. output targets and volume production; simple vehicle design — Absence of integrated product planning Heavy influence of Leyland and ex-Ford marketing — Immense changes and risks implied — Target date for production Aug '75			— Jaguar and Rover-Triumph separate Oct '72, rejoin June '73, separate Feb '74 — Feb '72 'programme approval' given to SD1 vehicle design by BL board — Difficulties in developing vehicle engineering; tension between Rover and other BL engineers. Dilution of Rover design capacity — Negligible personnel or production input — Use of suppliers and skilled assemblers		
PRODUCTION FACILITY		— Feb '71 Rover plan for P8 North Works rejected. Aug '71–Nov '72 15 alternative schemes prepared, leading to: — Nov '72 decision to build Specialist Car Centre costing £95m with flexible capacity — Design brief includes QWL features (group assembly, long cycles, casino line)		— Underdeveloped capacity of facility engineers for radical innovation. Dominance of traditional recipes. Little systems use. — Rejection of Volvo methods after May '73 Tension with design brief QWL features — Jan–Jun '73 debate on plant form after ID certificate granted Jan '73 — Competing interests of: Plant engineers, labour, community, contractors — Major interpretive role of architects. Diplomatic role of consultant engineers. — Special Projects Cttee Oct '73 — Jul '73 construction starts		
	— Corporate IR problem set: poor communication, fragmented authority, obscure corporate strategy — Decision to move to MDW 1970 and procedure agreement signed May '72. 'Pragmatic' view of QWL					
WORK ORGANIZATION	— Rover. New IR Director prime mover. 'Quiet revolution' to formal bargaining and MDW — Consultant's reports Jun '70			1971–73 creation of corporate IR Reforms based on MDW, mutuality, participation. QWL–established practices tension unresolved		
	— PEP agreement signed Mar '71 Phase I Phase II					

Figure 8   The SD1 Design Process 1968–1982. Reproduced with permission from R. Whipp and P.A. Clark (1986). *Innovation and the Auto Industry: Product, Process and Work Organization*, Frances Pinter, London.)

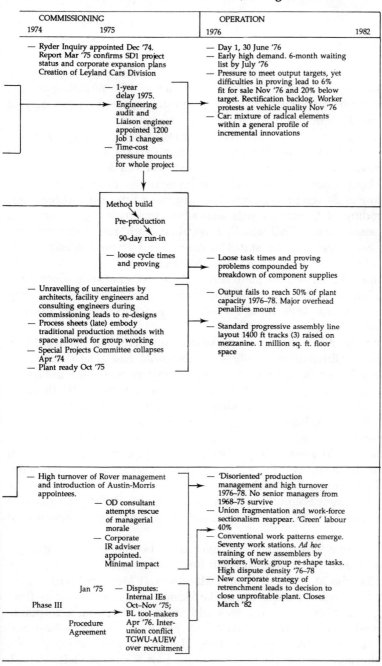

COMMISSIONING		OPERATION	
1974	1975	1976	1982

— Ryder Inquiry appointed Dec '74. Report Mar '75 confirms SD1 project status and corporate expansion plans Creation of Leyland Cars Division

— 1-year delay 1975. Engineering audit and Liaison engineer appointed 1200 Job 1 changes
— Time-cost pressure mounts for whole project

— Day 1, 30 June '76
— Early high demand. 6-month waiting list by July '76
— Pressure to meet output targets, yet difficulties in proving lead to 6% fit for sale Nov '76 and 20% below target. Rectification backlog. Worker protests at vehicle quality Nov '76
— Car: mixture of radical elements within a general profile of incremental innovations

Method build

Pre-production

90-day run-in

— loose cycle times and proving

— Loose task times and proving problems compounded by breakdown of component supplies

— Unravelling of uncertainties by architects, facility engineers and consulting engineers during commissioning leads to re-designs
— Process sheets (late) embody traditional production methods with space allowed for group working
— Special Projects Committee collapses Apr '74
— Plant ready Oct '75

— Output fails to reach 50% of plant capacity 1976–78. Major overhead penalities mount

— Standard progressive assembly line layout 1400 ft tracks (3) raised on mezzanine. 1 million sq. ft. floor space

— High turnover of Rover management and introduction of Austin-Morris appointees.

— OD consultant attempts rescue of managerial morale
— Corporate IR adviser appointed. Minimal impact

— 'Disoriented' production management and high turnover 1976–78. No senior managers from 1968–75 survive
— Union fragmentation and work-force sectionalism reappear. 'Green' labour 40%
— Conventional work patterns emerge. Seventy work stations. Ad hoc training of new assemblers by workers. Work group re-shape tasks. High dispute density '76–78
— New corporate strategy of retrenchment leads to decision to close unprofitable plant. Closes March '82

Phase III

Jan '75

Procedure Agreement

— Disputes: Internal IEs Oct–Nov '75; BL tool-makers Apr '76. Inter-union conflict TGWU-AUEW over recruitment

246 Peter Clark and Chris DeBresson

events—for example, the development of the Kalmar plant for Volvo. The information shown in Figure 8 is specific to Rover during the 1970s when its design expertise was being reconfigured.

By 1976, despite the many problems, there was a prize-winning European design. Moreover, the new factory also won an award. Yet subsequently, sales of the car barely exceeded the original estimate. Two aspects may be noted: the production and the market.

Producing the new car involved a new design requirement for cooperation between formerly competing design specialists and engineers spread across the sprawling, loosely coordinated federalism that was British Leyland. Coordination could not be achieved through formalistic standardizing of design practices used by American firms such as Ford, nor by the blending of formalism with extensive consensus decision-making of the Japanese firms. The situation required both new expertise and close, fast exchanges. So the states of design and implementation were more convoluted and fragmented than in international rivals. Moreover, commissioning the car and the plant revealed a managerial style which must have contributed to the confusion and uncertainty. After these faltering beginnings there was little room for later manoeuvre, as delays severely reduced cash flows.

Marketing the new car proved more difficult than had been publicly acknowledged. Distribution involves a substantial cost. The distribution networks for Rover in Europe were thin, inexperienced and uncertain. Moreover, the market image of Rover had to contend with those of successful European speciality models from Germany, Italy, France and Sweden. In the event, European sales were negligible.

When British Leyland was brought into state ownership there were earnest attempts to revive its performance through the introduction of shopfloor equipment, but these efforts were concentrated in the older plants at Longbridge, Birmingham and at Cowley, Oxford. It was later decided that the new Solihull car plant should be closed; the production of Rover transferred to Oxford in 1982. Land Rover production remained at Solihull.

The Rover name was retained and eventually became the main marque of the state-owned enterprise. By 1988 Rover combined an ageing range of models (i.e. the M series of Metro, Maestro and Montego) with a new range based on collaboration using the design skills of a foreign firm. Thus, the problem of car design was addressed by retaining a corporate potential whilst establishing intimate linkages with the innovation and design capability of the Japanese firm, Honda. Honda's expertise has provided the architecture for the new range of Rover models, though features such as styling of interior trim have been contributed by Rover.

Honda has also supplied key components and expertise in the operation of the production units.

As a state owned firm Rover possessed both immense debts and also considerable assets in property. The plant was modernized and the potential for survival strengthened. In 1988 the Conservative government 'sold' Rover to British Aerospace as part of its privatization philosophy. The debts were largely reduced and so the property value of Rover to British Aerospace is considerable, according to knowledgeable sources in the firm. *What is much more questionable is whether Rover has re-acquired a design capability.*

It has been said that British Aerospace will supply Rover with expertise in new metals and in information processing. British Aerospace employs 30 000 graduate engineers of various types so there must be possibilities for their specialized professional skills to be deployed in Rover. However, the incorporation of expertise across three corporate boundaries—Rover, Honda and British Aerospace—provides an enormous challenge. Given the significance of the organizational problems and the tacit nature of developmental knowledge, the attempted recreation of an innovation pole for the automobile industry in Britain should be closely observed.

CONCLUSIONS

This chapter combines the analytically focused narrative of design processes in Rover, a speciality car firm, with an examination of the abstract meta-frameworks proposed by Utterback, Abernathy, K.B. Clark and their colleagues at MIT and Harvard for understanding and prescribing the relationships between strategy and technology. The choice of a European speciality car firm provides a useful comparison with Abernathy's (1978) reconstruction of management strategy for technological innovation in Ford.

First, the case study; the existence of a number of speciality car firms illustrates the relevance of the DeBresson and Lampel (1985a,b) contention that generalist and speciality car firms can co-exist. Rover managed to survive beyond the shake-out of the early 1930s and to demonstrate that some adaptation of the original repertoire is possible by superimposing design capability on top of traditional work organization. However, that form of adaptation left certain organizational problems for later generations of management to face in the 1960s. In the 1930s Rover searched for and found market niches in the domestic markets of Britain and its Commonwealth. After 1950 the character of the speciality car segment was transformed by the growing market of wealthy executives—especially in Europe—and by developments in all areas of raw materials and of equipment. By the 1950s the speciality car segment faced pressures to combine expensive technological investments with increasing volume of production to amortize costs. There were constantly evolving demands to update expertise and to introduce novel expertise to handle raw materials, processes and equipment.

These requirements for new problem solving regimes in the technostructure were not fully grasped by Rover or by some of the other firms in their innovation pole in the UK. Rover also failed to control the distribution problem and so failed to penetrate the rich niches of continental Europe and of North America. Rover's output volume was nearer that of Porsche, the small German elite firm, than the larger outputs of Saab, Volvo, BMW and Mercedes. Moreover, whilst leading European speciality firms re-developed their innovation poles and recruited graduate scientific skills, Rover's design capability remained narrowly based, slow in pace, loose in heuristics and anchored in personal styles which were path-dependent and could not be readily transferred into British Leyland. The demise of a design capability had to be corrected through the importation of a modern design capability from Honda.

The Utterback and Abernathy (1975) framework drew attention to the relationship between technological innovation and strategy. The framework was tested and later refined by incorporating design and by raising the productivity versus innovation dilemma. American car firms moved strongly from fluid to specific innovation in most of their productive units up to the 1970s, but these processes were partially reversed during the 1970s. The pursuit of efficiency was achieved through a combination of skilful marketing of images whilst factoring the components beneath the skin of the car into packages which could often be developed incrementally. All this was achieved by a progressively elaborated technostructure containing many functionally specialized professionals recruited from the universities and trained to work within highly specific problem-solving regimes. Abernathy correctly recognized that these organizational structures possessed rigidities, but probably understated the problems of inertia.

Abernathy (1978) revealed limitations in the U–A model. However, the absence of international comparisons until Abernathy, Clark, K.B. and Kantrow (1983), obscured the hidden developments in the European speciality markets and the brilliant usage of design and development by the Japanese automobile assemblers. Equally significant was the lack of attention to innovation poles and their constituting networks.

The current position of Rover, given its dependence on Honda for key design inputs and on its new owner, British Aerospace, provides a fascinating challenge to its employees and to academic research into the strategic management of technological innovation.

REFERENCES

Abernathy, W.J. (1978). *The Productivity Dilemma. Roadblock to Innovation in the Automobile Industry*, Johns Hopkins, Baltimore.

Abernathy, W.J., Clark, K.B., and Kantrow, A.M. (1983). *Industrial Renaissance: Producing a Positive Future for America*, MIT Press, Boston.

Abernathy, W.J., and Clark, K.B. (1985). 'Innovation: mapping the winds of creative destruction', *Research Policy*, **14**, 3–22.

Boeker, W.P. (1988). 'Organizational origins: entrepreneurial and environmental imprinting at the time of foundation', in G.R. Carroll, (ed.), *Ecological Models of Organizations*. Ballinger, 33–52.

Clark, K.B. (1985). 'The interaction of design hierarchies and market concepts in technological evolution', *Research Policy* **14**, 235–51.

Clark, P.A. (1972). *Organizational Design: Theory and Practice*, Tavistock, London.

Clark, P.A. (1979). 'Cultural context as a determinant of organizational rationality: a comparison of the tobacco industries in Britain and France', in L.J. Lammers and D.J. Hickson (eds), *Organizations Alike and Unlike: International and Inter-Institutional Studies in the Sociology of Organizations*, Routledge & Kegan Paul, London.

Clark, P.A. (1986). 'The economy of time and managerial division of labour in the British construction industry 1965–1985', *Proceedings of the 7th Bartlett International Summer School*, University College, London University, London.

Clark, P.A. (1987). *Anglo-American Innovation*, de Gruyter, New York.

Clark, P.A. (1988). 'Computer aided management technologies and the British construction sector: problems of interpretation'. Paper to Colloque on 'Europe et Chantiers', Ecole Polytechnique, Paris.

Clark, P.A. (1989). 'Organization analysis and corporate chronologies', in D. Pymm and J. Hassard, *The Theory and Philosophy of Organizations: Critical Issues and New Perspectives*, Croom Helm.

Clark, P.A. and Starkey, K.P. (1988). *Organization Transitions and Innovation Design*, Pinter, London.

Clark, P.A., and Staunton, N. (1989). *Innovation in Technology and Organization*, Routledge, London.

Clark, P.A., and Windebank, J. (1985). 'Innovation and Renault, 1900–1982. Product, process and work organization', *ESRC Work Organization Research Centre Working Paper Series*, No. 13, Aston University.

Cusumano, M.A. (1985). *The Japanese Automobile Industry. Technology and Management at Nissan and Toyota*, Harvard University Press, Boston.

Daft, R. (1986). *Organization Theory and Design*, Gulf.

DeBresson, C. (1989). 'Breeding innovation clusters. A source of dynamic development', *World Development*, Vol. 17, No. 1.

DeBresson, C., and Murray, B. (1982). *Innovation in Canada*. Cooperation Research Unit on Science & Technology, British Columbia.

DeBresson, C., and Lampel, J. (1985a). 'Beyond the life cycle: organizational and technological design. I. An alternative Perspective', *Journal of Product Innovation Management*, 3, 170–87.

DeBresson, C., and Lampel, J. (1985b). 'Beyond the life cycle. II. An illustration', *Journal of Product Innovation Management*, 3 188–95.

Dill, W.R. (1962). 'The impact of environment on organizational development', in S.

Mailick and E. H. Van Ness (eds), *Concepts and Issues in Administrative Behavior*, Prentice-Hall, New York.

Freeman, C. (1982). *Unemployment and Technical Innovation*, Frances Pinter, London.

Galbraith, J.R. (1977). *Organization Design*, Addison-Wesley, Reading.

Hannan, M.T., and Freeman, J. (1984). 'Structural inertia and organizational change', *American Sociological Review*, **49**, 149–64.

Hickson, D.J., Butler, R.J., Cray, D., Mallory, G.R., and Wilson, D.N. (1986). *Top Decision Making in Organizations*, Blackwell, Oxford.

Hickson, D.J., Pugh, D.S., and Pheysey D. (1969). 'Operations technology and organization structure: an empirical reappraisal', *Administrative Science Quarterly*, Vol. 14, 378–97.

Kantrow, A.M. (1980). 'The strategy technology connection', *Harvard Business Review*, July/August, 6–21.

Kinch, N. (1984). 'The longterm development of a supplier-buyer relationship: the case of the Olofstrom-Volvo relationship. Paper: Centre for International Business Studies Upsala University, Sweden.

Larson, M.S. (1977). *The Rise of Professionalism: A Sociological Analysis*, University of California Press.

Lawrence, P.R., and Dyer, R. (1983). *Renewing American Industry*, Free Press, New York.

Mintzberg, H. (1983). *Structure in Fives: Designing Effective Organizations*, Prentice-Hall, Englewood Cliffs.

Nelson, R., and Winter, S. (1982). *An Evolutionary Theory of Economic Change*, Harvard University Press, Cambridge MA.

Pavitt, K. (1984). 'Sectoral patterns of technical change: towards a taxonomy and theory', *Research Policy*, **13**, 343–73.

Perez, C. (1983). 'Structural change and assimilation of new technologies in the economic and social systems, *Futures*, October, 357–75.

Perrow, C.A. (1967). 'A framework for the comparative analysis of organizations', *American Sociological Review*, **32**, 194–208.

Tushman, M.L., and Anderson, P. (1986). 'Technological discontinuities and organizational environments', *Administrative Science Quarterly*, **31**, September, 439–65.

Utterback, J.M., and Abernathy, W.J. (1975). 'A dynamic model of process and product innovation', *Omega*, **3**(6), 639–56.

Whipp, R., and Clark, P.A. (1986). *Innovation and the Auto Industry: Product, Process and Work Organization*, Frances Pinter, London.

Woodward, J. (1965). *Industrial Organization. Theory and Practice*, Oxford University Press, London.

Part 3

Crisis, Learning and Adaptation

11

Crisis Modes of Strategic Transformation: a New Metaphor for Managing Technological Innovation

Martyn Pitt

University of Bath

Metaphors used to describe the processes of strategic management include 'planning', 'adaptation', 'power politics' and 'interpretive sense making'. In the particular context of technological innovation, crisis is also a relevant metaphor to explain and prescribe action of a fundamental, transformational kind. Applying the crisis metaphor helps us understand *why* and *how* these transformations take place and so—at least in principle—offers insights into managing strategic technological change more effectively.

Strategic transformations, then, can be likened to coping with crisis. However, there are several crisis modes of organizational behaviour and in the complex and uncertain world of technological innovation it is suggested that a *proactive* mode, termed *constructed crisis*, offers managers the best prospect of stimulating effective strategic change.

The Strategic Management of Technological Innovation. Edited by R. Loveridge and M. Pitt
© 1990 by John Wiley & Sons Ltd

ADAPTIVE TECHNOLOGICAL TRANSFORMATION

Strategic development is sometimes portrayed as a comprehensive, planned managerial process that, when effective, ensures the survival and prosperity of the firm (Hofer and Schendel, 1978); less ambitiously, as *adaptive* behaviour involving purposive, albeit constrained, opportunistic and sometimes flawed decision making (Lindblom, 1959).

Long-run development is variously characterized as 'normal' periods of incremental change (relative continuity) disrupted by infrequent, major transformational changes or discontinuities. The latter have variously been labelled *global* (Mintzberg, 1978), *quantum* (Miller, 1982), *revolutionary* (Jonsson and Lundin, 1977; Ramaprasad, 1982) and *fundamental* (Johnson, 1987, p.248). Transformational change drastically alters a firm's market and competitive standing (Porter, 1980), its engineering and technological competences (Miles and Snow, 1978) and its administrative routines and standard operating procedures (Nelson and Winter, 1982). Radical changes arising from technological innovation are evidently of this general form. Coping with them effectively is a highly problematic strategic task.

Thus, radical technological shifts are essentially organizational crises, though a crisis can be a more positive transitional state than might initially be supposed: crises engender adaptive potentialities as well as threats. It follows that strategic management of innovation can be regarded as instigating and resolving transformational crises. The central theme of this chapter is that although managers can never fully control the direction, pace and outcomes of fundamental organizational change, they are closest to doing so when they behave proactively and adaptively.

Evolving, inherently ambiguous technological contexts present a continuing stream of novel circumstances with which managers must cope. In an adaptive firm (De Greene, 1982) technological crises should not be viewed as undesirable but as fundamental elements of the continuity-and-change syndrome. In a genuinely dynamic mode of behaviour, change is natural: technological momentum, not inertia, creates security. Strategists need to tolerate the uncertain and be disposed to learn experimentally. 'Constructing' crises is a means of opportunistic learning for able managers, a valuable facilitator of transformation, an antidote to inertia.

CRISIS CONCEPTS

As we know anecdotally, crises are common phenomena in organizations. Yet crisis is an unpopular, largely negative word in managerial vocabulary. Its use, with few exceptions, is synonymous with imminent, unfolding disaster, mandating reactive short-run, 'turnaround' management, arguably *the* heroic saga of our times. It conjures images of indomitable leaders facing unexpected problems, battling against formidable odds, yet emerging triumphant! The literature on organizational crisis rarely dwells on antecedents or uses the metaphor systematically and constructively (albeit with honourable exceptions, e.g. Shrivastava, 1988; Starbuck, Greve and Hedberg, 1978). For many, crisis challenges the 'illusion of control', signifying management failure and loss of face. Naturally enough, managers avoid these associations, so crisis-oriented learning is discouraged.

Yet organizational crisis is a multi-faceted concept. Hermann (1969, Ch. 2) offered two definitions (i) crisis-as-turning-point, implying abrupt and largely irreversible change of trend or status and (ii) a situation or episode characterized by surprise, high threat to organizational goals and a short time-frame for decision-making. These features suggest unusual instability and transition, endangering patterns of continuity. Decision-makers in crisis are greatly stressed, often showing dysfunctional or pathological behaviour (Daneke, 1985; Dunbar, Dutton and Torbert, 1982; Fink, Beak and Taddeo, 1971; Hermann, 1963).

Crisis can also be a *galvanizing* event with organization-enhancing outcomes (Meyer, 1982a; Pettigrew, 1985; Tjosvold, 1984), resulting from a shared, convergent interpretation and enactment of events (Ford and Baucus, 1987). Leadership style in crisis is important: crisis management requires decisiveness and leads to centralization of power as leaders struggle to control events (Dutton, 1986; Mulder, Ritsema van Eck and de Jong, 1971). Crises frequently involve the transformation of organizational beliefs, values and cultural patterns (Brunsson, 1982; Starbuck, 1982).

The precise causes and outcomes of a crisis are hard to appreciate until after the event, if at all. Firms are frequently affected by evolving situations beyond their control. In an admittedly extreme interpretation, Piore and Sabel (1984) posited an economic and technologically derived 'crisis of the industrial system'. Society at large has also been criticized for striking a 'Faustian bargain... (with high-risk technologies)... whose downside risks are not fully known to us' (Shrivastava, 1988, p.283). Governments intervene to influence the pace and direction of industrial development, creating environmental turbulence for industrial firms. But important as such influences are, it is their impact on the task of management at the *firm-in-sector* level with which the following arguments are concerned.

An optimistic view of technologically-driven crisis, then, is of a bridge between antecedent and future organizational states of the firm: its interpretation as an adaptive transformational *management* task implies (i) positive concern for long-run as well as short-run viability, (ii) periodic re-establishment of a new and better match between a firm's purposes and contextual potentialities, and (iii) proactive handling of transitional organizational episodes.

ANTECEDENTS OF ORGANIZATIONAL CRISIS

Firm-level crisis usually has organizational and environmental antecedents. The latter are often perceived as more common causes of crises especially by those preoccupied with organization-environment contingencies and resource dependencies. Unexpected events beyond the control of the focal organization create conditions of potential disaster from many sources, social, cultural, economic, political, regulatory as well as technological. Events and trends impinge directly on the firm via specific market or competitive pressures in its industrial sector. The multiplicity, complexity and unpredictability of events in the *emerging phase* makes their interpretation highly problematic (Duncan, 1972).

Conventional wisdom suggests comprehensive environmental scanning to facilitate the interpretation of external events and minimize the chance of surprise (Ansoff, 1976; Dutton and Ottensmeyer, 1987; Smart and Vertinsky, 1977). In various empirical studies of organizational responses to environmental crises (e.g. Dunbar and Goldberg, 1978; Grinyer and Spender, 1979; Meyer, 1982a) it is clear that crises test a firm's technical resilience along with its managerial competence.

Technological changes initiated or appropriated by competitors (including new entrants to the sector) are a common problem which many organizations prove ill-equipped to handle (Cooper and Schendel, 1976; McGee and Thomas, 1985). When a major innovation occurs in product design or production processes, well-established firms committed to obsolete technology often fail to stay the distance.

Situations become crises, however, when they prompt responses outside habituated routines. So perceiving crisis antecedents as wholly or largely environmental in origin almost certainly understates the degree to which the firm, as a contributor to its environment, *realizes* the crisis, as well as generating *responses* to it (Maidique and Hayes, 1984; Tushman and Anderson, 1986; Weick, 1988). There is little doubt that many firms nurture potential crisis-antecedents by virtue of their own actions. Managerial failures of leadership, organizational performance, control, structuring and so forth contribute directly to such crisis antecedents. Traumatic events focus actors' attention inwards on problem diagnosis (Billings, Milburn and Schaalman, 1980) and on testing personal and collective beliefs about appropriate responses (Meyer, 1982b; Starbuck, Greve and Hedberg, 1978).

Crises of *leadership* style and competence are common, yet amongst the most painful to resolve. Indeed, crisis realization frequently requires the rejection of senior members of the dominant coalition as the first step in crisis resolution. Crises of *performance* can be defined in many ways: inadequate rates of revenue or profits growth, failure to improve product specifications or quality, lack of new product introductions, market rejection etc. Inadequate economic returns are an unequivocal pointer to impending crises (Altman, 1971; Argenti, 1976); paradoxically, firms in high-risk situations do not always seek, let alone sustain, commensurately high rates of financial return (Bowman, 1980). Crises of *control* are failures to keep performance within acceptable limits. Crises of *structure* raise questions of how best to organize at particular stages of a firm's development (Chandler, 1962; Greiner, 1972; Mintzberg, 1979). Evolving structural problems particularly affect fast-growing technologically-advanced firms, leading to crises of *resilience* (Lippitt and Schmidt, 1967)

Particularly problematic situations arise from situations featuring *concurrent* threat and opportunity, such as innovations involving new products or production methods beyond the scope of the originating firm to exploit. The strategic dilemma is the realization of irreversible, but potentially imitable change in the firm's *technical* knowledge not matched by relevant *resources* and managerial *competence* to exploit this change. For a variety of reasons, the time-frame for first-mover exploitation of the opportunity is—or at least is perceived to be—limited, pressuring managers to act promptly to preempt competitive response.

Innovative opportunities become crises when firms implement inappropriate exploitation strategies including, of course, doing nothing. Inappropriate strategies arise from (i) predicting changes that fail commercially because they offer no net real benefits to users: e.g. rotary internal combustion engines that failed, not so much for technical reasons, but because piston engines improved far beyond what was envisaged in the mid-1960s (ii) not predicting the need for, or merit of, novel concepts, thereby failing to secure a strong competitive position in a changing technological context: e.g. plate-glass producers who ignored the possibility of enhanced process technology such as that which led to float glass, and (iii) initiating or responding to a technological trend, but misinterpreting the dominant form it will ultimately take: e.g. videotape rather than videodisc for home-view movies.

Some common antecedents of organizational crises are listed in Table 1. The list is intended to be illustrative, rather than comprehensive.

A PROCESS PERSPECTIVE ON CRISIS TRANSFORMATIONS

Thus far, the origins and effects of organizational crises have been discussed largely as if they were objectively indisputable events or trends. But crises are in many respects perceptual phenomena; events emerge as crises when organizational actors *interpret* them as such. Dutton (1986, p.502), for example, stated: 'Crisis can be seen as a composite perception based on several different dimensions of an issue. In particular, the perceived

Table 1 Types of organizational crises

(1) Cross-sectoral or global change creating new possibilities and threats for existing firms, such as:

- Non sector-specific technological change, e.g. automation; flexible manufacturing systems; communications; computer control
- Global changes and trends: e.g. trend to healthier diets and living; concern for the natural environment
- Imposed constraints: e.g. legislation constraining or proscribing classes of activity or eliminating quasi-monopolies
- Natural constraints and disasters: i.e. developments beyond the control of the firm having a major bearing on its potentialities and/or survivability

(2) Sectoral change, fundamentally (and differentially) affecting the competitive standing of firms in sector, for example:

- Product/process obsolescence or substitution
- Secular decline in demand for the products of the firm(s)
- Legislative constraints on freedom of action in the sector, e.g. controls on advertising
- Aggressive competitive actions aimed at fundamentally destabilizing the status quo in the sector
- Resource constraints affecting processes particular to the value chain of firms in the sector

(3) Major problems specific to the focal organization such as:

- Inappropriate competitive strategies
- Performance inadequacies
- Lack of control of activities
- Inadequate local resources, including finance
- Structural deficiencies
- Power conflicts between stakeholder groups in and around the firm
- Leadership deficiencies, including problems and conflicts of ideology, vision, culture etc.
- Unforeseen disasters such as the death of key executives
- Loss of key customers or suppliers arising from one or more of the above shortcomings

importance, immediacy and uncertainty of an issue'... (and, Milburn, Schuler and Watman (1983) would add—the lack of managerial control over it)... 'all contribute to how threatening an issue is perceived to be'. So crucial aspects are *whether*, to what *degree* and in what *ways* a crisis is recognized and responded to by managers in the firm (or more broadly, by the *internal coalition*—Mintzberg, 1983).

An important contributory influence on managerial perception of crisis, however, is what will be termed 'external validation', the recognition by legitimate 'external observers' of a highly problematic state affecting the focal organization. Who are these observers and what is validation? Three categories clearly exist. First, the *external coalition* of stakeholder interests (Mintzberg, 1983) such as shareholders, trading networks of customers and suppliers, non-trading contacts like banks, consultants and so on. Crisis 'validation' takes many forms here but, typically, is advice about and calls for enhanced performance, frequently accompanied by the enforced departure of the chief executive and other members of the dominant internal coalition associated with perceived failures.

A second influential class of 'observer' is, of course, competition; 'validation' here is competitive action aimed directly at appropriating the benefits of technological innovation otherwise available to the focal firm.

A third category of observer is the 'independent' expert such as the banker, stockbroker or industry commentator whose credibility resides in his/her ability to mobilize opinion via informed commentary (overt or otherwise). The latter may be genuinely impartial, based on more-or-less objective criteria, or grounded possibly in financial self-interest with a veneer of impartiality.

To pursue the analogy of crisis and strategic technological transformation, one must acknowledge the parts played by both the internal and external coalitions in these processes. The relative power of the various coalitions in and around the firm will influence the way that a crisis is acted out and therefore how it will be resolved (Figure 1).

Incipient crises

Actors who overlook or unwittingly suppress from the internal coalition externally-derived weak signals (Ansoff, 1976) of latent crisis are behaving predictably, if not advisedly. In any event, members of the external coalition may be poorly placed to recognize internal signs of emerging crisis. As long as crisis antecedents go unrecognized, this phase can be conceptualized as 'incipient' crisis. There can be no certainty that incipience will evolve into maturity, but this possibility clearly exists. Impending 'natural' disasters as well as those enabled by inadequate control of risk-laden technologies are obvious examples. However, it seems likely that all realized crises pass through an incipient phase during which the dimensions of the problem and its implications clarify and the crisis matures. This process is sometimes exceedingly rapid; conversely, it is sometimes protracted, meriting the title of 'ignored' crisis.

Ignored crises

Ignoring antecedents in the tacit hope they will dissipate is a possible, if dubious organizational response. It is feasible when the internal coalition

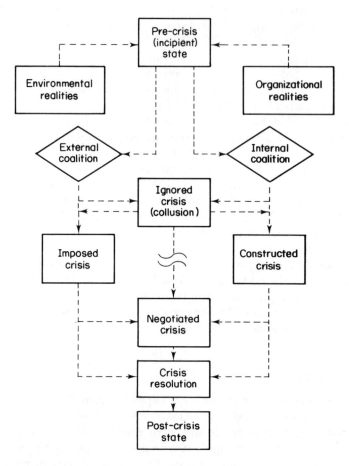

Figure 1   Modes of crisis transformation

dominates or when the external coalition colludes in feigning ignorance. The most common reason for ignoring antecedents, aside from complacency or oversight, is a reluctance to face the possibility that present policies and technologies are becoming inappropriate. There may also be a strong element of wishful thinking, particularly when failing to respond to competitive initiatives, born of a desire to avoid destabilizing change. A classic example was the position of the traditional Swiss mechanical watchmakers, facing technological innovation in the form of electronic wristwatches from U.S. firms in the early 1970s.

An equally common response to crisis antecedents is to make minor, incremental adjustments, preserving the apparent continuity of existing technical and administrative routines. This leads to the debilitating phenomenon of 'strategic drift' (Johnson, 1988). If the crisis antecedents do not prove temporary, strategic drift ultimately evolves into another and more urgent form of crisis.

Outright disasters, whether technological or economic, clearly cannot be ignored for long. In contrast, many environmental and organizational situations are complex, confusing and genuinely hard to interpret. Whilst in theory actors should diagnose antecedents promptly, this is far from straightforward, as General Motors, for example, found out during the oil shocks of the 1970s (Kanter, 1983; McCaskey, 1982; Quinn, 1980). There are various reasons why firms respond inappropriately, if at all.

First, industry sector environments are characterized by diversity, complexity, dynamism (rate of medium/long-run change), turbulence (rate of short-run change) and competitive hostility. As each increases in salience the sheer variety of issues for consideration creates confusion and makes assessment of weak signals highly problematic. Because strong technological positions are normally based on 'bundles' of technologies (Ford, 1988), early signals of impending change may take on significance only when they are recognized as forming a *convergent pattern*, for which there is rarely a precedent. In uncertain, hostile situations managers may know the *source* of the technological uncertainty, but still find its *implications*, construable in terms of viable, sector or firm-specific trajectories (Dosi, 1982), fundamentally unknowable for some considerable time.

Second, Smart and Vertinsky (1984), for example, noted that environmental characteristics condition a firm's organizational repertoire of strategic responses to external events. Firms in relatively stable, certain environments adopt standard operating procedures and systems on which they still rely in novel situations; contra-indications are frequently ignored because placid environmental conditioning reinforces cautious, incremental behaviour and creates susceptibility to emerging crisis.

Third, situations are construed as crises via intuitive as well as analytic processes. 'Problem formulation and decision analysis' (Lyles and Thomas, 1988; Thomas and Samson, 1986) and 'strategic issue diagnosis' (Dutton and Duncan, 1987; Dutton and Ottensmeyer, 1987) have been proposed as means of analysing complex, ill-structured problems, but evidence of their effectiveness is, so far, limited. In fact, there are various cognitive-psychological mechanisms that explain why human actors will have very considerable difficulty in being this analytical (Janis and Mann, 1977; Schwenk, 1984; 1988; Tversky and Kahnmann, 1984). Particular traps include biases, overconfidence, reliance on small samples of evidence, escalating commitment to ill-judged courses of action and premature closure.

Other factors that may hinder dispassionate issue identification and resolution are, notably, the politics of factional interests and cultural and ideological forces creating barriers to organizational change (Mangham, 1979; Meyer, 1982b; Pettigrew, 1973; Tichy, 1983). For all the above reasons actors may ignore emerging crisis signals or be pathologically incapable of reaching an appropriate and timely interpretation of them.

Imposed crises

Incipient and/or ignored crisis antecedents may reach a point where validatory pressures from the external coalition, as previously defined, effectively *impose crisis* on the focal organization, irrespective of whether its managers fully accept the diagnosis at this stage (Grinyer and Spender, 1979; Starbuck, 1982; Smith and Child, this volume; Whipp and Clark, 1986). This condition often arises when the internal coalition is weak or divided. It is characterized by a loss of managerial initiative and control over events: the internal coalition is forced to *react* not only to the emerging realities of the situation, it must also accept the technical diagnosis and dictates of the external coalition. For individual managers, imposed crisis is traumatic and frequently terminal. In general, the external coalition will not be in a sufficiently well-informed position to envision realistic *strategic* options for the firm, so the emphasis is likely to be on short-term palliatives. An important exception arises when a widely-networked external coalition is capable of diffusing knowledge inwards (Boisot, 1986). The means of achieving this quickly, however, are limited, save possibly through takeover.

Tushman and Anderson (1986) noted that technologically innovative actions by one competitor which destroy existing competences of the focal firm are another major form of imposed crisis. Since firms do not monopolize innovativeness, the set of competing firms, in effect, impose crises on one another. Revolutionary, frame-breaking technological change can entirely alter the shape of the sector over time. Harrigan (1980) noted the role of secular demand decline in imposing crises on entire industrial sectors. Faced with preemptive, imposed crisis, probably all the firm's internal coalition can do is to seek to 'negotiate' with the external coalition(s) to mitigate the adverse consequences.

Negotiated crises

In practice, external 'validation' by some form of exogenous action is often a catalyst for, or a reinforcement of, changing managerial perceptions in the focal firm. When problematic circumstances and appropriate responses are agreed unambiguously among internal and external coalitions, the firm can be said to experience a fully-owned or *negotiated crisis*. Shared ownership develops through dialogue during a period of increasing awareness.

Such crises typically become 'turnaround' situations (Hofer, 1980; Slatter, 1984). A convergent diagnosis soon directs efforts towards resolution; indeed, the pressure for prompt remedial action is likely to be irresistible. However, as Hofer argued, responses should be geared not only to short-run, resource-oriented technical considerations, but also to relevant longer-term strategies acceptable to major stakeholders. A strong focus on the short run means that negotiated crises, like imposed crises, are typically resolved by a largely *reactive* process. Alleviating crisis in the short run, ironically, exposes the firm to longer-term strategic pressures that will—in all probability—overwhelm it.

Clearly, then, resolving a negotiated crisis is not without its problems. Many classic 'turnarounds' began as imposed crises, requiring a change of leadership as part of the 'negotiation', prior to implementing other radical or unpleasant solutions. Radical responses do not guarantee success, indeed may convert a difficult condition into a terminal one; so managerial judgement is still a critical factor in problem resolution.

Crises arising from technologically *innovative* possibilities require negotiations that balance the risks of implementing a new technology against the risks of not doing so. Whilst such crises may be less urgent than other forms of crisis, the threat of competitive action can impose a sense of urgency on negotiations. As Quinn (1980) argued, effective managers handle these situations logically and incrementally, avoiding premature action until possibilities crystallize more clearly.

Constructed crises

Smart and Vertinsky (1984) concluded that environmental uncertainty leads to responsive or adaptive firm behaviour. Adaptive firms have dominant managerial coalitions disposed to investigate novel events and managers who seek environments for which their firms' capabilities are well-matched. Adaptive behaviour is necessary because comprehensiveness in strategic decision processes cannot, of itself, guarantee good economic performance when the firm operates in an unstable, unpredictable environment, the kind most likely to harbour antecedents of organizational crisis (Fredrickson, 1984: Fredrickson and Mitchell, 1984). Adaptive behaviour also reinforces the power of initiative-seeking internal coalitions to shape the agenda of crisis transformation.

But from a strategic choice perspective (Child, 1972), managers implement purposive choices which affect firms' behaviour far *beyond* a purely reactive response to events. This is the problem-finding and solving role of the dominant coalition. In so doing, managers act out a shared, albeit incomplete or flawed perception of their particular firm-in-sector context.

Further, in trying to clarify the significance of events, they act in ways that change the situation of the firm, as well as how it is subsequently interpreted; thus they create or *construct* crisis from equivocal antecedents (Weick, 1988).

That executives deliberately *champion* a perception of crisis within the internal coalition, when the indications are at best equivocal, finds support elsewhere (Harvey-Jones, 1988; Maidique, 1980; Pettigrew, 1985, p.447; Pitt, 1989; Schon, 1963; 1967). Such crises become a vehicle for stimulating innovative organizational responses. To the degree that constructed crisis is imposed on a passive or divided external coalition, it is a conscious managerial effort to dictate the transformational agenda. Not surprisingly, then, Dutton (1986) found that 'crisis issues' were obviously politicized, as key actors sought to avoid challenges to the dominance of their viewpoints.

For Dutton and Duncan (1987), convergent assessment within the dominant coalition of issue importance, urgency and response feasibility creates momentum for radical change. Shared views are initially change-averse: radical thinking on the part of more influential members begins the destabilization process. Articulating a radical reinterpretation of events creates the triggering conditions for crisis construction and innovative responses aimed at crisis resolution. Schon (1967, p.127) sums it up as:

> 'the (managerial) ability to create a sense of crisis around events that need not be interpreted in this way. . . . Only the strongest incentives can lead an organization to effect deliberate change. For an organization to increase significantly its capacity for innovation, it is not enough that a man or a few men—even at the top—understand that it would be advantageous for the organization to change. Something like a *state of crisis* must arise. The organization must come to feel that its survival, or at any rate, its survival as it has been, is threatened.'
> (emphasis added)

Thus, while tangible stimuli may form rational indicators of change, what is suggested is that constructed crises are *created* by managers determined to isolate or preempt opposition to change. Further, if Mintzberg and Waters' (1982) comments are correct, more generally, the construction and resolution of crisis are characteristically grounded in a prophetic, single-minded, *vision* of future direction. This, it is also suggested, is especially likely when state-of-art technologies are involved.

A vision, of course, can be idiosyncratic, strategically inappropriate, distorted, even pathological in its outcomes. Once crystallized, however, it is *highly resistant* to disconfirming factual data. External validation plays little or no part in the process. Constructed crises derive from proactive managers' personal priorities, beliefs (and fantasies), from synthetic rather than from analytic thinking. Crisis construction is an evocative and galvanizing device in the personal repertoires of many top managers.

Proactive coalitions foster change-receptive cultural norms to legitimize such visions, buttressed by simple, but potent symbols and slogans grounded in myths and past achievements (Martin and Powers, 1983; Smircich and Morgan, 1982). For example, Texas Instruments has used slogans like 'Tomorrow's Technology Today' and 'Creators of Change' to epitomize simple truths about what the firm stands for and what its future priorities should be.

CRISIS RESOLUTION

Transformational changes in firms, then, particularly those involving technological innovation, can arise in one of several crisis modes. Indeed, a crisis may evolve from one through another. The mode clearly affects the direction that crisis resolution takes, in particular, the degree of proactivity achieved by managers and the control they enjoy over the transformational process.

Crisis is a transitional state: in the management of *constructed* crisis several steps are involved in the process of resolution (i) cognitive construction and articulation of a state of crisis around specific technical issues and circumstances, (ii) postulation and evaluation of viable post-crisis states, (iii) postulation and evaluation of potential technical transformations between pre- and post-crisis states in the light of known organizational competences and resources, and (iv) selection (albeit intuitively) of an appropriate moment to trigger the transformation process.

Organizational states and transformations

Miller (1982), Miller and Friesen (1977; 1984) and others have conceptualized organizational archetypes or configurations as explanatory and predictive states in the study of strategic transformation. They argued that structural, technological, processual and contextual characteristics of firms allow only a limited set of configurations and that some of these are *non-viable in the long run*.

Strategic management, therefore, is future-oriented *pattern recognition* leading to real-time technological reconfiguration. Timing is crucial, because crisis-driven opportunism is feasible only at certain moments of technological development, analogous to Abell's (1978) 'windows' of opportunity. Specific transitions become actionable given 'a critical state of incongruence (lack of "fit") with the external environment' (Miller and Friesen, 1984, p.208; Venkatraman and Camillus, 1984); then, and perhaps only then, is reconfiguration possible *and* indicated. As Morgan (1986, p.255) speculated, a state of fit probably contains the seeds of tension from which transformation ultimately springs.

Generalized theorizing, however, offers little help to practitioners. Besides, 'fit' abstractions undervalue the autonomous contributions made by individual actors in the transformational process. Again, there is much anecdotal evidence that able strategists envision technological reconfigurations and how they may be realized. This reflective expertise (Schon, 1983) requires skill and judgement of a high order and in the final analysis, may be unamenable to systematic examination.

Nonetheless, within the *constructed crisis* mode, enabling and confounding organizational factors affect the probability of successful outcomes. They predispose actors to be receptive, hostile, or simply ignore emerging possibilities. These include the history of the sector, including firm-specific structures and interrelationships, technological trajectories (Dosi, 1982; Pavitt, 1986) and the convergence of industry-wide experience (Huff, 1982), all of which contribute to shared organizational beliefs (recipes or scripts) about appropriate behaviour in context (Gioia and Poole, 1984; Lord and Kernan, 1987).

Within the organizational repertoire of capabilities, managers should consider the extent of intended or serendipitous *slack resources*. The affects of slack are still disputed (Yasai-Ardekani, 1986): slack resources allegedly stabilize performance levels, *inhibiting* crisis and radical change (Cyert and March, 1963), whilst benign tolerance of slack resources is also said to be a potent *enabling* factor for innovative change (Peters and Waterman, 1982). Then again, poor economic performance tends to reduce resource slack, restricting the firm's capacity to achieve and build on experimentally-derived lessons (Singh, 1986). Thus, it would seem, tolerating some level of innovating resources when times are good may be an engine of subsequent transformational change. Failure to be tolerant may foreclose later options, escalating commitment to potentially high-risk residual courses of action (Schwenk, 1986; Staw, 1976, Whyte, 1986).

Patently, transformation is the product of action. Quasi-autonomous innovatory behaviour by 'change champions' has been widely noted as a key role in technological innovation. Champions set the agenda for crisis as well as its means of resolution; they are not only the 'strategic visionaries' who shape the internal environment in which new realities are sensed, amplified, legitimated and acted on, but also the highly motivated, if sometimes idiosyncratic technologists whose creative innovating capacity frequently leads to exciting, challenging opportunities (Maidique, 1980). Neither should one overlook those who catalyse administrative change, personally or by using others from outside to do so (Bennis, 1969; Mangham, 1978; Tichy, 1972). All such individuals are strong-willed, persistent and often intuitive thinkers with the courage to risk personal rejection in pursuit of their goals.

Still, organizations are complex, dynamic systems rarely understood *in toto* by their actors. Deliberate attempts to jolt the organization out of equilibrium frequently have unexpected outcomes, including outright failure. For transformational change must not merely disturb the status quo, it must alter it fundamentally and irreversibly. Only by stepping outside established administrative routines and technical repertoires in the pursuit of real learning can organizations change adaptively (Argyris and Schon, 1978; Chakravarthy, 1982; Hedberg and Jonsson, 1977). An insightful, if whimsical insight into the true nature of the adaptive firm was offered by Hedberg, Nystrom and Starbuck (1976) in their prescription for the 'self-designing' organization.

Finally, resolving constructed crises must take account of 'political' realities. Technologists in innovative firms, particularly at the cutting edge of innovative possibilities, increasingly recognize the power conferred by their expert knowledge. Still, many remain ill-equipped by (lack of) training, outlook or prior experience to exploit this power and to cope with the pressures of constructing and resolving crises in the face of hostile commercial uncertainties.

SUMMARY

Crisis has been proposed as an appropriate metaphor for strategic, technologically-oriented transformations of the firm-in-sector. More specifically, that the effective *management* of such transformations is a process of *constructing* and then *resolving* organizational crises adaptively and opportunistically. This is probably not a comforting message for many managers. There is after all, an accumulation of literature that portrays the act of management as essentially planning and controlling, not as coping with—or promoting—instability.

To see managing technological innovation in this light is not to advocate irresponsible behaviour. Rather, that managers must embrace a continuing dynamic of adaptive learning in the search for competitive advantage and profitable survival. They are encouraged to question (and to insist that others question) the status quo critically and more often. Then, as perceived necessary from time to time, to promulgate crisis constructs and visionary solutions among the influential coalitions in and around the firm.

There are few, well-researched prescriptions for practical success in this venture. Much of what we know is speculative, anecdotal and unsystematic. Still, the evidence so far is that for some executives crisis construction is a natural, intuitive, *expert* activity.

From a research perspective, the concept of states-and-transitions within a crisis framework needs further theoretical and practical development to understand how to recognize the need for constructed crises and how to implement innovative technological change in this mode. Thus it is argued, the 'crisis management' metaphor may well hold more promise for explicating the strategic management of technological innovation (and more specifically, the *basis* of management expertise) than traditional, normative analytic-planning models with which we are presently more familiar.

REFERENCES

Abell, D.F. (1978). 'Strategic windows', *Journal of Marketing*, July, 21–25.

Altman, E.I. (1971). *Corporate Bankruptcy in America*, Heath, Lexington.

Ansoff, H.I. (1976). 'Managing strategic surprise by response to weak signals', *California Management Review*, **19**(2), Winter, 21–33.

Argenti, J. (1976). *Corporate Collapse—Causes and Symptoms*, McGraw-Hill, New York.

Argyris, C., and Schon, D.A. (1978). *Organizational Learning: A Theory of Action Perspective*, Addison-Wesley, Reading, Mass.

Bennis, W.G. (1969). *Organization Development: Its Nature, Origins and Prospects*, Addison-Wesley, Reading, Mass.

Billings, R.S., Milburn, T.W., and Schaalman, M.L. (1980). 'Crisis perception: a theoretical and empirical analysis', *Administrative Science Quarterly*, **25**, 300–315.

Boisot, M. (1986). 'Markets and hierarchies in a cultural perspective', *Organization Studies*, **7**, 135–58.

Bowman, E. (1980), 'A risk/return paradox for strategic management', *Sloan Management Review*, **21**(3), 17–31.

Brunsson, N. (1982). 'The irrationality of action and action irrationality: decision ideologies and organization actions', *Journal of Management Studies*, **19**, 1, 29–44.

Chakravarthy, B.S. (1982). 'Adaptation: a promising metaphor for strategic management', *Academy of Management Review*, **7**, 1, 35–44.

Chandler, A.D. (1962). *Strategy and Structure*, M.I.T. Press, Cambridge, Mass.

Child, J. (1972). 'Organizational structure, environment and performance: the role of strategic choice', *Sociology*, **6**, 1–22.

Cooper, A.C., and Schendel, D. (1976). 'Strategic responses to technological threats', *Business Horizons*, Feb.

Cyert, R., and March, J.G. (1963). *A Behavioural Theory of the Firm*, Prentice-Hall, Englewood Cliffs, N.J.

Daneke, G. (1985). 'Regulation and the sociopathic firm', *Academy of Management Review*, **10**, 1, 15–20.

De Greene, K.B. (1982). *The Adaptive Organization: Anticipation and Management of Crisis*, Wiley, New York.

Dosi, G. (1982). 'Technological paradigms and technological trajectories', *Research Policy*, **11**, 147–62.

Dunbar, R.L., Dutton, J.M., and Torbert, W.R. (1982). 'Crossing mother: ideological constraints on organizational improvements', *Journal of Management Studies*, **19**(1), 91–108.

Dunbar. R.L., and Goldberg, W.H. (1978). 'Crisis development and strategic response in European corporations', in C.F. Smart and W.T. Stanbury (eds) *Studies on Crisis Management*, Butterworth, Toronto.

Duncan, R.B. (1972). 'Characteristics of organizational environments and perceived environmental uncertainty', *Administrative Science Quarterly*, **17**(3), 313–27.

Dutton, J.E. (1986). 'The processing of crisis and non-crisis strategic issues', *Journal of Management Studies*, **23**(5), 501–15.

Dutton, J.E., and Duncan, R.B. (1987). 'The creation of momentum for change through the process of strategic issue diagnosis', *Strategic Management Journal*, **3**(3) 279–95.

Dutton, J.E. and Ottensmeyer, E. (1987). 'Strategic issue management—systems—forms—functions and contexts', *Academy of Management Review*, **12**(2), 355–65.

Fink, S.L., Beak, J., and Taddeo, K. (1971). 'Organizational crisis and change', *Journal of Applied Behaviour*, **7**(1) 15–37.

Ford, J.D. (1988). 'Develop your technology strategy', *Long Range Planning*, **21**(5), 85–95.

Ford, J.D. and Baucus, D.A. (1987). 'Organizational adaptation to performance downturns: an interpretation-based perspective', *Academy of Management Review*, **12**(2), 366–80.

Fredrickson, J.W. (1984). 'The comprehensiveness of strategic decision processes: extension, observations, future directions', *Academy of Management Journal*, **27**(3), 445–66.

Fredrickson, J.W., and Mitchell, T.R. (1984). 'Strategic decision processes: comprehensiveness and performance in an industry with an unstable environment', *Academy of Management Journal*, **27**(2), 399–423.

Gioia, D.A, and Poole, P.P. (1984). 'Scripts in organizational behavior', *Academy of Management Review*, **9**(3), 449–59.

Greiner, L.E. (1972). 'Evolution and revolution as organizations grow', *Harvard Business Review*, July/Aug, 37–46.

Grinyer, P., and Spender, J-C. (1979). 'Recipes, crises, adaptation in mature businesses', *International Studies in Management and Organizations*, **9**(3), 113–23.

Harrigan, K.R. (1980). 'Strategies for declining industries', *Journal of Business Strategy*, **1**, 2.

Harvey-Jones, J. (1988). *Making it Happen: Reflections on Leadership*, Collins, London.

Hedberg, B., and Jonsson, S. (1977). 'Strategy making as a discontinuous process', *International Studies of Management and Organization*, **7**(2), 89–109.

Hedberg, B., Nystrom, P.C., and Starbuck, W.H. (1976). 'Camping on seesaws: prescriptions for a self-designing organization', *Administrative Science Quarterly* **21**(1), 41–65.

Hermann, C.F. (1963). 'Some consequences of crisis which limit the viability of organizations', *Administrative Science Quarterly* **8**(1), 61–82.

Hermann, C.F. (1969). *Crises in Foreign Policy*, Bobbs-Merrill, New York.

Hofer, C.W. (1980). 'Turnaround strategies', *Journal of Business Strategy*, **1**, 2.

Hofer, C.W., and Schendel, D.E. (1978). *Strategy Formulation: Analytical Concepts*, West, St. Paul.

Huff, A.S. (1982). 'Industry influences on strategy reformulation', *Strategic Management Journal*, **3**, 119–31.

Janis, I.L., and Mann, L. (1977). *Decision Making*, The Free Press, New York.

Johnson, G.N. (1987). *Strategic Change and the Management Process*, Blackwell, Oxford.

Johnson, G.N. (1988). 'Rethinking incrementalism', *Strategic Management Journal*, **9**, 75–91.

Jonsson, S.A., and Lundin, R.A. (1977). 'Myths and wishful thinking as management tools', in P.C. Nystrom, and W.H. Starbuck (eds), *Prescriptive Models of Organization*, North-Holland, Amsterdam.

Kanter, R.M. (1983). *The Change Masters*, Simon & Schuster, New York.

Lindblom, C.E. (1959). 'The science of muddling through', *Public Administration Review*, **19**, 79–88.

Lippett, G.L., and Schmidt, W. (1967). 'Crisis in a developing organization', *Harvard Business Review*, **45**, Nov./Dec., 103–12.

Lord, R.G., and Kernan, M.C. (1987). 'Scripts as determinants of purposeful behavior in organizations', *Academy of Management Review*, **12**(2), 265–77.

Lyles, M.A., and Thomas, H. (1988). 'Strategic problem formulation: biases and assumptions embedded in alternative decision-making models', *Journal of Management Studies*, **25**(2), 131–45.

Maidique, M.A. (1980). 'Entrepreneurs, champions and technological innovation', *Sloan Management Review*, **21**(2), Winter.

Maidique, M.A., and Hayes, R.H. (1984). 'The art of high technology management', *Sloan Management Review*, **25**(2), Winter.

Mangham, I.L. (1978). *Interactions and Interventions in Organizations*, Wiley, Chichester.

Mangham, I.L. (1979). *The Politics of Organisational Change*, Associated Business Press, London.

Martin, J., and Powers, M.E. (1983). 'Organizational stories: more vivid and persuasive than quantitative data', in, B. Staw (ed.), *Psychological Foundations of Organizational Behavior*, Scott Foresman, 161–8.

McCaskey, M.B. (1982). *The Executive Challenge: Managing Change and Ambiguity*, Pitman, Boston.

McGee, J., and Thomas, H. (1985). 'Making sense of complex industries', in D.E. Schendel, N. Hood, and J.E. Vahlne, (eds) *Global Strategies*, Wiley, New York.

Meyer, A.D. (1982a). 'Adapting to environmental jolts', *Administrative Science Quarterly*, **27**, 515–53.

Meyer, A.D. (1982b). 'How ideologies supplant formal structures and shape responses to environments', *Journal of Management Studies*, **19**(1), 45–61.

Milburn, T.W.,, Schuler, R.S., and Watman, K.H. (1983). 'Organizational crisis. Part One: Definition and conceptualization', *Human Relations*, **36**, 1141–60.

Miles, R., and Snow, C.C. (1978). *Organization Strategy, Structure and Process*, McGraw-Hill, New York.

Miller, D. (1982). 'Evolution and revolution: a quantum view of structural change in organizations, *Journal of Management Studies*, **19**(2), 131–51.

Miller, D., and Friesen, P. (1978). 'Archetypes of strategy formulation', *Management Science*, **24**, 921–33.

Miller, D., and Friesen, P. (1984). *Organizations: A Quantum View*, Prentice-Hall, Englewood Cliffs, N.J.

Mintzberg, H. (1978). 'Patterns in strategy formation', *Management Science*, May, 934–48.

Mintzberg, H. (1979). *The Structuring of Organizations*, Prentice-Hall, Englewood Cliffs, N.J.

Mintzberg, H., and Waters, J.A. (1982). 'Tracking strategy in an entrepreneurial firm', *Academy of Management Journal*, **25**, 465–99.

Morgan, G. (1986). *Images of Organization*, Sage, Beverly Hills.

Mulder, M., Ritsema van Eck, J.R., and de Jong, R.D. (1971). 'An organization in crisis and non-crisis conditions', *Human Relations*, **24**(1), 19–41.

Nelson, R.R., and Winter, S.G. (1982). *An Evolutionary Theory of Economic Change*, Harvard University, Belknap.

Pavitt, K. (1986). 'Technology, innovation and strategic management', in J. McGee, and H. Thomas, (eds) *Strategic Management Research: A European Perspective*, Wiley, Chichester.

Peters, T., and Waterman, R. (1982). *In Search of Excellence*, Harper & Row, New York.

Pettigrew, A. (1973). *The Politics of Organizational Decision Making*, Tavistock, London.

Pettigrew, A. (1985). *The Awakening Giant*, Basil Blackwell, Oxford.
Piore, M.J., and Sabel, C.F. (1984). *The Second Industrial Divide: Possibilities for Prosperity*, Basic Books, New York.
Pitt, M.R. (1989). 'Corporate birth, crisis and rebirth: the emergence of four small U.K. service firms', in R. Mansfield, *The Frontiers of Management*, Routledge, London.
Porter, M.E. (1980). *Competitive Strategy*, The Free Press, New York.
Quinn, J.B. (1980). *Strategies for Change: Logical Incrementalism*, Irwin, Homewood, Illinois.
Ramaprasad, A. (1982). 'Revolutionary change and strategic management', *Behavioral Science*, **27**, 387–92.
Schon, D.A. (1963). 'Champions for radical new inventions', *Harvard Business Review*, March–April.
Schon, D.A. (1967). *Technology and Change: The New Heraclitus*, Pergamon, Oxford.
Schon, D.A. (1983). *The Reflective Practitioner*, Basic Books, New York.
Schwenk, C.R. (1984). 'Cognitive simplification processes in strategic decision making', *Strategic Management Journal*, **5**, 111–28.
Schwenk, C. (1986). 'Information, cognitive biases and commitment to a course of action', *Academy of Management Review*, **11**(2), 298–310.
Schwenk, C. (1988). 'The cognitive perspective on strategic decision making', *Journal of Management Studies*, **1**, 41–55.
Shrivastava, P. (1988). 'Industrial crisis management: learning from organizational failures', *Journal of Management Studies*, **25**(4), 283–4.
Singh, J.V. (1986). 'Performance, slack, and risk taking in organizational decision making', *Academy of Management Journal*, **29**(3), 562–85.
Slatter, S. (1984). *Corporate Recovery: Successful Turnaround Strategies and their Implementation*, Penguin, London.
Smart, C., and Vertinsky, I. (1977). 'Designs for crisis decision units', *Administrative Science Quarterly*, **22**, 640–57.
Smart, C., and Vertinsky. I. (1984). 'Strategy and the environment: a study of corporate responses to crisis', *Strategic Management Journal*, **5**, 199–213.
Smircich, L., and Morgan, G. (1982). 'Leadership: the management of meaning', *Journal of Applied Behavioral Studies*, **18**, 257–73.
Starbuck, W.H. (1982). 'Congealing oil: inventing ideologies to justify acting ideologies out', *Journal of Management Studies* **19**(1), 3–27.
Starbuck, W.H., Greve, A., and Hedberg, B.L. (1978). 'Responding to crises', *Journal of Business Administration*, **9**(2), 111–37.
Staw, B.M. (1976). 'Knee-deep in big muddy: a study of escalating commitment to a chosen course of action', *Organizational Behaviour and Human Performance*, **16**, 27–44.
Thomas, H., and Samson, D. (1986). 'Subjective aspects of the art of decision analysis: exploring the role of decision analysis in decision structuring, decision support and policy dialogue', *Journal of the Operational Research Society*, **17**(3), 249–65.
Tichy, N.M. (1972). 'Agents of planned social change: congruence of values, cognitions and actions', *Administrative Science Quarterly*, **17**, 164–82.
Tichy, N.M. (1983). *Managing Strategic Change: Technical, Political and Cultural Dynamics*, Wiley, Chichester.
Tjosvold, D. (1984). 'Effects of crisis orientation on managers' approach to controversy in decision making', *Academy of Management Journal*, **27**, 130–38.
Tushman, M.L., and Anderson, P. (1986). 'Technological discontinuities and organizational environments', *Administrative Science Quarterly*, **31**, 439–65.

Tversky, A., and Kahnmann, D. (1984). 'Judgement under uncertainty: heuristics and biases', *Science*, **185**, 1124–31.

Venkatraman, N., and Camillus, J.C. (1984). 'Exploring the concept of 'Fit' in strategic management', *Academy of Management Review*, **9**, 513–25.

Weick, K. (1988). 'Enacted sensemaking in crisis situations', *Journal of Management Studies*, **25**(4), 305-17.

Whipp, R., and Clark, P. (1986). *Innovation and the Auto Industry*, Pinter, London.

Whyte, G. (1986). 'Escalating commitment to a course of action: a reinterpretation', *Academy of Management Review*, **11**(2), 311–21.

Yasai-Ardekani, M. (1986). 'Structural adaptations to environments', *Academy of Management Review*, **11**, 9–21.

12

Technological Change and Strategy Formulation

Ken Clarke

University of Bath

and

Howard Thomas

University of Illinois

Technological change is a central feature of an organization's product, process and system innovations. The increasing rate of technological development allied to the 'invasion' of technology into the area of information systems has created new threats for firms and industries, with new opportunities for gaining competitive advantage (Porter and Millar, 1985). That greater attention to technological change is needed by those who formulate strategies is well illustrated by the decline in the relative importance of Western economies over the last quarter century (Abernathy, Clark and Kantrow, 1983). Among many such commentators, Rothwell (1980) has suggested that poor managerial skills in technological development and innovation have contributed to Britain's declining competitive performance. The extent of the challenge is demonstrated by

The Strategic Management of Technological Innovation. Edited by R. Loveridge and M. Pitt
ⓒ1990 by John Wiley & Sons Ltd

Tushman and Anderson (1986) who investigated the transformation brought about by technological development in industries as diverse as locomotives, fountain pens, cement and airlines.

Historically the link between technology and the organization has been interpreted as a stable ratio between capital and labour which changed rarely and only in quantum leaps. The study of innovation and of R&D has been a specialized field somewhat remote from business economics. Whilst intrinsically useful, these approaches offer limited guidance when the issues of *strategic management* are addressed. Studies of strategic management emphasize its nature as a continuous process of identifying and resolving the strategic *dilemmas* that face the organization. Some of these dilemmas arise from the nature of the organization's environment. Therefore it is useful to examine how changes in technology affect both new and existing industries. Mature industries can be threatened by technological development that undermines the continued existence of their boundaries. New technologies can also create new boundaries as in the communications and computing industries. Technological development is multifaceted in its ability to destroy or to enhance existing productive competences and to affect consumer tastes and requirements.

There is a clear need for an enhanced framework to explicate the strategy–technology link; only moderate progress has been made since Kantrow (1980) pointed out:

'The overwhelming fact of the matter is that the most basic categories and terminology of technological strategy have not yet been satisfactorily determined.'

Technological innovation can be considered from at least three perspectives (Figure 1). Whilst these perspectives are related, the *nature* of the links are relatively unexplored. Most studies have adopted one primary perspective, with a predominance of work in (A) and (B), which are, however, largely deterministic in the way they assume technological change influences firms. Such a view is limited, since the application of technologies and the development of technological positions are, in the final analysis, *managerial choices*. This poses the question of how these choices are reached—process perspective (C).

Technology and technological development create pervasive pressures for change. This has always been so, but recent experience indicates more acute problems in coping with their effects on organizational design. However, much research examining technological change has adopted the viewpoint of the 'external' analyst rather than that of the practising manager, and has been largely contextual in nature. Much research has also focused on the

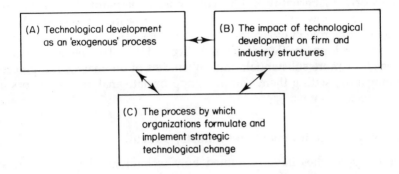

Figure 1

aggregated process (e.g. at the industry level) of diffusion rather than at the level of the adopting and implementing firm.

A central argument of this chapter is that studies of the strategic context *alone* are an insufficient basis for an understanding of the process of innovation. Strategic management is a complex, substantially organization-specific phenomenon. Whilst the contextual analysis throws light on the decision-making situation, it is important also to examine the issues from the viewpoint of the actors, focusing on the management processes within the technological context of the firm. Context and process interact, although *how* they interact and affect the subsequent *content* of strategy remains problematic for strategic analysts (Pettigrew, 1985a).

Technological development as the management of innovation

The attention given to technology as a firm-specific, strategic variable, then, is relatively recent. Economists have typically studied technological change as an aggregate phenomenon, largely treating the firm as a black box (for an excellent review see Dosi, 1988). They have only recently focused on the firm-level processes underpinning significant strategic technological decisions (Teece, 1984; Nelson and Winter, 1982). One strand of research has assumed an implicit link between technological change and the firm's innovation opportunities. Through the production function other research has focused on the factors that lead to successful technological innovation whilst largely disregarding other elements of strategy (Rothwell *et al.*, 1974).

Many studies have treated innovations as discrete, isolated events, with each innovation analysed in an episodic framework. This approach tends to assume that studying innovations is to study innovating. However, 'snapshots' of innovations may say little about the *continuous* dynamic of innovating. Indeed, only a partial view of the process will be obtained. The *dynamic* interaction between product and process innovation is found in the work of Abernathy and his colleagues (Abernathy, 1978; Abernathy, Clark and Kantrow, 1983). More recently some researchers have emphasized management's role in overcoming gaps and obstacles in the process of new product development (Maidique and Zirger, 1984). This research indicates the need for more substantial, 'in-firm' studies of the managerial processes of innovating, setting them within an appropriate analytic framework and firm-in-sector context.

Technological environment and industry structure

Various approaches have examined how technology shapes an industry sector and thereby affects the strategic alternatives open to the firms competing in it. Porter (1983), in an extension of the basic five-forces model of industry structure (Porter, 1980), offered a framework for understanding the technological forces that shape an industry. He discussed how technology influences the five competitive forces (that is, competitive rivalry, threat of entry, threat of substitution, bargaining power of suppliers, bargaining power of customers). For example, entry barriers may be raised by a technology's impact on economies of scale or switching costs. Alternatively, a novel technology can enable competitors to overcome existing entry barriers. Michelin's superior radial tyre technology is an example that allowed it to leap-frog competition and overcome distribution and advertising scale barriers in the U.S.A. But technology has probably had its strongest influence in the area of potential and actual *substitution*. The use of calculators in place of slide rules is an obvious example.

Pavitt (1984) argued that industrial groupings have broad *sectoral regularities* which may be articulated in 'technological trajectories'. In his study, based on 2000 'successful' U.K. innovations, firms were classified into three types: supplier-dominated, production-intensive (sub-divided into scale intensive and specialized suppliers) and science-based. Systematic differences were shown to exist across sectors in terms of the *source* of technology, user *requirements* (such as price or product performance), and the *means* by which technological benefits are appropriated by designers or by their customers. Typical sectors for supplier-dominated firms are housing, agriculture, private services, and 'traditional' manufacture. Scale-intensive firms are usually located in bulk materials industries such as steel, glass and basic chemicals. Specialized suppliers exist in the machinery and instruments industries, and science-based types are commonly found in electronics/electricals and also in chemicals. Firms in each sector, therefore, relate to a specific technological context.

Both Pavitt's and Porter's approaches imply that specific industries show systematic and consistent patterns of technological change. However neither approach explains convincingly *how* and *why* these particular changes develop over time and thus do not accommodate dynamic processual aspects of technological development.

Technology as continuity and change

The concept of technological trajectory (Nelson and Winter, 1982) posits an unfolding time-related pattern of technological change. Technology development and application within a particular industry setting may show relatively incremental, continuous progress for long periods, punctuated by infrequent but disruptive discontinuities.

The analogy of a smooth and continuous technological life cycle has been most extensively developed by Abernathy, Utterback and colleagues. An extensive study of the U.S. automobile industry elaborated the stages of growth and metamorphosis (Abernathy, 1978). There may be a myriad of technological possibilities and productive units in the early stages of industry development. Frequent, novel product innovations take place in the context of flexible, small scale, inefficient production processes (see also Clark and DeBresson—this volume). With the emergence of a dominant design, the pace of product innovation slackens and becomes incremental. Process innovation becomes more significant and is directed toward specialized, capital intensive, efficient systems. Eventually, products become more commodity like and are made in large scale production systems. The model is simple and has been applied to explain a number of phenomena, for example, to the evolution of the electric light bulb and semi-conductor industries.

None the less, the analogy clearly has limitations both as a descriptive or analytic tool and as a managerial aid. Some industries *never* reach the mass production stage: in the case of aero-engines a key technological requirement may be the establishment of flexible 'robust' designs as the basis for families of products (Rothwell and Gardiner, 1988). Utterback and Kim (1983) claimed a relatively wide applicability of the basic model in the area of product rather than process innovation. They argued that extending the model to non-assembled products, i.e. the process industries, highlights the significance of an 'enabling' process technology, not the emergence of a dominant product design, as marking the shift toward an *incremental* phase of innovations. To apply the model to the still emerging personal computer industry one would need to forecast the potential emergence and timing of a dominant design, and further, how that dominant design would configure around existing or possible elements in operating systems and data input systems such as keyboard, voice, 'mice', touch screens etc. Whilst some historical knowledge exists of the influences that might lead to a dominant design, there is no general framework for reliably predicting outcomes in a specific industrial context.

Abernathy, Clark and Kantrow (1983) suggested that the basic life-cycle model was no longer a valid description of the automobile industry in the 1980s, because it has experienced 'dematurity' and subsequent architectural renaissance through product and process innovations. Jones (1983) argued that dematurity had, itself, been broken into a series of disjunctures or staged sequences of product and process transformations. More futuristically, developments in information technology as applied to processes such as flexible manufacturing systems may reduce the deterministic link between products and processes.

More prosaically Abernathy and Clark (1985) emphasized that understanding the eventual *outcomes* of technological innovations requires a wider examination of their long-run impact on the firm, both on its production systems and its market linkages. This proposition goes beyond the regular, incremental innovation position implied by 'maturity'. They utilized the term 'transilience' (first coined in Abernathy, Clark and Kantrow, 1983) to characterize this more complex, disruptive capacity of major innovations.

Indeed, technological development is often disruptive to the point of discontinuity. Radical breakthroughs via technological innovation can disrupt the otherwise smooth workings of existing industry structures. In a study which included the U.S. cement, microcomputers and airlines industries, Tushman and Anderson (1986) demonstrated that periods of incremental technological change are punctuated by radical discontinuities. Such discontinuities produce both uncertainty and opportunities. A discontinuity which destroys existing competences is often linked to a new entrant to the sector; a 'competence enhancing' discontinuity is often linked to existing firms.

Discontinuity was addressed directly by Dosi (1982) who posited the existence of technological paradigms. He defined a technological paradigm '... as an 'outlook', a set of procedures and definitions of the 'relevant' problems and of the specific knowledge related to their solution' (Dosi, 1982, p.148). Paradigms 'direct' the trajectory of technological development by legitimizing and accommodating some solutions at the expense of others. By inference, major discontinuities are linked to paradigm shifts, in contrast to incremental developments within the existing paradigm. Little is known about what defines a paradigm, or the conditions under which a paradigm develops or undergoes complete revision. A 'brand-new' paradigm explains some aspects of Schumpeterian technological development, such as the switch to oxygen-based steel making. It seems reasonable to suppose that the temporal bunching of innovations is linked to the emergence of a new paradigm.

TECHNOLOGY AND STRATEGIC MANAGEMENT

Strategic management as a framework

Rosenbloom (1978) argued that the elements of corporate strategy and strategic management offer a broad framework within which more comprehensive 'explanations' of the process of technology change in firms and industries can be generated. Jemison (1981) proposed an approach which integrates the major perspectives of industrial organization, marketing and administrative behaviour. The first two are concerned with the content of strategy and the third with the processes of strategic decision making and managerial action in the context of the organization and its environment. Pettigrew (1977) similarly emphasized the need to see strategic management as a continuous, contextually based process, arguing that strategic choices address environmental and intra-organisational dilemmas, the resolution of which evolves into strategy. In this perspective, then, making sense of particular managerial decisions (or non-decisions) is central to the analysis of technological innovation within and across industry sectors.

Managerial dilemmas in technological innovation

The arguments above provide a useful framework within which to examine the relationship between technological change and strategy formulation. To what extent is an organization's technological environment traceable and predictable? What dilemmas does the technological environment produce for an organization? By what processes and/or structural arrangements are such dilemmas perceived, interpreted and resolved within it?

An effective technological strategy both creates and renews an organization's technological base and appropriates the benefits deriving from that base. But how? A number of authors characterize effective technological strategies in terms such as offensive-defensive (Freeman, 1982), technological leadership-followership (Porter, 1983), and first-to-market, second-to-market (Maidique and Patch, 1982).

Porter (1983) explored the conditions under which technological leadership is preferable to followership (and vice versa) by focusing on the competitive advantage to be gained via these alternative strategies. Recent studies linking competitive advantage and technology include Flaherty (1983), who suggested that technological leadership has played a significant role in determining the market shares of U.S. based semiconductor firms in international markets, and Spital (1983) who studied the U.S. semiconductor industry and concluded that innovation lead-time did *not* give long-term market share advantage to the first design at a new chip density or bit level. However, where the innovator was able to build up an image of technical excellence and reliability with customers, market share advantages did accrue. Spital concluded that the precise conditions for which innovation lead time provides advantages are not yet clear.

The product/process dimension also produces managerial dilemmas. One evident question concerns entry: specifically, whether, when and how an organization can and should break into the industry/technology life-cycle. Utterback and Kim (1983) argued that entry is required *before* the emergence of a dominant design if the organization is to be successful. However, if an industry such as automobiles is now in 'dematurity', the entry (and exit) choice could arise at several stages of an industry's development. Williams (1983) suggested that firms should examine the nature of the organizational risk and vulnerability at such transitional stages. The choice of product or process based strategies will depend on the investment content, incentive structures and managerial preferences over risk and return. However, technological uncertainty must be seen as qualitatively different from financial risk.

Much of the information on which to base product choices will be external to the firm—for example, which operating system to adopt in the case of microcomputers. The vulnerability is clearly embedded in an inappropriate choice of product technology. For the firm this is typically non-diversifiable—only rarely will it invest in the development of different, competing products unless it engages in joint collaborative ventures. An example is provided by the Philips Company which was forced to abandon its own video recorder design in the face of market acceptance of the VHS standard. More recently, when developing compact disc players, Philips chose to collaborate with Sony on development and adopted a strategy of rapid licensing to others.

The choice of new product technology is often uncertain and subjective, requiring answers to questions such as 'what kinds of needs' (Williams, 1983). If the alternative technologies are well understood, then the effects of substitution may be far more predictable (Harrigan, 1980). Process technology decisions may also be less problematic since a number of major decision factors are within the firm's control, notably experience curve effects, though process-related choices frequently depend on product choices. Product design choices for the firm typically relate to issues of, and information on, customers' 'learning by using'; process design choices to firms' 'learning by *doing*'.

To recap, important dimensions of constructive technological strategic choice include timing, understanding the product-process relationship, envisaging the role of various parts of the industry sector as sources of technological change, and predicting the creation and impact of radical technological changes. The basic proposition that the technological evolution of an industry sector, though complex, is historically traceable and potentially predictable for the future, offers a rich if controversial focus for future research. But, in practice, do firms which 'read' and pursue the central technological tendencies of their industry perform better in the long run? For example, do firms which synchronize their strategies to the evolving conditions of product/process life-cycles grow faster and achieve higher profitability? Under what conditions do significant radical shifts of industry-wide technology take place and how are they stimulated? Thus, further work should focus on whether firm-level competitive advantage is sustainable by a 'correct' reading of the technological forces at work in the sector.

FIRM LEVEL MANAGERIAL PROCESSES

The industry structure approach is of little value in understanding how each organization formulates its strategy. Generally, it adopts an economic paradigm to explain firms' behaviour, makes implicit assumptions that industry forces have broadly systematic effects on all firms in an industry, and that firms react predictably to those effects. It generally takes little account of the specific actors involved. Hence it is difficult to explain variability in firms and their behaviour. Yet firms clearly differ in their responses. The strategic groups concept (Porter, 1980; McGee and Thomas, 1988) has been posited as an intermediate level of analysis between firm and sector. However its theoretical justification and empirical utility are still contentious.

Therefore, for a complete analysis it is important to consider the intra-firm level processes of strategic management. This entails a somewhat different paradigm from the rational-economic model underlying the 'traditional' approach to strategic management. Recent developments have described the strategy of the organization as 'a pattern in a stream of decisions' (Mintzberg, 1978) and characterized the process as incremental and iterative (Quinn, 1980). Strategic management is portrayed as a process of building awareness, momentum and commitment to change amongst managers and groups within the organization. Strategy emerges from a complex political process of demand generation and mobilization of power around these demands (Pettigrew, 1977) within the organization. Hence it becomes necessary to add a socio-political element to the paradigm for the purpose of analysis and explanation.

It is useful to see the strategic behaviour of the firm rooted in its historical context, to understand its potential as emerging from the accumulated position and knowledge of the organization and more specifically, its actors, rather than as a 'free floating' open future. Studying technological change in the U.K. reprographics industry, Ghazanfar, McGee and Thomas (1986) showed how a number of firms failed to redefine their business in the face of changing technologies. They suggested rigidity of managerial values and perceptions, i.e. inertia around the current way of doing things, as the prime cause of this failure. Cooper and Schendel (1976) catalogued similar strategic responses to technolgical threats. As Pavitt (1986, p.174) pointed out:

> 'Firms do not "search" for innovations in a general "pool" or 'stock' of knowledge, all of which is equally accessible and assimilable by them. Instead, they search in zones that are closely related to their existing skills and technologies. What specific firms can hope to do in technology and innovation in the future is strongly conditioned by what they have been able to do in the past.'

Technological aspects of organizational strategy are amongst the most unpredictable both in significance and in timing, yet their effects will also be a function of the style of management; whether passive/reactive or proactive/innovative (Miles and Snow, 1978; Tushman and Anderson, 1986). To develop these arguments, it is clearly important to examine the ways managerial processes influence changes in technological strategies, specifically (i) the role played by formal analytical and structural processes, (ii) the effect of management's accumulated beliefs and values and the constraints these produce, and (iii) the nature of the internal political process.

The rational/formal process

All firms exhibit some level of formality in evaluating strategic options. However, what are the roles and effects of such processes in the context of technology?

Thomas (1984; 1985) emphasized that 'rational choice' paradigms, whilst needing modification to handle complex, ill-structured problems, may provide the basis for decision analysis as an aid to policy debate, questioning and dialogue. Bahrami (1983), in a study of 'high-tech' firms, acknowledged disillusionment with strategic planning, but illustrated how such firms develop systematic managerial and administrative processes as the firm proceeds into its 'adolescent' stage; that is, when the firm turns 'inward' to digest the organizational implications of growth.

It remains unclear, however, what the impact of restructuring the decision process will be. Child and Francis (1977) posed the question in examining strategic innovation:

> 'When a "great leap forward" is required strategically, does this not need some kind of extra-structural process?'

To support the view that it *does*, they cited two examples of significant strategic technological change by-passing formal decision processes. Peters (1983) further argued that the rational model of decision making has largely been ignored by the so-called 'excellent' U.S. companies. More specifically, formal evaluation techniques such as DCF and payback methods tend to have a short-term focus. They may even distort the strategic evalution of technological decisions (Senker, 1984). Furthermore, managers using these justifications for CAD/CAM implementation admitted they were based on unreliable information and probably spurious. Hayes and Garvin (1982) argued that a focus on return on investment and payback criteria produces incentives to disinvest and promotes short-term profit taking. Gold (1983) described the great difficulties of estimating the costs and benefits of technological innovation, arguing for a shift from capital budgeting evaluation methods to a 'continuing horizons' approach, which better recognizes longer-term benefits and contributions to the organization's competitive position.

A purely rational/economic approach via formal processes and structures therefore constrains evaluation and selection of technological strategy options. Whilst the impressionistic evidence is strong, to date there appears to be little systematic evidence of the resulting distortions of a firm's technological capabilities and position.

Managerial beliefs, attitudes and perceptions

It was argued earlier that an organization's accumulated historical experience conditions its approach to strategic management and indirectly, therefore, particular choices made. It has been said that managers' experience confers on them a firm or industry-specific set of substantially fixed beliefs—so called recipes—about their organization's strategic position and strategy to be pursued (e.g. Grinyer and Spender, 1979; Spender, 1983). Earl (1984) saw the 'recipe' as the way managers deal with choice in a programmed way. From their extensive background in studying technology and innovation, Nelson and Winter (1982) introduced the notion of managerial 'routine', which parallels the 'recipe', particularly the role of internal routines and rules in determining firm-level behaviour.

Recipes and routines facilitate the way managers handle complex situations, but limit their perceptions and interpretations of a situation. Explicating the content and nature of these 'recipes' is useful in highlighting several aspects of the strategy–technology link. For example, many firms engage in 'routine' innovation at the operational level without fully understanding its strategic implications (Senker, 1984). Hence potential strategic advantages may not be envisaged, much less achieved (Boddy and Buchanan, 1984).

Embedded recipes and routines may explain why managers select particular technological innovations and opportunities for attention. 'Routines' influence the 'patterning of particular ways to innovate' (Nelson and Winter, 1982). The work of Abernathy, Clark and Kantrow (1983) implied that in the 1960s and early 1970s the American auto industry became stuck in a process-oriented mass production recipe which became inappropriate in the subsequent and world-wide context of more complex product innovation and more rapid technological change.

After a study of innovation in U.S. corporations, Kanter (1983) concluded that strategic change derives not from environmental pressures producing 'automatic' change, but by *perceptions* of that environment, particularly perceptions of key actors. In a similar study of innovation in the U.S. food processing industry, Ettlie (1983) found that performance gaps were less of an innovative trigger than *perceived* environmental uncertainty. This distinction is crucial in understanding technological development and organizational choices. Downs and Mohr (1976) differentiated the intrinsic or primary properties of an innovation, e.g. cost, from the secondary properties that organizations and managers *attribute* to innovations, e.g. radicalness, (lack of) compatibility with existing resources and competences. We believe that the failure to distinguish such firm-specific secondary attributes has restricted progress in studying the management of innovation.

For example it is important to define what is meant by 'new' technologies. A particular technology may be 'new' in relation to the pre-existing scientific base. But, whether managers *ascribe* the quality 'new' to the technological development is also extremely relevant. The advent of a novel technology, e.g. laser based surface treating, and its effect on the firm's strategy will be seen as radical by some, whereas others, perhaps with more relevant technological experience, will see it as less radical. In a study of technology diffusion, Hayward (1977) suggested that:

> 'Innovations outside the normal practice and understanding of respondents take longer to diffuse than those innovations which are much in line with existing methods and do not require a great deal of understanding.' ·

Recipes and routines represent deep-rooted beliefs that are not easily displaced. However, the process of changing their technological content has not yet been extensively researched in the management literature. Johnson (1987) suggested that recipes, or in his terms paradigms, may change when confronted and challenged by 'outsiders' or those in positions of power. This position is corroborated by the finding that much successful innovation and adoption of innovations is a function of a 'champion' figure (Rothwell *et al.*, 1974; Maidique, 1980).

Political and cultural influences on managerial processes

Attitudes and values are significant in the strategic management process because they are embedded in organization *cultures* and articulated through political structures. Kanter (1983) identified 'integrative' companies which are essentially wholistic, open, entrepreneurial firms and 'segmentalist' companies which are compartmentalized, anti-change, fragmented, inhibited firms. The former are more successful at innovation. Kanter highlighted two aspects of a successful innovation-based strategy: (i) the necessity of integrating the various functions involved, and (ii) the significance of entrepreneurial roles in a successful technological innovation strategy. Her observation supports earlier work on the determinants of successful innovation reported from Project SAPPHO in the U.K. (Rothwell *et al.*, 1974) and by Maidique and Zirger (1984) in the USA.

Miller and Friesen (1982) outlined two models of the innovating firm: entrepreneurial and conservative. In the former, innovation is a 'natural state of affairs', in the latter innovation is only a response to challenges, threats and instabilities. Their central finding was that 'the determinants of product innovation are to a very great extent a function of the strategy that is being pursued'. What is not clear is whether the entrepreneurial and conservative orientations are the result of the strategies pursued, or vice versa.

Miles and Snow (1978) developed a framework which links propensity to adapt with organization structure and the strategic management process. The role which technology plays in shaping the organization strategy and the process by which technology becomes articulated varied systematically in different types of organization. *Defenders* established a single core technology to maintain efficiency. Financial and production experts in such firms were very powerful. For those organizations described as *prospectors*, a considerable portion of the technological core was frequently engaged in the production of prototype products and the firm developed multiple technologies embedded in people rather than in routine or mechanical operations. In such firms the dominant coalition comprised marketing and research and development. *Analysers* had a hybrid approach with a dual technological core, the dominant coalition built around marketing, applied research and production. Reactors had an unstable, poorly articulated strategy with no clear technological posture.

The implication of Miles and Snow's argument is clear: different functional groups (e.g. finance, marketing, R&D and production) have different attitudes toward technology. The politically dominant attitudes thus define the preferred strategic adaptation process for the organization. Not only do specific groups within organizations have a 'political' role but so too do individuals. The political role of the individual as entrepreneur has been revealed in a number of studies. Maidique (1980) identified Gatekeepers, Product Champions and Executive Champions as significant technological change roles. He hypothesized that the nature and importance of such roles may vary systematically with the industry life-cycle and the development stages of the firm.

When examining the political roles that influence strategy several issues were raised. The precise nature of the ideologies held by groups within the organization needs exploration. For example, is it possible to identify a technocratic ideology, how widely might it be shared and what is its impact on the firm's strategy? Connor and Becker (1983) reported that systematic differences existed between the values of engineers and the population at large. Engineers tended towards competence rather than morality, to be individually rather than interpersonally orientated, and ranked respect above love.

How do technologists articulate and exercise their influence? An extensive study of ICI by Pettigrew (1985b) indicated how technologists became powerful in the organization in the 1970s and apparently contributed to a reduction in its entrepreneurial culture. In a study which he claimed links power and innovation directly, Witte (1977) reported how a powerful coalition of 'promoters by power' and 'promoters by know-how' led to a greater of mobilization of organization members toward innovating.

SUMMARY AND CONCLUSIONS

Technological change, as an exogenous variable, produces certain kinds of technological opportunities around which firms and managers make choices. Yet at the same time, firms' technological competences are reshaping those opportunities and the managerial decision-making processes by which those opportunities are evaluated and exploited. So it is important to understand more clearly how firm-level differences such as organization culture (Kanter, 1983) and firms' accumulated technological positions (Pavitt, 1986) influence the translation of technological potentialities into organization strategies and ultimately, technological realities. Technological change interacts with organizational culture and managerial ideology to shape the firms' specific technological development and opportunities. The firm needs more than the economic/rational model to explain and guide its choice of an adaptive technological strategy.

Context and process interactions need further investigation. How does the context influence what enters the strategy formulation process? This is in part a question of individual predispositions and frameworks (Huff, 1982) and of more complex organizational processes (Lyles and Mitroff, 1980). It remains for explicit links to be made between contextual change and what can be described as strategic triggers. For example, to what extent do recipes, paradigms and routines derive from contextual technological change? Abernathy's (1978) work implied that prevailing *beliefs* about appropriate product and process development in a industry are strong. Dosi's (1982; 1984) technological paradigms also suggested strong, widely shared technological 'recipes' in an industry. It could be valuable for managers to use external agents to trace the various attitudes and perceptions toward the technological issues that confront their firm and how these influence strategic decision making. Do these attitudes vary systematically between the product development function, where market links and new product creativity may be key, and the process development function, where engineering and maufacturing skills are central?

It seems that innovatory cultures can be identified and are linked to the dominating influences in the organization's power structure. But how valuable is an innovatory culture? If the prevailing context is one of rapid change in product technology, an entrepreneurial culture may be highly appropriate. When the context indicates that the key need is process development, an analytical or conserving culture may be more appropriate. Much work remains to be done to trace how specific groups or individuals act to contribute to or block the emergence of new culture.

Fundamentally, how should managers in organizations perceive, evaluate, initiate and/or respond to technological change, in order to (i) achieve and exploit competitive advantage and (ii) improve the organization's technological base for the future.

To express these problems as specific propositions, we suggest the following areas for future investigation:

(1) There is need to unravel the complex *patterns* by which industry technologies unfold. The product/process dimension and notions of maturity/dematurity are over-simplistic; they need revisiting, refining and probably revising.
(2) The technological *context* within which firms take strategic decisions is not divorced from the *way* in which those decisions are made. It is necessary to integrate studies of context and process to understand better how firms' technological positions and strategies develop *over time*.
(3) The strategic *competence* of a firm depends on a wide range of technologies which cluster, develop and accumulate in complex and often firm-specific ways. This process of accumulation and the way it can lead to unique recipes and competitive advantages needs unravelling.
(4) Notions of technological discontinuity or radical technological change within industry sectors are well established, but how do discontinuities derive from or otherwise amend firm-in-sector competences and the

perceptual maps of managers? Studying these links systematically would help explain why some firms successfully manage the organization transitions contingent on radical discontinuities and some do not.

(5) Effective organizational integration is central to a technologically coherent strategy. Integration has both cultural and structural implications. There is need to define better the roles of individuals and groups and how they can best be organized to achieve an effective strategic technological strategy which is more than the management of R&D.

Thus, it is suggested, there is still much to be learned about the nature and process of the technological development of organizations. To secure answers to the foregoing questions appears to require two important conditions to be satisfied. First, a convergence of the research paradigms and methods of those whose prime interest is technological innovation with those whose interests are the strategic management of the firm. Second, recognizing that present knowledge of the processes by which managers perceive, evaluate and implement technological change is limited, to rectify these shortcomings may, itself, require innovative multidisciplinary approaches.

REFERENCES

Abernathy, W.J. (1978). *The Productivity Dilemma: Roadblock to Innovation in the Automobile Industry*, The Johns Hopkins University Press, Baltimore.

Abernathy, W.J., Clark, K.B., and Kantrow, A.M. (1983). *Industrial Renaissance: Producing a Competitive Future for America*, Basic Books, New York.

Abernathy, W.J., and Clark, K.B. (1985). 'Innovation: mapping and winds of creative destruction', *Research Policy*, **14**,3–22.

Bahrami, H. (1983). *Strategic Management in High Technology Firms: An Exploratory Study*, Research Paper No 697, Graduate School of Business, Stanford University.

Boddy, D., and Buchanan, D.A. (1984). 'Information technology and productivity: myths and realities', *Omega*, **12**, 233–40.

Child, J., and Francis, A. (1977). 'Strategy formulation as a structured process', *International Studies of Management and Organization*, **7**(2).

Connor, P.E., and Becker, B.W. (1983). Value Systems of Engineers: Students, Academics and Practitioners, Paper to Conference 'Management of Technological Innovation', Worcester Polytechnic Institute, May, Washington DC.

Cooper, A.C., and Schendel, D. (1976). 'Strategic responses to technological threats', *Business Horizons*, 61–79, February.

Dosi, G. (1982). 'Technological paradigms and technological trajectories: a suggested interpretation of the determinants and directions of technological change', *Research Policy*, **11**, 147–62.

Dosi, G. (1984). *Technological Change and Industrial Transformation: The Theory and Application to the Semiconductor Industry*, Macmillan, London.

Dosi, G. (1988). 'Institutions and markets in a dynamic world', *Manchester School of Economics and Social Sciences*, **56**, 119–46.

Downs, G.W., and Mohr, L.B. (1976). 'Conceptual studies in the study of innovation', *Administrative Science Quarterly*, **21**, 700–714.

Earl, P. (1984). *The Conceptual Imagination: How Big Companies Make Mistakes*, Wheatsheaf Books, Brighton.

Ettlie, J.E. (1983).'Performance gap theories of innovation', *IEE Transactions on Engineering Management*, **30**(2), May.

Flaherty, M.T. (1983). 'Market share, technology leadership and competition in international semi-conductor markets', in R.S. Rosenbloom (ed.), *Research on Technological Innovation, Management and Policy*, Vol 1, JAI Press, Greenwich, Conn.

Freeman, C. (1982). The Economics of Industrial Innovation, Second edition, Frances Pinter, London.

Ghazanfar, A., McGee, J., and Thomas, H. (1986). 'The impact of technological change on industry structure and corporate strategy: the case of the reprographics industry in the United Kingdom', in A. Pettigrew (ed.), *The Management of Strategic Change*, Basil Blackwell, Oxford.

Gold, B. (1983). 'Strengthening managerial approaches to improving technological capabilities', *Strategic Management Journal*, **4**, 209–20.

Grinyer, P., and Spender, J.C. (1979). *Turnaround: Managerial Recipes for Strategic Success*, Associated Business Press, London.

Harrigan, K.R. (1980). *Strategies for Declining Industries*, D.C. Heath, Lexington, Mass.

Hayes, R.H., and Garvin, G.A. (1982). 'Managing as if tomorrow mattered', *Harvard Business Review*, May/June, 71–9.

Hayward, G. (1977). 'Characteristics and diffusion of technological innovations', *R&D Management*, **7**(1).

Huff, A.S. (1982). 'Industry influences on strategy reformulation', *Strategic Management Journal*, **3**, 119–31.

Jemison, D.B. (1981). 'The importance of an integrative approach to strategic management research', *Academy of Management Review*, **6**(4), 601–8.

Johnson, G.N. (1987). *Strategic Change and the Management Process*, Basil Blackwell, Oxford.

Jones D.T. (1983). 'Technology and the UK automobile industry', *Lloyds Bank Review*, No.148, April.

Kanter, R.M. (1983). *The Change Masters: Corporate Entrepreneurs at Work*, Unwin Paperbacks, London.

Kantrow, A.M. (1980). 'The strategy technology connection', *Harvard Business Review*, July–August.

Lyles, M.A., and Mitroff, I.I. (1980). 'Organizational problem solving: an empirical study', *Administrative Science Quarterly*, **28**, 102–9.

Maidique, M. (1980). 'Entrepreneurs, champions and technological innovation', *Sloan Management Review*, Winter.

Maidique, M., and Patch, P. (1982). 'Corporate strategy and technological policy', in M.L. Tushman and W.L. Moore (eds), *Readings in the Management of Innovation*, Pitman, Boston.

Maidique, M., and Zirger, B.J. (1984). 'A study of success and failure in product innovation: the case of the US electronics industry', *IEE Transactions on Engineering Management*, **EM31**, 192–203, November.

McGee, J., and Thomas, H. (1988). Technology Innovation and Strategy: Progress and Future Directions, Paper to the State of the Art in R&D Management Conference, Manchester Business School, July.

Miles, R., and Snow, C. (1978). *Organization Strategy, Structure and Process*, McGraw-Hill.

Miller, D., and Friesen, P.H. (1982). 'Innovation in conservative and entrepreneurial firms: two models of strategic momentum', *Strategic Management Journal*, 3, 1–25.

Mintzberg, H. (1978). 'Patterns in strategy formulation', *Management Science*, 24, 934–48.

Nelson, R.R., and Winter, S.G. (1982). *An Evolutionary Theory of Economic Change*, Harvard University Press.

Pavitt, K. (1984). 'Sectoral patterns of technical change: towards a taxonomy and a theory', *Research Policy*, 13, 343–73.

Pavitt, K. (1986). 'Technology, innovation and strategic management', in J. McGee and H. Thomas (eds), *Strategic Management Research: A European Perspective*, Wiley, Chichester

Peters, T.J. (1983). 'The rational model has led us astray', *Planning Review*, 10, 16–23.

Pettigrew, A. (1985a). 'Contextualist research and the study of organizational change processes', in E. Lawler (ed.), *Doing Research That is Useful for Theory and Practice*, Jossey Bass, San Francisco.

Pettigrew, A. (1985b). *The Awakening Giant: Continuity and Change at ICI*, Blackwell, London.

Pettigrew, A.M., (1977). 'Strategy formulation as a political process', *International Studies of Management and Organization*, 7(2), 78-87.

Porter, M.E. (1980). *Competitive Strategy.: Techniques for Analysing Industries and Competitors*, The Free Press, New York.

Porter, M.E. (1983). 'The technological dimension of competitive strategy', in R. Rosenbloom (ed.), *Research on Technological Innovation Management and Policy*, Vol. 1, pp. 1–33, JAI Press, Greenwich, Conn.

Porter, M.E., and Millar, V.E. (1985). 'How information gives you competitive advantage', *Harvard Business Review*, July/Aug., 149– 60.

Quinn, J.B. (1980). *Strategies for Change: Logical Incrementalism*, Irwin, Homewood, Illinois.

Rosenbloom, R.S. (1978). 'Technological innovation in firms and industries: an assessment of the state of the art', in P. Kelly, and M. Kranzberg (eds), *Technological Innovation*, San Francisco Press.

Rothwell, R. (1980) 'Policies in industry', in K. Pavitt (ed.), *Technical Innovation and British Economic Performance*, Macmillan, London.

Rothwell, R. *et al.* (1974). 'SAPPHO updated—Project SAPPHO Phase 2', *Research Policy*, 3, 258–91.

Rothwell, R., and Gardiner, P. (1988). The Strategic Management of Re-innovation, Paper to the 'State of the Art in R&D Management' Conference, Manchester Business School, July 11th– 13th.

Senker, P. (1984). 'Implications of CAD/CAM for management', *Omega*, 12, 225–31.

Spender, J.-C. (1983). 'The business policy problem and industry recipes', in R.B. Lamb, *Advances in Strategic Management*, Vol. 2, pp.211–29, JAI Press, Greenwich, Conn.

Spital, F.C. (1983). 'Gaining market share in the semi-conductor industry by lead time in innovation', in R.S. Rosenbloom (ed.), *Research on Technological Innovation, Management and Policy*, Vol. 1, JAI Press, Greenwich, Conn.

Teece, D.J. (1984). 'Economic analysis and strategic management', *California Management Review*, 26, 81–110.

Thomas, H. (1984). 'Strategic decision analysis: applied decision analysis and its role in the strategic management process', *Strategic Management Journal*, 5, 139–56.

Thomas, H. (1985). 'Decision analysis and strategic management of research and development: a comparison between applications in electronics and ethical pharmaceuticals', *R&D Management*, **15**, 3–22.

Tushman, M.L., and Anderson, P. (1986). 'Technological discontinuities and organizational environments', *Administrative Science Quarterly*, **31**, 439–65.

Utterback, J.M., and Kim, L. (1983). Innovation and the evolving structure of the firm: a framework for technology policy, Paper to Conference 'Management of Technological Innovation', Worcester Polytechnic Institute, May, Washington DC.

Williams, J.R. (1983). 'Technological evolution and competitive response', *Strategic Management Journal*, **4**, 55–65.

Witte, E. (1977). 'Power and innovation: a two centre theory', *International Studies of Management and Organization*, **7**(1).

13

New Production Technology and Work Roles: a Paradox of Flexibility versus Strategic Control?

Bryn Jones

School of Social Sciences, University of Bath, England

Various perspectives are now converging on a solution to a longstanding tension between stable productive efficiency and continuing innovation. The growing belief is that this tension can be resolved at the operations level by combining reprogrammable production technologies with more flexible work organisation. 'Japanisation', 'flexible specialisation', 'new techno-economic paradigms', 'Post-Fordism' etc. have all been used as labels for the new manufacturing model. Despite this diversity of terminology, most interpretations give central emphasis to the nexus of product innovation, flexible automation, reorganised work tasks. However, the diversity of current practice amongst firms upgrading their operations technology indicates an uncertainty as to whether more decentralised and diffuse decision making for fully flexible operating capacity is compatible

The Strategic Management of Technological Innovation. Edited by R. Loveridge and M. Pitt
© 1990 by John Wiley & Sons Ltd

with secure strategic control from the top. Greater operating flexibility may weaken or usurp the control of higher-level managers and the implementation of detailed strategic programmes.

A review of some evidence from different national contexts supports the view that change is needed across a range of the firm's social and managerial dimensions to optimise the new technological and organisational possibilities. The analysis also suggests that managements are seeking not 'one best way' but a variety of approaches which amount to practical versions of such academic paradigms as technological and cultural determinism.

FORDISM AND FLEXIBLE SPECIALISATION

It was Abernathy who first identified a growing incompatibility between stable forms of productivity growth and the increasing need for constant innovation in U.S. industry (Abernathy, 1977). From his thesis a number of further claims have been made about the emergence of a new operations paradigm for Western manufacturing operations. For Clarke, Hayes and Lorenz (1985, p. 9) the productivity-innovation tension can be overcome by developing management styles, philosophies and personnel and the overall reorganisation of all of these in such a way that they are responsive to technological innovation for long-term rather than short-term productivity gains.

The organisational diagnosis offered by proponents of the U.S. Quality of Worklife movement has also been associated with such changes. Walton (1985, pp. 237–265) claimed that more devolved decision-making, broader work roles and less hierarchy in management systems are the most rational model for the future. (Although Walton carefully observed that new operating technologies have often not been associated with shifts towards more participative work roles.) A more specific theory for overcoming the productivity–innovation dichotomy, one which centres on the role of new technology, was Piore and Sabel's (1984) distinction between the declining mass manufacturing regime of Fordism and the innovative smaller batch systems of 'flexible specialisation'.

'Fordism' signifies the marketing of a narrow range of standard products manufactured in large volumes using special purpose technology that cannot be easily modified to make different parts and products. The classic case is, of course, the initial mass production paradigm of the Ford Motor Company. It may be debatable whether this paradigm has been uniformly and universally applied to Western manufacturing up until the 1970s. It is also unclear whether core mass-manufacturing enterprises, such as automobiles, have now abandoned this paradigm, or, as with Ford itself, whether their attempt to incorporate Japanese policies and techniques has changed the basic Fordist approach into a quite different form of mass-manufacturing. These issues are beyond the scope of this chapter and their detailed analysis can be found elsewhere (Tolliday and Zeitlin, 1987; Wood, 1988). For present purposes Fordism is a convenient term to describe the manufacturing philosophy just described, and the pattern of specialised, hierarchical labour organisation and 'low trust' employee-management relations that invariably accompany it.

In many respects the polar opposite of Fordism is an operations and marketing strategy of 'flexible specialisation'. Unlike Fordism flexible specialisation involves a portfolio of distinct products aimed at market niches, specialist customers and quality-seeking consumers. Production runs tend to be short and done by versatile production machinery which is frequently reset by skilled production staff who work on a cooperative and 'high trust' basis with design and technical staff and even customers. These organisational relationships enable the enterprise to switch rapidly from making one type of product to another. The intellectual origins of flexible specialisation stem from the discovery of industrial districts of innovative small firms combining craft skills with computerised production technology in the regions of the Third Italy (Brusco, 1982; Sabel, 1984; and Jones, 1989 for a discussion of flexible automation and Italian small firms). Piore and Sabel have since gone on to claim that variants of the core elements of flexible specialisation distinguish successful large and small firms in Europe, Japan and the U.S.A. from less innovative companies with the narrow job responsibilities, centralised decision making and slow-changing products and processes of the mass production era (Piore and Sabel, 1984).

With its emphasis upon 'high trust', cooperative labour relations and technological adaptability, the flexible specialisation hypothesis crystallises, at the operations level, a number of the prescriptions and predictions about the strategic, technological, organisational and employee relations features of the new model manufacturing firm of the late twentieth century. The best-selling management pundit, Tom Peters recently made flexible specialisation a special case of the need for the kind of strategic management raised by Clarke et al. (Peters, 1988). Walton's prescription of high-trust work organisation was re-interpreted as a flexible specialisation phenomenon in the 'new industrial relations' thesis of Kochan, Katz and McKersie; they associated innovations in work regulation with both participative labour relations and successful use of advanced technology (Kochan, Katz and McKersie, 1986). At a higher level of abstraction the implicit links between the re-programmability of computer-based technologies makes the model of the flexible specialisation enterprise consistent with the new 'techno-economic paradigm', which commentators such as Perez (1985) and Freeman (1987) predicted as the successor to the mass production economies of the period 1945–1975. It also provides an organisational context and rationale for claims that the future of work with new production technology will be one of less specialised and more responsible jobs (Hirschhorn, 1985; Kern and Schumann, 1987; Coriat, 1987).

Other commentaries see versions of flexible specialisation in the model of operations management and labour organisation associated with Japanese manufacturing (Murray, 1988; Oliver and Wilkinson, 1988). The basic elements therefore exist for academic consensus around a new paradigm for analysing the organisational, technological and human aspects of changes in manufacturing operations. There is however a danger that concepts such as flexible specialisation may be taken too readily as summarising an all-inclusive model of new forms of operations management and organisation; or alternatively, diverse changes taken as convergent with a single interpretation. On these kinds of assumptions commentators can question the extent to which any serious change from previous practices is occurring (Lane, 1988; Wood, 1989).

Three important issues need to be clarified before a new universal paradigm is taken as the yardstick for measuring and promoting specific changes. Firstly a clearer line needs to be drawn between the prescriptive and descriptive elements in recent theories. Secondly, as Wood has argued, consideration needs to be given to the possibility that the innovation-averse mass production paradigm may have several successors, rather than one essential form. Thirdly, as a guide for practitioners and policy-makers, there needs to be more detailed identification of the areas in which change must take place to secure the overall organisational character associated with flexible, innovative and cooperative factories and firms.

This chapter offers a preliminary clarification of some of these problems by focusing on one aspect of the productivity-innovation contradiction on which considerable data exist. That is, the implicit paradox of more detailed managerial control through advanced computer-integrated technologies being accompanied by greater autonomy for the direct workforce. Flexible automation may bind *what* is finally produced and *how* it is produced more closely to strategic enterprise decisions; but the degree and quality of production flexibility depends upon a reduction of direct control over work at the point of production.

After outlining the relevance of the boundary between the prescription and description of new developments in manufacturing organisation of work and technology, some studies of flexible automation will be analysed using a classificatory model of labour–management relationships, the WEARI scheme, which indicates the dimensions along which change must take place in order to achieve a systematic reorganisation of work roles (Rose and Jones, 1985).

PRESCRIPTION, DESCRIPTION AND DIVERSITY

Flexible specialisation is counterposed to the mass production of standardised products by multinational corporations. Mass production is characterised as an undifferentiated entity based on standard products and special purpose, 'dedicated' technologies. On the other hand, Piore and Sabel were careful to identify a variety of organisational forms that flexible specialisation can take. Regional conglomerations, as in the industrial districts of the Third Italy or the New York garment trade are one type. Federated enterprises, as with inter-locking Japanese firms, are a second. The third and fourth variants are the 'solar' firms and their networks of orbiting suppliers and the internally segmented workshop factories of complex capital goods producers such as the Boeing aircraft company (Piore and Sabel, 1984, ch. 10, especially pp. 265–268).

The potential weakness in this analysis stems from its mixture of description with prescription and the corollary of allowing only a single successor to mass production, rather than a variety of organisational types. For the only common denominators of the four flexible specialisation variants is that they 'do not produce long runs of standardized products' and that they require inter-organisational forms of cooperation. So Piore and Sabel's description of the present range of flexible specialisation organisations (including as 'solar' organisations divisions of General Electric) is already broad. This diversity in the current incidence of flexible specialisation supports the prospects for a generalisation of the model, but it raises the question of whether description is being guided by prescription. Are definitions being over-stretched in order to strengthen the case for the generalisability of flexible specialisation? Is the term, as used by Piore and Sabel, being used to encompass quite distinct types of organisation?

Let us assume, in order to progress the argument, that a mass production paradigm—Fordism for short—has dominated the standards of manufacturing culture, if not the practices of all firms, until the economic crisis of the early 1970s. What have been the significant changes in the organisation of operations that have broken with that paradigm? A shortlist would include: (i) more rapid design changes and shorter lead times between design and production; (ii) decentralisation rather than integration of production and related activities; (iii) a shortening of the production runs of any given product; and (iv) the replacement of specialised work roles by multi-skilled, flexible work groups. From the flexible specialisation and similar perspectives, more especially Perez's (1985) 'techno-economic paradigm', the re-programmable technologies of CNC machine tools, Computer Aided Design/Computer Aided Manufacturing (CAD/CAM) and production and inventory control systems are the means for achieving these kinds of change. However closer inspection suggests that the combinations of technology and methods are either consistent with Fordist philosophy, or are clustering into separate and distinct types of organisation, rather than all converging into a single system of operations.

One indicator of possible diversity is, as Williams *et al.* (1987) have shown, that Japanese firms have combined more rapid design changes and shorter lead times to continue with mass productions runs, which simply have a shorter life-cycle. The decentralisation of production has, at the very least, been variable between countries. In Britain most sub-contracting has been of ancillary activities such as cleaning, catering and transport (Marginson and Edwards, 1988), rather than the kinds of manufacturing operations which apparently gave such a stimulus to small-firm flexible specialisation in Italy during the 1970s. Many large British firms are using computer technologies to attain higher levels of integration of processes throughout their main sites (Jones, 1988b): a thoroughly Fordist strategy.

Where the length of production runs has fallen significantly this is not necessarily because firms are seeking to move away from standardised mass production to more customised batch production. Perhaps the most distinctive development in this respect is the emulation of Japanese 'just-in-time' methods which are aimed at reducing, or even eliminating, the holding of stocks of parts and finished products. At least as practised in the automobile components industries this strategy of shortening production runs does not appear to lead to craft-like work roles among line workers. Existing reports suggest that work flexibility in this context principally means an intensification of effort, as operators switch rapidly amongst machines of a similar type (Turnbull, 1988). In Britain, more specific initiatives for flexible work roles do not, on the whole, appear to have led to multi-skilling, or polyvalence. Apart from a consistent but only sporadically successful campaign to integrate electrical and mechanical maintenance roles, most change seems to have involved adding on minor product inspection and equipment checking tasks to the same routine production jobs (Pollert, 1988). More fundamental regroupings of tasks into semi-independent work teams has been a minor trend but there is, as yet, no evidence that such reorganisations are associated with a switch from mass production, or specifically with the computerisation of production (Jones, 1988a).

All of these caveats and complications do not, of course, mean that a Fordist mass production paradigm still reigns unchallenged. Rather they suggest two other alternatives. Either that the changes are so dispersed and specific that, at least in a country such as Britain, eclecticism is more likely than systematic reorganisation. Or alternatively, that there is less of a uniform convergence on a post-Fordist type than the development of a limited number of organisational types; perhaps ranging from largely unreconstructed, low-innovation, mass production operations to fully developed, flexible specialists. Companies can change, and have been changing, the character of their operations by discrete, but radical changes to product and process routines and equipment. However, if they are to transcend the highly circumscribed role of most employees under Fordism, and if they are to create the capacity for continuing innovation, then the social structure of the enterprise must be changed. Indeed, the more radical the operations, methods and goals envisaged, especially where the latter also encompass strategic business objectives, then the more likely it is that this social sphere must be overhauled rather than modified. Many firms, perhaps the majority, will consciously or unconsciously prefer partial, piecemeal or zero changes to the risks and upheavals of comprehensive restructuring. However, it is useful to have as a yardstick a specification of the social structural dimensions of the organisation which would need reform to achieve systematic change.

A CAMPAIGN ON FIVE FRONTS?

British debates in the early 1980s on the extension and limits of managerial control over work and workers led to the identification of five dimensions to the management of labour. These are: *Work*—the organisation and performance of work tasks; *Employment*—the contractual and customary terms and conditions under which workers are hired, fired and rewarded; *Authority*—the relationships and procedures involved in the direction and supervision of workers; *Representation*—the institutions, e.g. collective bargaining, consultative schemes and so on, for articulating and conveying workers' interests and opinions to management; *Interpretative frameworks*—the discourse of managers, workers and unionists in which they express and communicate all aspects of the other dimensions—in academic argot: managerial 'ideologies' and 'corporate cultures'.

The application of this WEARI model in the depths of the 1980s recession to case studies of British firms that were reorganising operations and work roles suggested a paradoxical limitation to managerial initiatives (Rose and Jones, 1985; Rose, 1988). In these unionised firms the implementation of radical reforms to working practices and operating methods depended upon winning subsidiary changes in each dimension of the social structure. Elevating the responsibility expected of specific manual occupations required changes to their terms of employment; most obviously, fatter wage packets, but in some instances expectations of regrading to staff status.

More responsible or more complex work roles called into question existing methods of supervision: fewer direct checks on workers' performance, perhaps even the redundancy of some existing supervisory roles. A need to broaden the scope of meetings with union representatives without getting change schemes bogged down in the rituals and confrontations of bargaining; perhaps setting up less formal consultative arrangements in parallel. Finally, the need to break with existing norms and attitudes of what managers and employees could and ought to expect from each other: the need for new ideas and symbols, either to express new intentions or to obscure their full implications.

The element of paradox about these initiatives lies in the need to plan broad and detailed reforms in the adjacent social dimensions summarised in the WEARI model, if the change sought in just one area (here reorganised work roles) is to be comprehensive. But there is then always the risk that the overall scheme will have to be modified precisely because its scale increases the probability of unanticipated obstacles or resistance. The contrast is with schemes which are deliberately restricted to the achievement of only marginal change; for example, automating a machine operator's tool replacement task, while leaving the rest of the job intact. In these latter types of case minor changes in a highly specific area may be achieved, but they remain limited, because they are contained within unreformed networks of occupational definitions and work practices, authority and representational structures and interpretive discourses.

An illustration of the importance and fragility of transdimensional change schemes is provided by the adoption of JIT methods in British automotive components producers. In at least one of the largest of these firms, a major reorganisation of work roles for greater task flexibility accompanied a shift to a Group Technology layout of machines, production 'on demand' rather than for stocks and periodic recruitment of temporary workers. However, as Turnbull (1988), pointed out, this seeming textbook recreation of Japanese practices has been rendered highly unstable at key points because, as yet, management has not achieved or not felt able to seek a complementary 'Japanese' remodelling of industrial relations: the 'representation' dimension in the WEARI model. It might be added that new practices could be more coherent if they were also accompanied by changes in the interpretative frameworks of worker-management and industrial relations. Of course, whether such a further line of initiative is practicable, or even thinkable, for the managements of traditionalist British manufacturing firms is another matter (Loveridge 1981).

Let us now turn this social dimensions model to the specific topic of the successful use of flexible manufacturing technologies.

COMPUTER-INTEGRATED MANUFACTURING: FLEXIBILITY VERSUS CONTROL?

In an early outline of the routes between Fordism and flexible specialisation Sabel (1984, pp. 209–219) speculated that large traditional manufacturing corporations of the Anglo-American type might attempt to raise productivity and increase the flexibility of their operations in a limited fashion, by pursuing what French economists of the 'regulationist' school have called neo-Fordism (Aglietta, 1979; Palloix, 1976). This consists of minor expansions of work roles which actually tighten managerial control over performance while seeking greater flexibility of production lines and product modifications from 'technological fixes' of computerised systems. Such strategies constitute an attempt to remain within the reassuring parameters of Fordist practice, 'to meet or create new demands for more varied products while holding fast to familiar principles of command and organisation' (Sabel, 1984, p. 210), without the risks and uncertainties of delegating more autonomy and discretion to less specialised shopfloor work-roles.

Flexible Manufacturing Systems (FMSs) are groups of workstations, normally metalcutting machine tools, whose operations are controlled by on-line computers. These controls comprise the 'downloading' of the part-programs that command the movements of the cutting tools and materials in relation to each other, the rates at which they interact, and the sequence of operations. Other software can control the selection and setting up of the tooling, with the movement of jobs between the workstations accomplished by automatic transfer devices, rail-mounted shuttles or carts powered by electrified pathways beneath the shopfloor. The movement of jobs and the downloading of the part-programs is decided by a master program for the scheduling of work through the system. There are as yet very few cases of FMS being applied to final assembly operations.

Two distinctive features are claimed for such FMSs. Firstly, they bring labour-saving automation, previously restricted to the mass producers, to the domain of small and medium batch production. Secondly, they provide the option of increasing the range of parts that can be made within economic time limits without costly alterations to equipment. A further potential advantage is that FMS is an essential building block for the more advanced automation stage of Computer Integrated Manufacturing (CIM) in which different functional activities such as design, planning, inventory and sales would be controlled and integrated by inter-linked computer systems. CIM would mean the automatic transmission of data and instructions between these different stages of the manufacturing process (Kaplinsky,

1984). However, even software linkages between the CAD/CAM design and planning systems and FMS are difficult to find at present (see Jones, 1988b).

Thus FMS could be used to expand the range of parts and products made within the same unit, what might be termed operational flexibility. This kind of usage could be sought along flexible specialisation lines: by changing work roles and social organisation to complement and support the technological potential to introduce new designs. Or it could be the core of a neo-Fordist strategy which aimed to achieve more limited innovations from the technology while freezing the socio-organisational framework to avoid the risks of delegating responsibility to production workers. Alternatively this dilemma could be avoided entirely. FMSs could be used in a much more restricted fashion simply to achieve savings in labour costs by eliminating workers' tasks, and inventory costs, *via* associated reductions in the lead times between scheduling, machine setting and producing given parts and sub-assemblies. There is evidence that firms pursuing this latter approach will prefer to reduce or simplify product ranges in order to use the computer systems to secure greater central determination of operational processes. This might be called strategic flexibility.

Many firms having adopted FMSs are in sectors which seem to be well suited to something like flexible specialisation. Earth moving and agricultural equipment, aerospace, machine tools and diesel engine firms all operate in markets which have either been shrinking and fragmenting over the last decade, or which have always been difficult to supply with large standardised volumes. Relatedly, they have continued to employ relatively large numbers of workers with craft-type skills and experiences. Now shorter production runs, broader product mixes and constant involvement of skilled workers are all elements of the flexible specialisation recipe. It would therefore seem, at least at first sight, as though such enterprises would be predisposed to use automation technologies in a flexible specialisation manner. On the contrary: direct cost savings in labour and lead times tend to be the principal gains sought from these FMSs. It is as if the managers concerned had chafed for many years at their incapacity to match the cost economies and production flow principles of their mass production counterparts, but had now been handed instantaneous means of adopting the Fordist paradigm.

Amongst FMS users in the U.S.A., the majority of FMS lines are relatively 'dedicated' to the manufacture of small 'families' of related parts. They make extensive use of automatic monitoring and setting devices responding to powerful central computers normally housed away from the shopfloor. Experimentation with the systems to extend the range of components they can produce is precluded or discouraged. There are rigid occupational specialisms, with skilled manual workers being excluded from any programming tasks (Jaikumar, 1986; Jones, 1989). Conventional supervisory controls persist and job descriptions and payment systems tend to be regulated by conventional industrial relations and personnel management procedures (Jones, 1988b, 1989).

The continuation of low trust in work roles and labour relations in most of these cases may be attributed to the extent to which Taylorism is still ingrained in North American management culture. The adversarial ethos and legal contractualism of the U.S. collective bargaining system also plays a part. As seniority rules underlie unions' influence on rate-for-the-job bargaining and dismissal procedures, they are keen to retain tightly demarcated job definitions. Managers for their part see manual workers involvement in programming tasks as an encroachment on managerial duties and prerogatives; they are able to use legal precedents to exclude computing work from unionised workers on those grounds (Jones, 1985a, 1985b). Why do U.S. managers in such firms choose to use flexible automation to lower particular direct operating costs for relatively non-flexible and non-innovative purposes? Why are they using the technology to make small to medium batches of a narrow range of products when they could increase the product range by more innovative experimentation with the new systems?

The response of operations management analysts to this question might be that experimentation means considerable downtime and a lengthening of the 'payback' period for recouping investment costs (cf. Jaikumar, 1984). The answer from an industrial relations, organisational and radical sociological perspective is more likely to be that the innovations required for a qualitatively more flexible use of the technology would mean allowing production staff more involvement in running different designs and programmes. Yet such local autonomy would loosen the extent to which managerial strategic plans could expect to encompass the detail of design targets, output deadlines and labour costs. Managements would, to put it bluntly, lose their control over labour and its span of discretion.

These possible explanations are dealt with in the following discussion of case study material from British, American and Japanese FMS plants. (Based upon fieldwork financed by the Economic and Social Research Council of the U.K.) However, it must be admitted that a full attack upon this problem would need more systematic investigation of the various links between companies' financial environments, market strategies and overall business policies. At the moment we can make only the safe, but relatively unenlightening judgement that here they have turned Sabel's prognosis on its head. There is another route to neo-Fordism which consists of firms outside the mass manufacturing sectors trying to use flexible automation technologies to get closer to textbook Fordism. If, as some believe, market trends are turning mass production and large batch categories into the need for medium and small batch output, then this hybrid of flexible technology and the Fordist cost paradigm will turn out to be a more generalised phenomenon.

There are exceptions to this Tayloristic neo-Fordism. One notable American example confirms that there is a link between the degree of innovativeness required of production operations and the degree of independence and expansiveness of work roles. A small mid-West tractor firm, 'Alpha', overshadowed in sales volumes by the superior distribution network of its larger competitors, drifted into a policy of niche marketing and frequent model improvements. Its FMS, purchased in the early 1970s as a collaborative experiment with a machine tool manufacturer, was frequently pressed into service as the most suitable production line to achieve the tight deadlines and rapid retooling involved. Through a mixture of miscalculations in management's labour relations strategy, the relaxed response of the plant union to changes in the job descriptions and the common craft background of the FMS operators, all recruited from the same jobbing shop, a *de facto* autonomous workgroup emerged which shared overlapping tasks and knowledge and was able to limit attempts by line managers to exercise authority over them in arbitrary, Tayloristic fashion (cf. Jones and Scott, 1987).

Across the Atlantic unplanned elements of flexible specialisation and de-Taylorised work roles have also been found in British FMS installations. However, as might be expected from the more pragmatic and less systematic management culture of this country, they have rarely gelled even into operations as effective as Alpha. In the U.K. the purpose of FMSs, which are often in similar sectors to their U.S. counterparts, has also been seen as bringing Fordist cost efficiencies to small and medium batch operations. There has, consequently, been little incentive to expand work roles, delegate discretion or try out cooperative methods of working. Jaikumar claimed that technicians and operators in the U.S. are discouraged from experimenting with the systems to increase innovative capacity and their own competences because the resulting downtime conflicts with senior managers' perceptions of their FMSs as 'dedicated' to narrow and well defined product ranges and the need for intensive working for short-term recouping of investment costs.

Similar, though not uniform, evidence of this development was found in Britain (Jones and Scott, 1985). However, the generally restricted ranges of products put through the British FMSs diminishes the need for radical revisions of work practices and work reorganisation for more flexibility and responsibility. Technological transformation and socio-organisational stagnation result from this and from the fact that FMSs are embedded in institutional sediments that have survived into the different commercial environment of today. No doubt realising the residual strength of these social complexes, many managers prefer instinctively to bypass rather than remould them. There are isolated examples of efforts by more determined managements' to redesign, rather than adapt work roles for more compatibility with the technological potentials. But as one of Peter Scott's British case studies suggests (Scott, 1987, pp. 212–20), without a comprehensive strategy and plan for change the Pandora's Box of complications suggested by the WEARI model is likely to deter changes.

In Scott's 'Turnco' study a fluid workteam were trained up to truly polyvalent skills for programming, inspection, scheduling, operating and maintenance functions. Unfortunately these skills and the FMS were developed on a separate experimental site. When the FMS was re-installed in the main plant personnel management objected to a continuation of the technician status and pay grades that had been awarded to the FMS team. They feared that it would further complicate the already contentious pay aspects of industrial relations for the rest of the workforce. The FMS crew were regraded individually and reverted to specialised occupational duties (Jones and Scott, 1987, pp. 32–4).

The Japanese utilisation of FMS, of course, suggests other possibilities. Jaikumar provided evidence that work roles are more expansive, with production staff undertaking programming and more time devoted to non-productive experimentation with the systems. The average range of component types produced is nine times greater than in comparable U.S. installations. Yet Jaikumar claims that overall utilisation rates are higher, presumably because cooperative work practices and more efficient engineering staff lead to more reliable running, which creates the time for more experimentation. My own more qualitative studies of a smaller number of Japanese firms did not suggest that the range of products was significantly higher than in North American FMSs. Cost minimisation was emphasised more than productive versatility. The most noticeable difference is that if the optimisation of flexible automation technologies *does* depend on the discretion, responsibilities and fluidities of production jobs, then the social organisation of the Japanese factory does not have to be *adapted* to the technology. The point is that unlike the USA, or to a lesser extent Britain, the Japanese do not have to adapt an inflexible social organisation to a (potentially) flexible technology. In Japanese plants the social and organisational relationships into which the FMSs are introduced are already 'flexible' in a number of relevant ways (Jones, 1989).

Thus in Britain and the U.S.A. the combination of small/medium batch production with a technology capable of considerable versatility in product range and craft skills amongst operating staff do *not* add up to flexible specialisation. In Abernathy's terminology the final roadblock to innovation was essentially the same in both countries. It was a failure to realise the potential flexibility of the technology by expanding responsibilities amongst technical and production staff. American managers were more constrained by the depth of their Tayloristic management culture and the strength of industrial relations institutions. Yet their British counterparts seemed less restricted by dogma and the web of rules than by their inability, or reluctance, to formulate strategies that would reform the institutional conditions for flexible work and technology combinations.

CONCLUSION

The developments in both Britain and the U.S.A. and the evidence from Japan suggests that the flexible automation alternative to mass production may comprise of a number of hybrid forms. It may, ironically, be facilitating the realisation of Fordist principles in sectors that were not mass producers. It may combine this flexible Fordism with non-Taylorist controls and work roles, as in Japan. It may even result in some innovative working arrangements of the flexible specialisation type, as instanced in the Alpha case. But in general the prospect of using these systems to tighten hierarchical control over final operations may prove more appealing to many managers than the surrender of detailed powers to the shopfloor *that is necessary* for versatile and innovative productive capability.

In operations management, as in social science analysis, a number of paradigms can be adopted. We ought not to be surprised at the appeal of a diversity of managerial approaches. Erudite and strategic executives may see their organisations and the path of change after the fashion of Talcot Parsons's complex systems model, which requires shifts in overarching cultural values to complement changes in local sub-systems and to make actors receptive to them. But equally possible are the 'technological determinists': managers who content themselves with breaching the institutional rigidities that they are able to perceive by a narrowly focused set of technological changes. In the British case the evidence suggests that the majority will practice eclecticism and settle for solutions somewhere in between.

Advanced computer technologies such as FMS can provide extra competitive advantage, by expanding market opportunities as well as operational ones. For the most part, however, they tend to be used in a predictable fashion in a more intensive pursuit of familiar forms of cost savings. The organisational changes necessary to follow a more innovative and imaginative use of the technology are substantial. From the perspective adopted in this chapter, shifts in this direction will require broader task ranges with more discretion in work roles, the need for cooperative rather than hierarchical decision making in authority relationships and reforms to the representative institutions of industrial relations. In the interpretative dimension grandiose schemes for new corporate cultures may not be important. These may anyway risk degenerating into contrived propaganda exercises. However, decisive shifts in the attitudes, beliefs and expectations of managers and workers underpin the other changes. Perhaps the very scale of so much change, in so many dimensions, deters consideration of more expansive uses of the technology. Campbell and Warner's (1988) study of the related technology of CAD/CAM raised some of the same issues from another angle. They proposed that more corporate emphasis on employee *training* may resolve many of the obstacles to more productive use of advanced technologies.

However, if operations managers are to face up to these issues a prior change of strategic orientation must link technology and organisation with marketing policies. If shifts in the operational and socio-organisational field do depend on such strategic linkages, the evidence of this chapter suggests that executive management is generally not yet willing to, or capable of formulating strategies of such scope and detail.

ACKNOWLEDGEMENT

I am grateful to Martyn Pitt for comments and helpful suggestions in the re-drafting of this chapter; the remaining flaws and errors remain my responsibility.

REFERENCES

Abernathy, J. (1977). *The Productivity Dilemma*, Johns Hopkins Press, Baltimore.

Aglietta, M. (1979). *A Theory of Capitalist Regulation*, New Left Books, London.

Allen, S., Purcell, K., Waton, A., and Wood, S. (1986). *The Changing Experience of Employment*, Macmillan, London.

Brusco, S. (1982). 'The Emilian Model: productive decentralisation and social integration, *Cambridge Journal of Economics*, **6**.

Campbell, A., and Warner, M. (1988). Organisation for New Forms of Manufacturing Operations, Cambridge University Engineering Dept. Management Studies Research Paper 2/88.

Clarke, K., Hayes, R.H., and Lorenz, C. (1985). *The Uneasy Alliance: Managing the Productivity-Technology Dilemma*, Harvard Business School Press, Boston, Mass.

Clegg, S., and Dunkerley, D. (1981). *International Yearbook of International Studies*, RKP, London.

Conference of Socialist Economics (1976). *The Labour Process and Class Strategies*, Stage One, London.

Coriat, B. (1987). Information Technologies, Productivity and New Job Content: Skill as a Competitive Issue, paper for the BRIE meeting on Comparative Production, Berkeley, September.

Freeman, C. (1987). *Technology Policy and Economic Performance: Lessons from Japan*, Pinter, London.

Hirschhorn, L. (1985). *Beyond Mechanisation*, MIT Press, Cambridge.

Hirst, P.Q., and Zeitlin, J. (1989). *Reversing Industrial Decline*, St Martin's Press/Berg Publishers, London.

Hyman, R. and Streeck, W. (eds) (1988). *New Technology and Industrial Relations: International Experiences*, Blackwell, Oxford.

Jaikumar, R. (1984). *Flexible Manufacturing Systems: A Managerial Perspective*, Working Paper 1–74–078, Harvard Business School: Division of Research.

Jaikumar, R. (1986). 'Post-industrial manufacturing', *Harvard Business Review*, November–December, 69–76.

Jones, B. (1985a). 'Flexible Technologies and Inflexible Jobs: Impossible Dreams and Missed Opportunities'. World Congress on Human Factors in Automation; Dearborn, Michigan: Society of Manufacturing Engineers.

Jones, B. (1985b). 'Controlling production on the shopfloor: state administration and regulation in the British and American aerospace industries', in S. Tolliday and J. Zeitlin (eds), *Shop Floor Bargaining and The State*, Cambridge University Press, Cambridge.

Jones, B. (1988a). Post-Fordist Production: A New Role for Labour? Bologna: Second International Colloquium on the Sociology of Work, 2–4 February 1988.

Jones, B. (1988b). 'Work and flexible automation in Britain: a review of developments and possibilities', *Work, Employment and Society*, **2**, 4 December.

Jones, B. (1989). 'Flexible automation and factory politics: Britain in comparative perspective, in P.Q. Hirst and J. Zeitlin (eds), *Reversing Industrial Decline*.

Jones, B. and Scott, P.J. (1985). FMS in Britain, Unpublished report, University of Bath.

Jones, B., and Scott, P.J. (1987). 'Working The System: flexible manufacturing systems in Britain and the USA', *New Technology Work and Employment* **2**(1), pp.27–36.

Kaplinsky, R. (1984). *Automation: the Technology and Society*, Longman, London.

Kern, H., and Schumann, M. (1987). 'Limits of the division of labour: new production and employment concepts in West German industry', *Economic and Industrial Democracy*, **8**.

Knights, D., Willmott, H., and Collinson, D. (1985). *Job Redesign: Organization and Control of the Labour Process*, Gower, Aldershot.

Kochan, T. A., Katz, H.C., and McKersie, R.B. (1986). *The Transformation of American Industrial Relations*, Basic Books, New York.

Lane, C. (1988). 'Industrial change in Europe: the pursuit of flexible specialisation in Britain and West Germany', *Work, Employment and Society*, **2**(2), 141–68.

Loveridge, R. (1981). 'Business strategy and community culture', in S. Clegg and D. Dunkerley (eds), *International Yearbook of International Studies*.

Marginson, P., and Edwards, P. (1988). *Beyond the Workplace*, Basil Blackwell, Oxford.

Murray, R. (1988). 'Life after Henry (Ford)', *Marxism Today*, October.

Oliver, N., and Wilkinson, B. (1988). *The Japanization of British Industry*, Blackwell, Oxford.

Palloix, C. (1976). 'The labour process: from Fordism to neo-Fordism', in Conference of Socialist Economists.

Perez, C. (1985). 'Micro-electronics, long waves and world structural change: new perspectives for developing countries', *World Development*, **13**(3), 441–63.

Peters, T. (1988). *Thriving on Chaos: A Handbook for Managerial Revolution*, Macmillan, London.

Piore, M.J., and Sabel, C. (1984). *The Second Industrial Divide*, Basic Books, New York.

Pollert, A. (1988). 'The flexible firm: fixation or fact', *Work, Employment and Society*, **2**, 3 September.

Rose, M. J. (1988). *Industrial Behaviour: Research and Control*, Penguin, London.

Rose, M. J., and Jones, B. (1985). 'Management strategy and trade union response in plant level reorganizations of work', in D. Knights *et al.*, *Job Redesign: Organization and Control of the Labour Process*.

Sabel, C. (1984). *Work and Politics*, Cambridge University Press, Cambridge.

Scott, P. J. (1987). Craft Skills in Flexible Manufacturing Systems, University of Bath, Ph.D. dissertation.

Tolliday, S., and Zeitlin, J. (eds) (1987). *Between Fordism and Flexibility*, Polity Press, London.

Turnbull, P. (1988). 'The limits to Japanisation', *New Technology, Work and Employment*, **3**, 1.

Walton, R. (1985). 'From control to commitment: transforming workforce management in the United States', in K. Clark, *The Uneasy Alliance: Managing the Productivity-Technology Dilemna*

Williams, K. Cutler, T. Williams, J., and Haslam, C. (1987). 'The end of mass production' *Economy and Society* **16**, 405–439.

Wood, S. (1988). 'Between Fordism and flexibility? The US car industry', in R. Hyman and W. Streeck (eds), *New Technology and Industrial Relations: International Experiences*.

Wood, S. (1989). *The Transformation of Work?* Hutchinson, London.

14

The Context and Process of Organizational Transformation: Cadbury Limited in its Sector*

John Child and Christopher Smith

Work Organization Research Centre, Aston University

Problems of industrial competitiveness have stimulated inquiry into the behaviour of firms within their sectors, namely the population of firms which provide similar goods and services. There is particular interest in the process whereby new or rejuvenated firms may emerge to compete more effectively within their sectors. The policy objective of achieving this industrial transformation now complements a long-standing academic interest in organizational change for the insight it affords into organizational dynamics and the relationship of organizations with their environments.

The chocolate confectionery firm, Cadbury Limited, operates within a well-defined sector and has recently undergone a significant

* This chapter originally appeared in the *Journal of Management Studies*, Volume 24, No.6, November 1987, under the title of 'The Context and Process of Organisational Transformation—Cadbury Limited in its Sector'. Reproduced with changes by permission of the publishers, Basil Blackwell Ltd.

The Strategic Management of Technological Innovation. Edited by R. Loveridge and M. Pitt
Published by John Wiley & Sons Ltd

transformation. This chapter analyses the empirical insights into organizational transformation which the Cadbury case offers, within a perspective that is sensitive to the significance of the firm's location within its sector. An outline of this perspective and its antecedents is presented first. The situation of Cadbury Limited within its particular sector is then examined. This provides the necessary context for a consideration of the company's transformation which is first described substantively and then addressed in terms of the process whereby it was accomplished.

THE 'FIRM-IN-SECTOR' PERSPECTIVE

Strategic innovation in firms refers to major systemic decisions and their implementation concerning the choice of products and markets, of production processes and technologies and of work organization. It therefore involves a significant transformation for the organizations concerned.

Three aspects of 'sector' are regarded as relevant to organizational transformation. First, the sector constitutes a set of *objective conditions* which can create pressures for transformation insofar as a firm's viability depends upon the extent to which its behaviour is appropriate to those environmental conditions. Second, the sector is a *cognitive arena* with which its members identify. In this role, the sector is the bearer of external exemplars against which a firm's current strategy and structure, and the ideology underlying these, can be compared and which may therefore serve to guide the new configurations to which a transformation is directed. Third, a sector does not only consist of product competitors; it is also a *network* of potential and actual collaborators. Lawrence and Dyer (1983) have encapsulated these sector characteristics into two 'environmental domains', namely those of resources for which organizations compete and of information in which the intensity of cognitive fields and networks generates a level of information complexity for firms to handle. These writers regard changes in the resource scarcity and information complexity within a sector as triggers for organizational change.

The sector as determinant of transformation

The deterministic view of transformation sees the critical link between sector and firm to lie in the criteria the sector imposes for the firm's survival. The sector is taken to constitute an *objective reality* possessing identifiable and measurable characteristics which are of consequence for corporate strategy and structure. The structure–conduct–performance model in economics, for example, posits that sector (especially market) structures influence the actions which firms take (Scherer, 1980). Relevant sector structural characteristics include market concentration, height of entry barriers such as capital intensity, R&D intensity and technological specificities, buyer and seller power, labour market conditions, and governmental actions towards the sector. The underlying assumption is that the sector, particularly when strongly competitive, largely determines the path a firm must take for future success, including distinct patterns of technical change (Pavitt, 1984). Poor performance, not just absolutely but also relative to competitors in the same sector, is a major trigger of organizational transformations (Bowman, 1985).

The population ecology school has applied a biological analogy in arguing that there is a survival path within a sector of competing organizations. Deviance from that path acts to trigger transformation in the wayward organization. Either managements learn to recognize the survival path and transform their organizations as and when it requires or those organizations will become transformed willy-nilly into extinction. Different sectors are seen to carry different organizational design requirements (Aldrich, McKelvey and Ulrich, 1984). Application of the biological life-cycle model to the sector raises the possibility that it may pass through several phases of development from its founding era through to maturity and potential demise. Abernathy and Utterback (1975) advanced a dynamic model of product and process innovation (and associated forms of work organization) in which these main dimensions are combined differently as the sector passes from an early 'fluid' pattern, through a 'transition' pattern to the 'specific' (i.e. dedicated plants, vertically integrated, formalized) pattern of maturity. It is implied that firms which do not transform themselves over time in line with the sector life-cycle trend will find it difficult to survive unless they can remove themselves from the mainstream of the sector by finding protected niches in which to operate. The ability of managements to *recognize, interpret and implement* the emergent requirements of their sector in the products, processes and organizational modes they adopt is seen to be crucial to their survival under competitive conditions.

Abernathy, Clarke and Kantrow (1983) added a further state to the sector model, that of 'de-maturity'. They argued the pressing need for American manufacturing industry to respond to the entry of foreign competition into mature sectors by undertaking a process of dematurization. In conditions of non-trivial environmental change the incremental adaptation which characterized mature firms becomes insufficient. They need instead to undergo a new learning process in which innovation is refocused from processes back to products and distinctive technical solutions are sought to the product attributes now favoured by the market. This is particularly the case if the market now places a premium on variety, state-of-the-art design, special applications, or is in other ways no longer responsive to standardization.

The concept of de-maturity has parallels with that of 'neo-Fordism' developed by French industrial economists to explain how mass production sectors have adapted to competitive changes and labour opposition by reorganizing assembly lines, utilizing new technology and employing new categories of labour (Palloix, 1976; Aglietta, 1979). More recently this concept has been supplemented by American pluralist models of production, such as 'flexible specialization' in which it is argued that in order to meet intensifying international competition large mature firms will need to disaggregate their inflexible and unwieldy systems of production and management (Piore and Sabel, 1984).

When the life-cycle model is applied to organizations themselves (cf. Kimberly, 1980), it postulates that as they become more mature they take on rigid attributes that make it increasingly difficult for them to adjust to changing contexts in an innovative manner. The analogy of organic development also implies that firms will be stamped with the character of their inheritance and early development. These may constitute a further hindrance to transformation. Organizations become encumbered with their founding *ideologies* (Miles, 1980), with sedimented *structures* which reflect the conditions and circumstances applying at the time of their birth (Stinchcombe, 1965), and with distinctive *competences* no longer suited to competitive requirements. This suggests that the legacy of a firm's history will bear heavily upon its ability to effect a permanent transformation. Competitive pressures to adopt a superior strategy and structural configuration are therefore likely to be mediated by an organization's inherited tradition, structured power distributions and particular inherited competences.

The biological analogy therefore raises some key issues for the present case study. One is whether the notion of product life-cycle applies to sectors such as the one to be considered here, processed food, where standardized brands may often appeal because of their unchanging classic 'original recipe' qualities. In such sectors transformation appears less likely to be led by product innovation. While there is differentiation between core and experimental brands, the former finance the latter and product continuity co-exists with product change within the same market segment.

A second issue concerns the possibility of more than one sector survival path around different types of product appeal which may be valued in particular social and regional segments of the market. The recognition of a sector survival path by a firm's management is likely to depend on its identifying other firms following that path as competitively relevant peers. In a case such as chocolate confectionery where giant firms dominate an oligopolistic sector, the significant other firm, rather than the sector in a classical market sense, may often provide the signals which direct managerial attention towards transformation.

Transformation as cognitive reorientations

These considerations identify the second aspect of sector relevant to organizational transformation, which is that of a *cognitive arena* with which its members identify. Research by Spender (1980) indicates that, at least within relatively well-established sectors, the senior managers of constituent firms hold very similar constructs of the sector's operational dynamics which effectively furnish the 'rules of the game' for the sector. These sector strategic 'recipes' can be expected to change only when there is a substantial shift in market or technological conditions, quite possibly emanating from beyond the previous pragmatic boundaries of the sector as, for example, with new entry from overseas. It has also been noted in this vein that 'sectors may be characterized by distinctive corporate languages, constructs and frameworks, all of which have an important influence on the evolvement of learning paths in the sector'. Spender's findings related to fairly homogeneous sectors and they do not rule out the possibility of multiple recipes corresponding to different sectoral segments.

The concept of strategic recipe introduces the significance of cognition for organizational transformation. Organization analysts point out that external conditions only become known through the perceptions of organizational members and that in this sense the sector is a mental construct. The human actor does not react to an environment, he enacts it (Weick, 1969, p. 64, emphasis in original). Some, however, pursue this argument to the point of maintaining that environments (sectors) are themselves merely perceptual phenomena (Feldman, 1986; Smircich and Stubbart, 1985). A less extreme position recognizes that sector environments have real properties distinct from the perceptions of particular actors within them, but that those perceptions and social constructions are of consequence in their own right if they determine the *behaviour* of firms. The sector *per se* is always liable to be greater than the sum of its present actors (even if they could somehow pool their perceptions and consequential actions) because of the potential for new entry among producers, suppliers, buyers, government legislators, inventors and so on. Pragmatically, it is suggested that the key issues concern matters of degree. The first concerns the degree of strategic choice available to organizational decision makers such that they enjoy the power to enact environments substantively, over and above their perceptual enactment (Child, 1972). This is clearly relevant to the ability of firms to avoid undertaking transformations which are contrary to their traditions or otherwise unwelcome. The second issue concerns the circumstances under which new subjective interpretations of objective conditions emerge to define a 'crisis', or at least the need to act, and thereby cast doubt on the validity of an existing recipe or approach. This is relevant to the triggering of transformations. A third issue concerns the continuing relation between objective conditions and subjective interpretations in shaping the course of transformation including its duration and scope.

Sectors as networks

Up to this point the sector has been discussed primarily as a competitive arena. However, the presence of a shared identity among sector members, often sustained by long-standing ties, also provides the basis for co-operative relationships. The third aspect of sector is therefore that of a *collaborative network* within which the firm is located. As Richardson (1972, p. 895) pointed out, 'firms are not islands but are linked together in patterns of co-operation and affiliation'. Collaborative sector networks can play a significant role in facilitating organizational transformations especially those involving technological change. The importance of collaboration between users and suppliers has been analysed with respect to the equipment and process innovations which may be central to transformation (Von Hippel, 1982). Joint ventures can not only transfer technology but also help to diffuse 'advanced' knowledge of techniques for labour control and productivity improvement where the collaboration involves one partner considered to be at the cutting-edge in this area. Channels of collaboration can become channels for the import of new management practice.

Collaboration is also evident in the role played by consultants in assessing the need for and advising on the process of transformation, including consulting engineers whose expertise is brought to bear in the construction of new process facilities. It is not unusual for organizational transformations to be accompanied by the inflow of new staff possessing knowledge of the new mode of operation that is sought and/or in the successful accomplishment of change. When pursued in the direction of organizational fragmentation, transformation will embed the remaining organizational core within a network of now externalized transactions undertaken with designers, suppliers, distributors and production sub-contractors.

It is in this respect important to recognize that management may be differentially immersed in sectoral information and knowledge through collaborative contacts. We could perhaps distinguish between sector boundary managers such as R&D engineers, marketing managers and corporate managers who are more conscious of innovations and new trends within a sector, and firm-specific (or core) managers such as line production managers and industrial engineers who are tied into the internal labour markets and the practices of the particular firm. Boundary managers may be a strong source of diffusion of sector developments through the firm and firm-specific management a major embodiment of tradition and resistance to change.

The three aspects of sector identified within the 'firm-in-sector' perspective—objective conditions, cognitive arena and collaborative network—furnish an array of propositions concerning the form and process of organizational transformation. The study of Cadbury Limited was informed by this perspective and it provides insights into a number of the issues which have been raised.

CADBURY LIMITED IN ITS SECTOR

The study of organizational transformation on which this chapter draws is reported at length in Smith, Child and Rowlinson (1990).

Cadbury Limited is the UK arm of the Confectionery Division of Cadbury-Schweppes, a diversified international food and drink corporation producing soft drinks, tea and other hot beverages, jam and biscuits as well as chocolate and sugar confectionery (in 1986 the company largely divested itself of tea and some foods through a management buy-out of Cadbury Typhoo). Our attention will be further focused on the chocolate confectionery business primarily located at Bournville, Birmingham.

Chocolate confectionery can be differentiated within the food industry in terms of several *objective conditions*. First, the origins of its products. Unlike traditional, indigenous products like dairy goods, meat and bread, which retained small-scale local markets and handicraft production, chocolate confectionery originated as a luxury product lacking a traditional local market. It therefore developed an international market, initially a luxury one which extended later into a mass market. Confectionery producers typically became capital intensive, concentrated and international, which led them to apply the latest organizational and technological developments drawn from an international rather than a local constituency. They applied a scientific management approach, including labour deskilling, much earlier than in areas such as bread and butchery.

Second, active marketing has been a prime factor in the development of Cadburys and other major food firms since the late nineteenth century. The absence of indigenous methods of processing created the need to generate demand for loyalty to exotic brands of commodities. The fact that confectionery is generally purchased in prepared form also widened the distribution opportunities open to manufacturers. They spread from local and regional to national markets and accordingly developed large-scale mass production. Most segments of the UK food industry have now become concentrated into a few giant firms, with chocolate confectionery being the most concentrated after sugar production, oils, fats and margarine, and canned soup.

During the inter-war years manufacturers like Cadbury advocated policies to restrict the power of wholesalers and to eliminate small inefficient retailers, both of which stood in the way of expanding sales through offering better value for money (Cadbury, 1947). More recently, and particularly since the 1960s, concentration in retailing began to affect the power of the large food manufacturers. The latter's control over conditions of sale was substantially weakened by the abolition of resale price maintenance, concentration in the ownership of shops and the advent of own brands. At the present time the introduction of computerized point-of-sale systems is enabling the large retailers to exercise market-sensitive discrimination between manufacturers to an even more refined degree.

Chocolate can be instantly consumed without additional preparation, it is compact, and enjoys longer storage properties than most foods and can therefore be sold in a wide variety of outlets including garages and places of entertainment. Because of this spread of outlets and the heavy advertising of long-accepted products with low or zero obsolescence, chocolate products have retained a greater degree of 'brand integrity' in the face of own labelling by the large supermarket companies. Nevertheless, by the late 1970s Cadburys was ascribing the need to rationalize-out smaller product lines to pressures emanating from the growing concentration and business sophistication of retailers.

Competition between the chocolate manufacturers intensified in the post-war period. Market saturation combined with the sophistication of taste and image associated with affluence; both were increased by TV advertising from the late 1950s. This was seen by Cadbury at the time to imply that it had to support its brands on a broader basis than just its straightforward pre-war appeal to 'good value' (Cadbury, 1964). Moreover, the heavy dependence on a major raw material, cocoa, the market price of which has fluctuated considerably over recent years, poses an area of high risk to profits.

The chocolate confectionery sector, in common with many other areas of the food industry, is not characterized by significant product life-cycles. Once a favoured recipe and process is discovered, a long history of consumer acceptance becomes a marketing virtue. The major contribution to chocolate sales is made by long-established products such as the Mars bar, Kit Kat or in Cadburys' case the Dairy Milk family and Flake. New products are introduced, particularly among count lines, but only a few are successful and these then normally remain in the product range for many years. Most new products involve little technical development in their make-up, though some like Cadburys' Wispa did depend upon new processing techniques. Though product development aims to find the occasional new winner and improve quality and storage characteristics, the main competitive initiatives within the sector are directed towards cost-reduction through process improvement, raw material substitution to decrease expensive cocoa content and advances in packaging.

There has been a strong sense of continuity within the sector and its constituent firms around their products and the traditional skills of their manufacture. Product stability was further supported by a long period of market share stability between the two major British chocolate manufacturers, Cadbury and Rowntree. These Quaker manufacturers also shared a particular ethos concerning welfarism, formalized employment policies and social responsibility. While these manufacturers were well-known for their interest in new methods of management and for their general spirit of enquiry, there was also an assumption that any change would be incremental and would not sever the continuity with the past. This meant that subsequent transformation took place within an institutionalized context that was unreceptive to fundamental change.

The Cadbury *cognitive model* of chocolate manufacture was to produce a large number and broad range of branded products. The limited market for some of these products meant they could only be produced in small quantities or at certain times of the year, so injecting discontinuities in production. Moreover, Cadburys' production was divided between different sites and indeed the company claimed as late as 1964 that 'there are great dangers in so far rationalizing production as to concentrate the whole of a factory's resources on a single line' (Cadbury, 1964, p. 41).

This broad product range, multi-site approach presents a contrast, at least in food, with the policy of American firms which located production in Britain. They produced single products or small product ranges, along standardized lines in single capital-intensive plants offering the maximum economies in production. This could be called a Fordist strategy, compared with the paradigm of hybrid mass and luxury production followed by British manufacturers. Mars provides the outstanding Fordist example in chocolate confectionery, from 1933, producing a single simply-wrapped product in one highly mechanized plant, a product outside the labour-intensive areas of boxed assortment and filled chocolate bar production and of elaborate wrapping. British manufacturers, particularly Cadburys, were moving along the path of capital intensification before the Second World War, but they shunned product rationalization. It was only when competition intensified particularly for the custom of large retailers operating on high throughput and low margins, who were prepared to stock only the most popular and competitively-priced lines, that the American strategic recipe was accepted as relevant.

Cadburys resisted a whole-hearted transfer to a rationalized approach until the 1970s. Just after the Second World War the company acknowledged the savings in labour, factory space and paper, achieved as a result of the enforced wartime reduction of lines from 237 in 1939 to only 29 in 1942. These savings much more than outweighed the slight reduction in total tonnage produced (Cadbury, 1947, pp 38–40). Yet by 1962 it had increased its product range to 60 (Cadbury, 1964, p. 25). Later, following the merger with

Schweppes which encouraged the enlarged corporation to expand itself into a general food company through product and geographical diversification, a policy of increased product proliferation was pursued by Cadburys for several years. It was not until the mid-1970s when Cadburys' market share experienced a dramatic fall despite this policy of brand proliferation that it began significantly to reduce its product range and to perceive the Mars model of concentrated mass manufacture and marketing as worth emulating.

Thus, under conditions of benign competition based on considerable market segmentation, more than one strategic recipe may be viable within a sector. This strategic variety was sustained in chocolate confectionery despite a keen awareness of the alternative approach being adopted by an oligopolistic rival. It is only when competition intensifies and demonstrates one approach to have economic superiorities over the other(s) that sufficient pressures on the performance of the losing firms force their managements to engage in active learning from the bearers of the alternative recipe, potentially stimulating radical transformation.

A number of writers have drawn attention to the way that downturns in organizational performance or other organizational 'crises' serve as triggers for major changes in top personnel, cultures and policies (e.g. Frost et al., 1985; Pettigrew, 1985). Cadburys' share of the total chocolate market fell from 31.1% to 26.2% between 1975 and 1977, with heavy falls in the product areas directly competitive with the Mars and Rowntree–Mackintosh Yorkie bars. Moreover, the company misread the cocoa market in 1976. As a result trading profit declined in two consecutive years, 1975 and 1976, a substantial fall at a time when inflation was running at around 20% per annum. This crisis was the trigger for transformation.

The intensification of competition affected the nature of the *sector network*. It increased the secrecy between the competing chocolate producers and so altered their channels of learning. While it proved possible for Cadbury managers to visit some continental European firms with which the company was not in substantial competition, this facility was no longer available within the UK. As a result, much of the learning required to shift from one sector *template* to another depended on other mediated means of knowledge transfer. One was to recruit staff from Mars ('the men from Mars') which by the late 1970s was adopted as a model for Cadbury's new production rationalization philosophy. Another was to rely on a transfer of advanced production technology through the medium of equipment suppliers, predominantly continental manufacturers, who had developed and tested new equipment in other chocolate making firms. These equipment suppliers became an important bridge between competing players in the sector because (i) those players were not prepared to trade know-how directly and (ii) the model of rationalization to which Cadbury was moving envisaged a concentration on the core chocolate-making function and a withdrawal from the substantial amount of equipment design that the company had itself previously undertaken. Indeed some new areas of process design, particularly that of microprocessor sensing and control, lay outside the company's sphere of experience altogether. This necessitated a dependence on system suppliers for the development of software that was quite central to the organization of its operations.

The role which mobile key staff and equipment suppliers played in technology transfer to the company illustrates the collaborative aspect of sectors as social networks forged by working contacts. The 'market' for transfer of design concepts and technical knowledge is paradoxically *imperfected* by competition, and networks can play an important part in transcending what would otherwise be significant barriers to change.

TRANSFORMATION AS CHANGE AND CONTINUITY

Long-term competitive trends in the confectionery sector created a general need for Cadburys to change, while the company's sharply deteriorating performance in the mid-1970s provided the specific trigger for the substantial transformation that has taken place since. This has so far been rather loosely described as a movement from the traditional UK food sector recipe to one more in accord with that pursued by successful American entrants. Since organizations are social institutions with cultures and histories, change is liable to be in dynamic tension with continuity, as Pettigrew (1985) noted in the case of ICI. What we call transformations typically combine the incremental extension of some existing policies and practices with other features that are more radically innovative. Transformations can build upon a history of calls for change as well as upon previous tangible demonstration projects. The Cadbury experience contains these interleaves of change and continuity, including within its tradition a combination of managerial receptivity to new ideas with a deeply embedded set of practices at its historically core Bournville plant.

Transformation nevertheless denotes metamorphosis or at least a movement away from a previous condition. In this case the change has been widely perceived as the rejection of something that both its managers and workers were aware made the company different, and which may justly be called Cadburyism. This was a tradition self-consciously articulated in writings and other public statements issuing both from within the company and externally. It gave the whole company a strong self-image.

Cadburyism had both its widely-known and somewhat less widely appreciated characteristics. It is well-known for its employment, welfare and community-related aspects. Cadburys developed an internal labour market at an early stage. Prior to the First World War the firm drew practically all of its labour force from local elementary schools. The main personnel differentiations then made and perpetuated until recently were (i) within production, between processing operations as areas of male employment and wrapping/packing as female areas and (ii) between male full-time continuing contracts and female full-time contracts terminated on marriage. A sharp distinction was drawn between general factory work, craft work, and engineering. During the inter-war period the company started to recruit from secondary and public schools and even a few university graduates. It formulated a policy for recruitment, training and advancement which recognized that entry through these various gates would lead to a differentiation of final position both by level and specialty (Cadbury, 1947, p. 66).

The Cadbury employment structure was therefore formalized at an early stage of Bournville's development as a factory. Even before the First World War, formalization in the personnel area extended to selection procedures, compulsory continuing education, apprenticeship schemes, health and safety provisions, and various welfare benefits including holiday schemes and pension funds (Cadbury, 1912). It was allied to the offer of secure continuing employment for men, recognition of trade unions along segmented lines that accorded with the factory's own workforce distinctions and after 1918, male and female Bournville Works Councils. Taken together and cemented by the strong ideology of the Bournville community which the Cadbury family expounded and most employees readily accepted, these various policies and practices clearly constituted a very firm basis for the development of a strong tradition.

It is somewhat less widely appreciated that the Cadburys did not simply regard their employment policies as reflective of a Quaker social conscience and community spirit but also as productive of business efficiency (Cadbury, 1912, Introduction). Rowlinson (1986) describes in detail how they were among the first British employers to engage with and selectively apply Taylorism. In so doing they established a 'Works Organization Department' in 1912 to deal with the organization of labour and wage payment systems and out of which a strong industrial engineering tradition developed. Piecework systems were introduced during 1913, their encouragement of exact task specification being recognized as a benefit in addition to their motivational effects. The company also hired American efficiency consultants to assist in the reorganization of work.

Together with other Quaker chocolate makers, especially Rowntrees, the Cadburys gave considerable thought to the contribution of management organization to business efficiency (Child, 1964). One of the early and enduring manifestations of this lay in the adoption of a functional organization, with major departments in the charge of directors coordinated by interlocking committees chaired by directors. For instance, functional departments allied to production included engineering, design, quality, production planning, cost, labour and statistics.

A consistent Cadbury priority was investment in mechanization and technical progress financed through retained income. (The company was under family control and enjoyed financial independence which permitted a long-term orientation towards profitability and planning.) Mechanization was introduced after the mid-1920s. Cadburys was the first British company to invest in flow-line production for moulded chocolate bars. Combined with falling raw material and transport costs, it permitted price cuts which in turn stimulated sales, enabling it to expand control over the moulded bar market, a segment it has continued to dominate. Various schemes were devised to provide for the employees displaced through mechanization, including alternative work in the factory and Bournville community, and benefits were provided for those actually made redundant including financial encouragement to retrain or set up in business (Cadbury, 1947). So while investment and work reorganization have a long history in Cadburys, attention was given to relieving the burden of their consequences for workers. Moreover, changes were introduced slowly and after consultation, or according to one manager, 'democratic consensus: slow persuasion and very mild coercion'.

The heavy investment programme which has been a main engine of recent transformation thus represents a *continuity* with the traditional priority given to mechanization. A *discontinuity*, however, appears in the rejection of the company's traditional concern to make the transition as smooth as possible in employment terms through consultation, and not shatter the community ethos.

Cadburyism, then, was a well thought-out body of ideology and practice. It was a set of distinctive competences which had built a successful food business and a source of legitimacy for managerial prerogatives especially when exercised by family-managers in accordance with its precepts. Its very strength and embeddedness has led to a keen awareness of established practice from which each period of change has been liable to be judged a retreat. At the same time, Cadburyism as an ideology was less inflexible, for it combined its concept of business in the community with a keen search for improved efficiency and profitable new investment. It was a managerial philosophy which therefore engaged actively with new currents of thought and technique.

The changes which constitute the transformation of Cadbury are described in detail by Smith *et al*. (1990). Prior to examining salient features in the process whereby the transformation was accomplished, there is space here only to provide a chronology of main events followed by a summary of the main discontinuities which took place:

1966–9   'Project Ambridge'. Relocation of cocoa bean processing to a new purpose-built factory with modern continuous processing units in a greenfield site (Chirk, opened November 1969). Programme of manufacturing rationalization announced. Product specialization for factories.

1969   March: merger with Schweppes. Schweppes men occupy chairmanship and key financial positions with greater orientation to short-term profitability. Followed by a period of brand proliferation.

1970   November: launch of 'Operation Profitability'. Significant loss of jobs involving the first redundancies at Bournville since the 1930s and which extended to middle management grades.

1971–3   'Achievements' at Chirk (four-shift system, single union, flexible working practices, close relationship between management and workers) presented to Confectionery Group and received 'an interested but unconvinced response' (Whitaker, 1982, p. 71). March 1973: Confectionery Group 'Pink Paper' on employee relations: endorses achievements at Chirk, stresses the need to improve efficiency and favours a participative route forward.

1974 on   Development among senior Bournville management of new industrial relations strategy aimed at alignment of participative bodies with operating divisions. Growing managerial impatience with the Factory (formerly Works) Council for standing in the way of change, duplicating trade union functions and criticizing management. Creation of Bournville Factory Management Team, a move away from functional organization. Quality, industrial engineering and management accounting operationally integrated into production.

1975–6   Sharp decline in Cadbury Limited market share and trading profit.

1976   Adrian Cadbury's Chairman's Statement: change of policy to concentration on core businesses (including chocolate confectionery) and improvement of operating performance.

1977   First major dispute at Bournville since 1953 and the first ever to involve the whole workforce. Four-and-a-half week strike over wages. Replacement of the personnel manager by a hard-liner.

1977   Special Industrial Engineering Report on factory rationalization, laying out a systematic, plant-by-plant programme of change based on the idea of 'stretching' performance on the best lines, consolidating production and labour into high performance plants with negligible capital investment.

1978   January: five-year Long Range Plan. Key elements were major new investment, four-shift working, end of craft demarcation,

	job flexibility, subcontracting, and a significant reduction in employment.
1978–80	Management withdrawal from the Factory Council, sacking and suspension of several senior shop stewards and the substitution of a new decentralized factory consultative structure operating to managerial initiative (information-giving, briefing groups, etc.).
1978–86	Large fixed capital investment in core product range. Closure of old plants, with three or four typically replaced by one new plant. From 1980 on, increasing fragmentation of employees' collective identity via close symbolic and social attachment (team concept) to each new facility coming on stream. Growing use of computer process control for greater flexibility of working practices.
1978–82	Cadbury Limited engineering functions decentralized to report to the Manufacturing Director and operationally contained within Bournville Assortments and Moulding 'Factories' in the form of 'Integrated Project Groups'. Simplification of line managerial hierarchies.
1979–80	January 1979: recruitment of labour to operate four-shift system. June: strike over shift working. Cadbury management climb-down under pressure from top corporate management. Shiftworking agreed with reduced working hours and an increase of payment through shift allowances in 1980. 'Old guard' TGWU stewards retire. January 1980: Dominic Cadbury becomes M.D. of Cadbury Limited. Headcount reduction reinforced as key target.
1982	Decline in Cadbury Limited trading profit and a substantial fall in its return on assets. November: 'Operation Fundamental Change' aimed at securing better return from new investments through work organization reform: compulsory redeployment, job flexibility, all-purpose operators, linking of pay deals with productivity improvements, cancellation of overtime guarantee in contracting-out agreement.
1983	August: First formal flexibility agreement of Bournville signed for the new Wispa plant.

Attempts to extend flexibility to other production lines have continued since and employment levels have continued to be cut. By 1986, however, the large investment programme was running out and union resistance was stiffening to further rationalization of demarcations between engineering trade and production roles, and to the reorganization of union branches to suit the now decentralized moulded and assortments factory units.

Whilst the redundancies incurred during 'Operation Profitability' in 1970 and the brusque way these were handled, administered the first major shock to the Cadbury culture, it is primarily the changes after 1978 that constitute the organization's transformation. Cadbury Limited manufacturing personnel were reduced from 8565 in 1978 to 4508 in 1985. Over the same period the number of products was cut from 60 to 32 (down to only 29 in 1983) and production lines from 142 to 52. From the start of the Long Range Plan in 1978 to 1985 there was an overall productivity gain of 75%. The long-established factories at Bournville and Somerdale were massively reconstructed. There were also considerable changes in work organization. Continuous shiftworking was introduced, a significant measure of flexibility implemented and many activities subcontracted out (including card box manufacture and multi-packaging, and much engineering support). The line managerial hierarchy was simplified as part of a fundamental reorganization of the management structure, including engineering, which integrated and decentralized activities around the two main product areas at Bournville—assortments and moulded chocolate. The Factory Council, which had been a highly publicized cornerstone of the Cadbury tradition since 1918, was abolished.

The main elements of discontinuity with past practice contained within these events were:

(1) A shift to a short-term profitability orientation;
(2) Labour elimination as a proclaimed objective;
(3) The substitution of managerially-initiated communication for partici-
    pation;
(4) An attack on functional differentiation at all levels of the organization
    and as between production and service functions;
(5) A simplification of organization structure and its focus onto the two
    core product areas;
(6) A transition from batch to continuous flow production aided by
    electronic process control and automation;
(7) Continuous shift working;
(8) An attack on traditional rigidities in labour deployment.

The mode and pace of change especially since 1978 have themselves been in sharp contrast with the past. The hardening of managerial attitudes and tactics in negotiation was much out of keeping with the Cadbury tradition although softened by the offer of high compensatory material benefits. The reduction of product and production lines after 1978 was also perceived by many participants as a further discontinuity, in contrast to the period of brand proliferation during the early 1970s. However, earlier precedents for this policy have been noted and it should be apparent that certain other aspects of the changes actually represent a return to earlier Cadbury policy—especially the commitment to substantial factory investment and the emphasis on the competitive appeal of established 'good value' brands.

There is ample evidence that the changes have been perceived by Cadbury employees as a major discontinuity. Attitude surveys conducted for the company by Imperial College in the early 1970s, after the merger with Schweppes and the first wave of redundancies, indicated that a move away from former Cadbury practices was perceived. Interviews with Bournville foremen in the mid-1970s showed that they shared the same perception (Child and Partridge, 1982). Interviews with all levels of employees conducted by other Aston University researchers in 1986 indicated that the programme of changes since 1978 was judged across the board to be a movement away from the old Cadburyism, a finding supported by interviews recorded for a recent TV documentary on the company's transformation (Central TV, 1985).

THE PROCESS OF TRANSFORMATION

Organizational transformation entails both an intellectual or cognitive re-framing and a material structural change. The process of transformation at Cadbury passed through several stages, of which its realization since 1978 represents only the visible manifestation. Many of the concepts on which it drew were already articulated by the mid-1960s in 'Project Ambridge' and in respect of capital intensification and mechanization had entered Cadbury thinking forty years earlier still. A tangible application of those concepts and a successful break with embedded Bournville practice, the Chirk factory, was operational by the end of 1969. The diffusion of that new conceptual application from an outlying plant to the centre of institutionalized traditional practice was subsequently resisted. It needed to be triggered by new economic pressures which brought the advocates of change into prominent positions and legitimized their claims of urgency.

Once adopted into the mainstream of company intentions, the new concepts were formalized into a plan and granted the hard currency of investment, which became the substantive vehicle for many of the elements in transformation such as employment reduction, flexible work organization and continuous shift working. For its actual application, the investment had to pass through another stage in the process, contained in the work of new plant project teams. These teams drew their design constructs both from precedent within Cadbury and from the external sector network, including team members' knowledge of other confectionery plants and inputs from equipment suppliers and consulting engineers. Their work was also strongly structured by key productivity and financial targets laid down by senior management. They in their turn established new product and process parameters for the subsequent process of work organization change.

Transformation thus proceeds through different phases to which there are not necessarily clearly defined beginnings and ends. Whilst in relative terms it may be correct to suggest that firms move between periods of stability punctuated by transformations (Miller and Friesen, 1984), the intense phases of transformation cannot be so sharply delimited from either their genesis or their legacy. In Cadburys, just as in ICI (Pettigrew, 1985), it took decades rather than years for the process to be achieved. This is partly because it transcends many levels and both the cognitive and political linkages between those levels must be active if the process is not to stall.

The origins of transformation in the case of the chocolate confectionery sector lay in the intensification of oligopolistic competition within a saturated home market, which was furthered by the growing dominance of large retailers. It became increasingly important to support key brands in terms of advertising and value. This placed a premium on a more effective marketing-manufacturing nexus to permit a limited range of products to be produced more efficiently.

In this way, the viable strategic recipe or 'path of survival' for a company within its sector can shift. That shift, however, must first be 'recognized': intellectually and conceptually learned and articulated. A cognitive reconstruction has to be made of sector conditions and their implications. The capacity for accomplishing this may initially reside only with certain organizational actors who are oriented to change by prior experience, close contact with external organizations and probably personality. Cadbury's Managing Director post-1980, Manufacturing Director, Group Technical Director and at a more philosophical level the Chairman, were prominent actors of this kind who were also supported by a number of middle and staff managers, mostly graduates. They provided the channels for a new outlook to infuse into the mainstream of organizational awareness: that is, a linkage between sector and organization.

At this point the process enters the realms of organizational politics and inherent conservatism. The relevance of linkage between levels remains. For example, in a large diversified firm such as Cadbury-Schweppes, organized into separate corporate, divisional and operating unit levels, the impetus for change could have arisen at any of these levels, but required the force of its rationale and perceived urgency to be conveyed to other levels. The political problematics around these vertical linkages required for the acceptance and support of change were illustrated in Cadburys by the management's failure to gain corporate backing for its initial attempt to introduce four-shift working, once this had resulted in the 1979 strike. It is reported that there was a particular lack of empathy over this issue between a key ex-Schweppes corporate executive and the Cadbury managers pressing for change in Bournville.

When the transformation process reaches the stage of specific projects, relations between key change advocates and project team members constitute another critical inter-level linkage. Within the project design process in Cadbury, embedded assumptions and practices on engineering design, staffing and consultation mediated more radical proposals. The composition of project teams was therefore critical, both because of their members' acquired perspectives and because change was facilitated when a project team included a particular 'champion' who enjoyed good links with the corporate hierarchy.

Cadburyism remained a strong ideology in the consciousness of family members, managers and workers. As a benchmark of change it provoked key change agents to set out their case against the fundamentals of the tradition. For example, the Cadbury (later Group) Technical Director twice put up proposals to evacuate production from Bournville, the centre and very essence of the Cadbury tradition. He argued that Bournville was too large and too institutionalized. He claimed that his intention on the second occasion was to spur the two leading family directors into a commitment to radical change. His rationale was one of mobile capital, that ideally a company should change its site of operations every fifteen years or so because 'bad practices' always set in over time. Less far-reaching but equally dramatic frontal challenges to the dominant ideology were made by some younger graduate line managers who argued against long-standing Cadbury arrangements such as a large central engineering function, the Factory Council and top management accessibility to worker representatives.

At the same time, the company was able to invoke elements of the ideology and remould these in different circumstances. The search for precedents for radical changes was apparent among those family managers who, in effect, act as official historians for the company. Thus, Sir Adrian Cadbury has in recent statements attached the family's traditional belief in community to the possibilities of close corporate identity and personal relationships offered by the smaller primary employment core created by the process of employment rationalization and restructuring pursued in his company (e.g. Cadbury, 1983).

In short, a dominant traditional corporate ideology should not necessarily be seen merely as an obstacle to transformation. For it may encourage a clearer articulation of alternatives, the more highly developed it is, and if reshaped or reapplied flexibly it may provide an important legitimating bridge for the transition from one organizational policy/structure configuration to another. If we invoke the Greenwood and

Hinings (1986) analogy of 'tracks' of organizational change, the Cadbury case indicates a transformation process in which the initial movement from inertia requires the stimulus provided by advocacy of a complete directional change, but which subsequently redirects the critical mass of the organization in a tangential direction to which the long-standing frame of meaning can be accommodated. To the extent that change was thus reconciled with organizationally-embedded constraints of continuity, transformation was accorded additional legitimacy.

The process of transformation therefore involved debate: the advocates of change had to pit their visions against more tangible established recipes. Their power, deriving from position and/or family shareholding plus a successful record of achieving change, was a necessary condition for ensuring that their visions prevailed. Adoption was also contingent upon the emergence of a declining competitive position. Acknowledging the part played by power and contingency, a role in Cadbury's transformation must, nevertheless, be accorded to 'vision' itself and the key symbols by which it was articulated. By vision we refer to a clear sense of general direction which is in advance of any specific planning. As the leader of Project Ambridge and later a main architect of reorganization at Bournville said, 'you do have to have an image of the future'.

Cadburys' had increasingly recruited graduates in the post-war period and its management was led by articulate professional careerists together with well-educated family-managers. So it is not surprising that several strategic visions for change were advanced during the 1970s, emanating to a large extent from the different cognitive frames held by those in varying positions (specialties and hierarchical level) and with contrasting pedigrees (Cadbury, Schweppes and recent recruits). Product market visions divided between *product proliferation* (held particularly by marketing specialists and alluding *inter alia* to the precedent set by the Fry's side of the company to resolve its business crisis in the 1930s) and *product reduction* with allusions to the American food manufacturing template. Visions for enhancing productivity divided between work *intensification* with only modest capital investment (mainly advocated by industrial engineers) and labour *elimination* through high capital investment in automation (advocated by higher level engineers, the new systems engineering specialty, and the key agents of change through the post-1978 Long Range Plan). Work organization visions divided between improvement by enlisting co-operation through *participation* (a traditional Cadbury approach), active transfer of *improved practices* from demonstration sites (Chirk) and radical exponents of removal to the *tabulae rasae* of *greenfield sites*. Different implications were drawn from the same preceding projects. The Chirk model was, for example, appropriated by the technical director to represent management's power to redesign fundamentally. By contrast, the manufacturing director drew from it a template for work organization to apply within the company's traditional sites.

There was competition between various advocates of these visions and power blocs as well as alliances across specialisms which incorporated mixed visions. The Long Range Plan of 1978 incorporated an amalgam of the visions of automation/labour elimination, product reduction and Chirk-inspired new working practices. It was therefore somewhat akin in origin and nature to what Abell (1975) termed a collective modified preferred outcome. However, unlike earlier strategic documents such as the 1973 Pink Paper, the 1978 Plan and its derivatives were focused and programmed by the phasing of capital investment. Whereas the Pink Paper had been somewhat of an intellectual think-piece couched in the indulgent ideology of Cadburyism, plans issued after 1978 betrayed a sense of competitive urgency and were organized tightly around mechanistic and unifying slogans.

The force of this new dominant vision, which guided Cadburys' organizational transformation, stemmed from a (sometimes uneasy) alliance of views which appeared to meet the needs of the company's problematic strategic position. Its bearers had come to occupy senior positions. In addition to their positional power, these men also possessed the significant ability to crystallize their visions into a few simple watchwords, such as 'decentralization', 'core businesses', 'rationalization', 'headcount reduction' and 'flexibility'. Such slogans served to focus attention and effort on a limited set of targets, which in the literature of management is widely assumed to be a requirement for effective action and change (cf. Peters and Waterman, 1982). A stated target like reduction of headcount then served as a lever to secure a more complex configuration of changes, such as the stretching and intensification of jobs, flexible team working and simplification of the managerial hierarchy. The translation of these watchwords into specific targets was a powerful, aspirational device. This was later admitted by the company's managing director with respect to the headcount target:

'I laid a lot of emphasis on headcounts—I said the numbers will come down, they are going to come down by 10% per annum. And frankly it was a bit of a ball park figure. It wasn't a very scientific figure but people got it into their heads that it was about 1000 a year . . . and that was pretty well how it turned out.'

This is not to imply that there was a randomness in Cadbury's design for change, but rather that it proceeded from and was driven by relatively simple, clear-cut criteria. They provided an effective basis for achieving and monitoring fundamental transformation.

The prior experience of the major agents of Cadbury's transformation gave them a point of reference, a source of corporate legitimacy, and a determination to stick by their visions, extolling both their relevance and urgency. The technical director, the main advocate of radical change through root-and-branch transfer of sites, had played a major role in the factory closures and rationalizations of production initiated in 1967 and 1971. The Cadbury manufacturing director, with immediate responsibility for the Long Range Plan, had been the Chirk project leader. The Managing Director, who intensified the rate of headcount reduction and work organization changes in the early 1980s, had been responsible for major rationalization in a rapid turnaround of the North American business.

The significant agents of change were family members or had been with Cadburys for most of their working lives. Cadburys had internal change agents who in this way bridged continuity with reform. This contrasts with the emphasis upon the role of external change agents to be found in much of the literature on the subject. It is not, of course, to state that Cadbury's transformation drew upon internal sources only; quite the contrary. In addition to the engagement of outside organizations in its implementation, noted earlier, the company's internal change agents drew important concepts and ideas from a wide range of outside sources. Within the sector network, for example, visits to continental European confectionery manufacturers promoted a realization of the productivity gains available from more efficient working practices, which was claimed to be a direct impetus for mangement's emphasis on work organization reform in recent years.

The key actors were also prepared to use their strategic positions within the company to force major breakthroughs. The group technical director employed his power of veto over new investment proposals to ensure that these incorporated computer-controlled automation. We have noted how the managing director forced through headcount reductions. The manufacturing director was willing to risk the launch of the Wispa Bar to insist on total union acceptance of that plant's new flexible working agreement. While the key change agents were long-term senior players in the company game, they also relied upon more junior agents to front up the realization of key steps in the transformation process. The author of the 1977 Special Report was brought in by the group technical director to articulate the specific principles by which plant rationalization and labour intensification could be realized. He directly confronted those managers who clung to the Cadbury tradition of maintaining employment and achieving change gradually out of a concern for people. Backed by the group technical director, his report fed directly into the Long Range Plan for manufacturing. Similarly, a personnel manager with experience in plant closures was brought in from elsewhere in the company to confront the shop stewards and Factory Council representatives in the process of demolishing the council and introducing new practices such as continuous shift working. Once they had done their job, such agents were quickly moved elsewhere and phases of consolidation began.

Overall, the implementation of change proceeded through a mixture of confrontation and incrementalism. Confrontation, as in the 1979 strike, tested the limits to change. Where those limits were restrictive they set a challenge to the change agents to proceed gradually and informally. The 1979 strike entailed a retreat from introducing continuous shift working. This was introduced more quietly eighteen months later, once a more attractive personnel policy had been devised and a more determined managing director installed, who as a family member was in a better position to ensure corporate support for the change. Flexible working became an issue of formal confrontation in the 1983 Wispa negotiations, but it had previously been introduced informally on other newly commissioned Bournville production lines by management determined to recruit only those workers who were willing to work flexibly. Selection 'tests' identified workers who fitted management's definition of a responsible attitude and who appeared willing to work on a team basis. So by the time that flexible working practices emerged as an issue of formal confrontation, management had already prepared the ground for their introduction.

Despite the radical nature of the changes and management's determination to implement them, the degree of overt conflict that ensued was relatively minor albeit set against a previous history of good industrial relations. The percentages of time lost through industrial action by non-management groups in Cadburys between 1977 and 1982 were: 1977: 2.93 1978: 0.14 1979: 4.51 1980: 0.01 1981: nil 1982: 0.01. The strikes of 1977 and 1979 account for the peaks in those years, but the amount of time lost through industrial action is remarkably small compared, for example, with the loss through sickness absence which remained above 5% throughout the period.

The low incidence of overt conflict during a period of transformation must be ascribed in part to the tradition of good industrial relations and to the balance management struck between taking a firm stand and a attempting to create consent through communications (including videos), joint discussions and low profile incremental adjustment. Probably the most influential component of this personnel policy, however, lay in the provision of material benefits to labour. Movements into new production plants were accompanied by improvements in grading for production workers, referred to as 'grade drift'. Guarantees of fixed overtime for tradesmen during the period of heavy investment were exchanged for the contracting-in of workers to install equipment and erect new buildings. Average overtime doubled in the late 1970s and early 1980s. The introduction of new shift systems reduced total working hours and increased earnings through shift allowances.

Wage rates increased, though not as fast as productivity. Value-added per worker increased by just under one half between 1979 and 1982, while average wages or salaries and pension cost per employee increased by just over one third. Output per employee rose dramatically through the reduction in the number of workers and a greater utilization of the working day, reversing declines in the mid-1970s. The redundancies themselves were mainly 'voluntary', legitimated by retirement of several 'old guard' senior shop stewards with large redundancy payments in the early 1980s. Relative to other firms in the sector, redundancy payments were quite high and two to three times greater than statutory redundancy pay. Such payments were conditional on 'full co-operation in achieving an orderly run-down' in employment during the period of capital investment and were part of a carefully packaged personnel policy designed to assist the transformation.

Central to this policy was a 'Security of Employment Agreement' signed with all unions in July 1981. This institutionalized retraining, transfers and redundancy payments for those displaced by automation and rationalization. The agreement gave management greater control over labour utilization in exchange for job security, but only for those currently in permanent jobs and not comparable to the security that had been traditional in the company up to the first wave of redundancies in the early 1970s. Following this agreement the company increased its numbers of temporary workers who enjoyed no security of employment, no entitlment to redundancy payments, no payment for illness unless through industrial accident and a generally marginal employment position. To some extent, then, the material benefits which played a part in buying workers' acceptance of transformation have been traded at the expense of those available to their successors.

CONCLUSIONS

The Cadbury case suggests the following conclusions:

It is possible to distinguish *objective* sector phenomena which are relevant to understanding the behaviour of firms and more specifically the conditions under which they experience transformation.

There was a co-existence of *strategic models* within the chocolate confectionery sector that persisted until new competitive conditions which emerged in the 1970s forced a shift towards the 'Mars model'. Although some aspects of this realignment had already been undertaken by the company, it was only embarked upon systematically after a marked deterioration in market share and profitability. Nor did Cadbury's transformation involve a complete break with its traditions. In contrast to Mars, for example, it remained unionized, multi-site and continued to produce a range of higher-price brands.

A lengthy process of *re-cognition* preceded transformation. This involved a combination of symbolic and power-relevant activities, namely reframing the definition of relevant contextual conditions and appropriate internal arrangements with the accession into powerful positions of those advocating new interpretations and solutions.

Competing *frames of meaning* and recipes for improvement were advanced by actors whose views were conditioned by their training, specialty and particularly their comfort with technological change and previous experience in the company.

The traditional and hitherto dominant *corporate ideology* was not simply a barrier to transformation. Rather it provided a clear position against which the case for change had to be developed. The ideology was itself malleable, stressing receptivity to new techniques and intellectual enquiry. Moreover, the ideology derived partly from the Cadbury owner-managers who returned after the mid-1970s to the leading corporate and subsidiary positions. This legitimation assisted in the implementation of change.

The process of transformation passed through several *stages* and did not exhibit a clear beginning or end. It depended upon the import of knowledge, particularly of new technological opportunities, through the sector network and it also required linkages between units at different levels to align higher corporate backing for the implementation of specific transformational projects.

The formulation of key *concepts* and *symbols* made an important contribution to the activation of change. These were specified into leading targets for change upon which managerial drive could be focused.

Change also relied on the *managerial power* of key agents. In addition to their *position* power, their *legitimacy* in a period of crisis was strengthened by their previous record of managing change successfully in other areas of the company, albeit where forces resistant to change had been less entrenched (greenfield sites and the recently acquired North American subsidiaries).

REFERENCES

Abell, P. (1975). *Organizations as Bargaining and Influence Systems*, Heinemann, London.
Abernathy, W.J., Clarke, K.B., and Kantrow, A.M. (1983). *Industrial Renaissance*, Basic Books, New York.
Abernathy, W.J., and Utterback, J. (1975). 'Dynamic model of product and process innovation', *Omega*, 3, 639–657.
Aglietta, M. (1979). *A Theory of Capitalist Regulation*, New Left Books, London.

Aldrich, H., McKelvey, B., and Ulrich, D. (1984). 'Design strategies from the population perspective', *Journal of Management*, **10**, 67–86.

Bowman, E.H. (1985). 'Generalizing about strategic change', in J.M. Pennings, and Associates, *Organizational Strategy and Change*, Jossey-Bass, San Francisco.

Cadbury, Sir A. (1983). 'Cadbury-Schweppes: more than chocolate and tonic', *Harvard Business Review*, Jan.-Feb., 134–44.

Cadbury, E. (1912). *Experiments in Industrial Organization*, Longmans, Green, London.

Cadbury Brothers Ltd (1947). *Industrial Record 1919–1939*, Cadbury Brothers Limited, Bournville.

Cadbury Brothers Ltd (1964). *Industrial Challenge*, Bournville, Cadbury Brothers Limited.

Central T.V. (1985). *'Venture' Programme; Report on Cadbury Limited* by Gareth Jones, 1 April.

Child, J. (1964). 'Quaker employers and industrial relations', *Sociological Review*, **12**, 293–315.

Child, J. (1972). 'Organization structure, environment and performance: the role of strategic choice', *Sociology*, **6**, 1–22.

Child, J. and Partridge, B.E. (1982). *Lost Managers*, Cambridge University Press, Cambridge.

Feldman, S.P. (1986). 'Management in context: an essay on the relevance of culture to the understanding of organizational change', *Journal of Management Studies*, **23**, 587–607.

Frost, P.J., Moore, L.F., Louis, M.R., Lundberg, C.C., and Martin, J. (1985). *Organizational Culture*, Sage, Beverly Hills.

Greenwood, R., and Hinings, C.R. (1986). 'Organizational design types, tracks and the dynamics of change', *Working Paper, Department of Organizational Analysis, University of Alberta*. February.

Kimberly, J.R. (1980). 'The life cycle analogy and the study of organizations', in J.R. Kimberly, R.H. Miles, and Associates, *The Organizational Life Cycle*, Jossey-Bass, San Francisco.

Lawrence, P.R., and Dyer, D. (1983). *Renewing American Industry*, Free Press, New York.

Miles, R.E., and Snow, C.C. (1978). *Organizational Strategy, Structure and Process*, McGraw-Hill, New York.

Miles, R.H. (1980). 'Findings and implications of organizational life cycle research', in J.R. Kimberly, R.H. Miles, and Associates, *The Organizational Life Cycle*, Jossey-Bass, New York.

Miller, D., and Friesen, P.H. (1984). *Organizations: A Quantum View*, Prentice-Hall, Englewood Cliffs.

Palloix, C. (1976). 'The labour process: from Fordism to Neo-Fordism', in Conference of Socialist Economists (eds), *The Labour Process and Class Strategies*, Stage One, London.

Pavitt, K. (1984). 'Sectoral patterns of technical change: towards a taxonomy and a theory', *Research Policy*, **13**, 343–73.

Peters, T.J., and Waterman, R.H. Jr. (1982). *In Search of Excellence*, Harper & Row, New York.

Pettigrew, A.M. (1985). *The Awakening Giant*, Blackwell, Oxford.

Piore, M.J., and Sabel, C.F. (1984). *The Second Industrial Divide*, Basic Books, New York.

Richardson, G.B. (1972). 'The organisation of industry', *Economic Journal*, **82**, 883–96.

Rowlinson, M. (1986). 'The Bournville factory: a greenfield development', unpublished paper, Work Organization Research Centre, Aston University.

Scherer, F.M. (1980). *Industrial Market Structure and Economic Performance*, Second edition, Rand McNally, Chicago.

Smircich, L., and Stubbart, C. (1985). 'Strategic management in an enacted world', *Academy of Management Review*, **10**, 724-36.

Smith, C., Child, J., and Rowlinson, M. (1990). *Innovations in Work Organization: Cadbury Limited 1900–1985*, Cambridge University Press, Cambridge (forthcoming).

Spender, J-C. (1980). Strategy-making in Business, unpublished Ph.D. Thesis, University of Manchester.

Stinchcombe, A.L. (1965). 'Social structure and organizations", in J.G. March, (ed.), *Handbook of Organizations*, Rand McNally, Chicago.

Von Hippel, E. (1982). 'Appropriability of innovation benefit as a predictor of the source of innovation', *Research Policy*, **11**, 95–115.

Weick, K.E. (1969). *The Social Psychology of Organizing*, Addison-Wesley, Reading, Mass.

Whitaker, A. (1982). *People, Tasks and Technology: a Study in Consensus*, Cadbury Limited, Bournville.

15

Incremental Innovation and Appropriative Learning Styles in Direct Services

Ray Loveridge

Aston University

Implicitly or explicitly much of current literature on the strategic impact of new technology accepts that information technology (IT) has brought about a generic change in both manufacturing processes and service delivery. Some writers have gone so far as to suggest that its effects on either one are such as to bring both closer together in a manner that blurs traditional boundaries between manual and non-manual occupations (Toffler, 1980). However, this metamorphosis is generally seen to be brought about by the ascendancy of service employment over that in manufacturing as the latter is reduced by automation. The importance of knowledge-based occupations is seen to be increasing as manual effort declines in value. More especially the value of knowledge provided in direct services is believed to be growing in significance (Bell, 1974; Kumar, 1978).

This interpretation is an optimistic one. Others suggest a 'self-service' society is a more likely outcome of technological change, in which consumers will come to depend on automated services. The growing

The Strategic Management of Technological Innovation. Edited by R. Loveridge and M. Pitt
ⓒ1990 by John Wiley & Sons Ltd

demand for services is seen to allow their 'commodification' in order to bring them to a mass market (Braverman, 1974). This self provision from a range of pre-programmed service options would reduce the direct involvement of knowledge workers and, eventually, the overall numbers involved in the design of the software required to store and analyse information. By contrast it would require much more from the consumer by way of collaboration in self-servicing (Lovelock and Young, 1979) and in the development of a domestic self-sufficiency (Gershuny and Miles, 1983). Thus the producers of 'self-tech' or easy to use 'intelligent' machines might well gain greater strategic influence within society at the expense of the former providers of professional services (Coombes and Green, 1989).

This debate is paralleled by conflicting interpretations of present trends in patterns of work organization. On the one hand the manner in which tasks which were formerly specialized by skill or function can be brought together in flexible bundles has been seen as providing the basis for a new form of craft-autonomy (Handy, 1984; Sabel, 1982). When complemented by the ability of strategic management to monitor operations without the intercession of layers of middle-management, this can be interpreted as offering an opportunity for dialogue and reduction in conflicts arising from lack of communication (Child, 1984). On the other hand the type of devolved control over operation that appears often to accompany the introduction of information technology has been interpreted as a means of fragmenting the work force and the consumer market in a manner that reinforces the central control of strategic managers. The ability to monitor both worker task performance and consumer reaction by means of IT allows a more pervasive central executive influence on both than was previously possible through the use of overt bureaucratic regulation (Pollert, 1988; Watanabe, 1988).

Like any tool IT offers few constraints that have not been designed into its operations. It has been argued by Gustavsen and Hetty (1985) that IT has removed the necessity for many of the administrative and logistical constraints left by past eras of work organization. The flexibility implies a recognition of the potential to 'enact' the socio-technical possibilities contained in the intrinsic nature of the new tool and its associated 'know-how'. It also implies a wide agenda for the enactment of conflicts between new and old vested interests within organizations. Child, Ganter and Kieser (1987) suggest that the intrinsic conservatism present in organizational activities has militated against an exploration of these possibilities. In this chapter it is proposed that many academic studies of technological innovation neglect the strategic complexity that attends periods of rapid transition simultaneously in several areas of business activity.

Decisions to invest in IT seem often to have been taken with little managerial comprehension of the technical and social implications for the business at large (Caulkin, 1989; Child and Loveridge, 1990). These implications are often discovered through a process of pragmatic learning which begins from conceiving IT in relation to a 'localized or *limited*' opportunity and moving outwards to its systemic meanings. This stumbling and revelatory form of operational incrementalism (Lindblom, 1959) gives way to a developing strategic logic in systemic applications of IT. This later, logical incrementalism (Quinn, 1978) has however been shaped by earlier 'problem' applications and, often, by the crises that triggered the search for earlier IT solutions. Hence a consideration of creative modes of IT deployment may have been effectively eradicated from the strategic agenda.

The evidence used in this chapter is drawn from a six nation study undertaken by the author and colleagues in all of these countries but focuses on the United Kingdom in detailed instances (Child and Loveridge, 1990). The intention of the study was to describe the effects of micro-electronics in three sectors; those of retailing, banking and hospital health care, at both organizational and inter-organizational levels.

Conjoining markets and technologies

The transformation in the business environment that has taken place over the last twenty-five years has been described as the confluence of two separate streams of development (Marshall, 1978; Freeman; 1988). One stream of events consisted of a series of market crises. The mass markets for standardized commodities were perceived by inventors to have reached a level of saturation towards the end of the 1960s. An increase in basic commodity prices in 1973–4 caused by an alliance of oil producers is seen to have triggered an already imminent withdrawal of capital from manufacturing. For almost another decade competition between Western producers was pursued even more intensely as the level of economic activity increased once more, using techniques appropriate to mass markets of the 1950s. By 1982 the better adapted manufacturers of South East Asian countries had made such inroads into global markets that a second crisis of confidence occurred with Western capital markets. The effects were particularly severe in the U.S.A and the U.K because of the stringent monetarist policies adopted by their governments. Since that time managements have sought to deal with the 'chaos' resulting from the disappearance of familiar domains and boundaries through their wide-scale adoption of lean, flat and flexible organizations modelled on selected features of Japanese manufacturers. The emergence of IT over the period of the latter crises has enabled the further development of loose-linked work-processes that may be geographically and administratively remote from their strategic orchestrate. This phenomenon is sometimes portrayed as constituting a well articulated paradigm or archetype of organization on a par with that enunciated by Henry Ford (Perez, 1985).

This widely accepted view of the confluence of market need and technological supply might, perhaps, be seen to be based on an optimism stemming from the relative versatility, the compactness and the capacity of the micro-circuit, or more particularly of the micro-processor. Taken with the cheapness of its supply the micro-processor was seen to provide the raw material for an irresistible revolution in the design of whole socio-technical systems. While not doubting the fundamental impact of IT its development can now be seen to be taking a variety of paths marked by varying degrees of complementarity or conflict between the aspirations and motives of both designers and users. The fusion or convergence of the computing and telecommunication elements comprising any IT system had been forecast for over a quarter of a century. Its 'delayed' emergence is, perhaps, a product of both the misplaced optimism of early technology forecasters fuelled by the early success of military applications of IT and of the more uncertain conditions accompanying its commercial applications. Until recently interfaces between system elements have been technically problematic, expensive to bring about (relative to the cost of individual items) and involving a high degree of operational risk. Pioneering suppliers were themselves undertaking extremely uncertain commitments in designing and servicing equipment and software that had, almost by definition, to be both specific to users' needs (with which they were unfamiliar) and to makers' use of equipment components whose design changed every few months. It was noticeable that the largest manufacturers such as IBM were, and are, often second, or even third, into new IT markets.

In these circumstances normal accountancy bench marks of cost/performance ratio and risk assessment were found to be relatively useless by user-innovators, including those in our study, except as political devices to be used by the sponsors of a new IT investment in convincing politically important others and justifying their determination to innovate (see also Rothwell, 1977). Furthermore the costs of implementation, maintenance and adaptation of new equipment has remained high as a result of lack of trained labour at all levels of operation, programming and analysis. Much of this problem might be seen as relating to the reluctance of users to rely on off-the-shelf packages and to insist on conveying the supposed idiosyncrasies of their own organizational operations into idiosyncratic software, and even hardware, configurations (Brady, 1989).

While this long drawn-out search for common standards might be regarded as a normal characteristic of an immature market it must also be seen to involve a higher level of innovative uncertainty than anything previously experienced. On the one hand computing machines and the use of accompanying software are nowhere as explicitly concrete in operation or accessibility as the socio-mechanical devices they replace. More importantly the procedures followed in computer programs most often represent substitutes for those previously administered through a hierarchy of social roles. At the extreme they may, as suggested by Braverman (1974), represent an automation of bureaucracy. The failures that have followed the adoption of advanced manufacturing technology by many Western firms have been attributed to the unsystematic nature of their previously existing modes of social organization and the failure of these systems to comply with the logic built into the new tools (Caulkin, 1989). Hence the perceived need for idiosyncratic programs and the specific knowledge of existing operational norms required of programmers and analysts.

Crisis and innovation

Barras (1986) has described the adoption of IT in the service sector as representing a form of 'reverse innovation'. This term implies that unlike manufacturing innovation, which is generally led by a new product design, executives in the service sector have applied IT to 'back-office' or indirect processes and later extended them through offering a similarly automated service to customers in the 'front-office'. Examples might be seen in what Rajan (1987) has described as the second wave of banking automation involving the introduction of cash dispensers or ATMs (Automatic Telling Machines) and other customer activated terminals in high street branches, lobbies or shopping centres in the late 1970s and 1980s. The first wave of mainframe computers had been introduced by the mid-1960s in order to process routine flows of deposits and loans. These, in due course, became passed to head office from the high street branches by way of on-line terminals. This in turn led to the accessing of terminals by the clients themselves either through ATMs or through home-based computer terminals.

One of the major reasons given for the introduction of mainframe computing in the late 1960s was the escalating scale of administrative costs as the level of indirect clerical and technical staff numbers increased more than proportionately to those of direct workers. This continued to be so in the 1970s but in an increasingly crisis-ridden atmosphere as product demand flattened or shifted drastically into new and unknown areas. The items occupying many strategic agendas were therefore those concerning the threat of take-over, merger and rationalization as rival organizations sought to survive through product diversification and keen price competition. Some observers have described this process as marking the late 'maturation' of an activity—the provision of personal services—that had been characterized by the survival of small, often family-controlled, traditional modes of selling and buying (Fuchs, 1968; Miles, 1987).

The formerly institutionalized boundaries between different types of direct service were in one way or another often endorsed by State regulation. In retailing for instance, small owner-run outlets were supported by price maintenance legislation which constrained large chains in taking advantage of economies of scale available to them; restrictions on opening laws had the same effect and even, in Italy for instance, ownership of more than one store within a given area was forbidden or, as in Belgium, limits were placed on the size of any one outlet. In banking State recognition derived, and derives, largely through authorized dealings with the central bank. But again area restrictions can be imposed (as in Italy and the U.S.A) and different forms of banking, for example commercial, trustee savings, building societies or cooperatives can be dealt with in quite different ways within tax laws etc. Only in the third sector, that of medicine, does the formal requirement for personal licensing for practice exist in all industrialized countries. The effect of this individualized system of State recognition will be explored later in the chapter but we may note in passing that fragmented or polyarchic nature of the multiple hierarchies through which health care was provided across all six European countries in our study.

Most organizational and technological innovation in these sectors tended to owe their origins to U.S practice and generally the diffusion to Europe was extremely lagged. Supermarkets, for instance, were pioneered in California by Safeway in the 1920s, but it was not until the 1950s that Tesco 'piled it high and sold it cheap'. In the United Kingdom even after the repeal of the Restrictive Price Maintenance Act in 1954, this mode of selling took a decade to become significant in the U.K. and longer elsewhere in Europe. Any explanation has to take into account the relative size of the market in the U.S.A and the newness and mobility of much of its consuming population. Not only were economies of scale obtainable but the consumer was encouraged to expect a standard service in all parts of that continent. For the far less mobile European consumer, expectations were more locally idiosyncratic. By the late 1950s, however, both geographical and social boundaries to consumer markets in Europe were being recognized around new and higher standards of mass consumption. Within a decade a wider and more variable range of products was offered across all income brackets, including the lowest, and, more particularly, those catering for young consumers.

This movement in tastes coincided with a gradual erosion of boundaries in all three sectors of personal service provision. These boundaries often carried a morally normative significance not only in respect to the consumer but also in the respect shown for the status of other traders or providers of specialist services by those in the field. Once weakened, voluntary constraints on trading provided little protection against the vicissitudes of the recession that followed the rise in commodity prices in 1974. Cross-product competition that had begun in a period of economic growth now became the norm in finance and some modes of distribution and health care. At the same time control over the value chain (or in the case of banks, the circuit of credit creation) became a complementary goal. Providers of final services became aware that their ability to supply a wide and rapidly varying range of products crucially depended on the control they exercised over their own sub-contractors. It would be simplistic to believe that they had, hitherto, been unaware of this fact of strategic life, but as Senker's chapter in this volume demonstrates, the nature of these dependencies had grown more complex, not only in the procedural forms required to encompass the variability in trading conditions of the 1970s but also in the technicality embodied in the substantive exchange of goods and services.

IT was portrayed by its designers and in-house champions as an aid to achieving just such control over the value-chain and as a means of cross-trading over new and complex product portfolios. It must be emphasized that, in itself, IT did not normally act to destroy the core competences of direct service providers or their markets in the all-embracing manner suggested by the transcillience mapping developed by Clarke (1983) and by Tushman and Anderson (1986). Rather it provided a means— a potentially revolutionary means—whereby the strategic environment might be reconfigured to secure the continued security and growth of the organization as defined by corporate management. This was seen largely in terms of the appropriation of areas of competence that had previously been regarded as complementary to those of their own organization or as valued inputs to its activities. It was, then, the shift in perceived market boundaries brought about by crises of confidence that provided the key ingredient for the development of an appropriate logic for strategic action. Often the withdrawal of State recognition and of their previous legitimation of these boundaries was both an enabling factor and trigger to predatory action.

For this reason we ought to distinguish the manner in which new technology *enhanced* strategic competences from the manner in which it enhanced operational performance (see Tushman and Anderson, 1986). In the latter, operational applications, strategic consent for investment in new technology was often based on a mixture of pragmaticism and internal politics of the kind traced in Pettigrew's (1973) classical study of the computerization of a retail chain between 1955 and the late 1960s. The author highlights the role of the O.M. manager (later systems analysis head) as a bridge between operational and strategic management in both facilitating the innovation and his own personal rise to authority in the organization. Members of the board are invited to presentations by computer suppliers but do not attend. The reasons seem to relate to both their own technical incomprehension, and resultant inhibition, combined with an awareness of threat to their hierarchical role and status. Later these feelings are assuaged, in part by a growing comprehension of the technology, but also because its operational implications become assigned to a senior member of the organization and are therefore capable of being handled in social interaction between peers. Emergent skills become formally evaluated and assigned job titles and hierarchical positions in the organization. In doing so they may take the place of tasks hitherto carried out by non-specialists who performed the transitory enabling roles. This process was approximated in several histories of the organizations studies in our study.

The development of a strategic logic may be precipitated by a desire on the part of the chief executive to be a pioneer or first mover in a systemic or radical introduction of an IT network such as that which often accompanies Epos. This intention appears to be often frustrated by unanticipated snags, hold-ups and lack of fit between the pre-prepared system and that required for effective operations. The most evident successes could be found in the 1970s among firms such as Citicorp and the Belgian retailing group GB-Inno who appeared able to marry totally innovative products such as ATMs, debit cards and smart cards with the long-term planning necessary for internal organizational change. The major obstacles to the successful integration of operational and strategic logics seemed often to rest in; (i) the incomprehension shown by strategic management towards the technological logics and languages of IT complemented by a closure of these disciplines around the early success of skills based on mainframe computing, (ii) the unwillingness of chief executives within competing firms to collaborate in the provision of a collectively provided infrastructure that would enable the more rapid realization of its private benefits. In the following paragraphs a brief description is attempted of this strategic development within the three sectors under analysis.

BANKING

Crisis and the origins of strategic logic

The status of the largest Clearing Banks derives from their ability to borrow from, and lend to, consumers and manufacturers on a short-term basis but, in aggregate, on a scale sufficient to affect the economy at large (Frankel, 1977). In the past they have distinguished themselves from other points on the credit-creating circuit such as building societies or trustee savings banks which borrowed and lent long to consumers, and from industrial, venture or merchant banking which provided these services for entrepreneurs and commerce. They had in Britain, up until the last 20 years, been only loosely connected to the creation of industrial credit through the issuing of stocks and shares and through the issuing of bonds and securities. Insurance and life assurance had also been regarded as the province of specialist (McCrae and Caincross, 1973) underwriters.

Over a period of economic decline during which the multinational manufacturer developed its own expert treasury management, clearing banks sought to compensate for declining profits on short-term commercial loans by entering all forms of consumer credit creation, whilst at the same time buying into merchant banking and securities trading (Strange, 1986). In both areas of credit creation business expanded rapidly in the mid-1980s as did the number of external contenders at all points of the circuit or credit chain. In the U.S.A., for example, Sears Roebuck, the mail-order business, cashed more cheques than any American bank while plastic credit cards issued by retailers of all kinds rivalled those of banks (Naisbitt, 1984).

By the mid-1980s most large British banks were able to offer a complete range of financial services through wholly or partially owned subsidiaries, whilst competing with the remaining specialist agencies in all areas. Much of this ability for cross-trading, as well as the enormous increase in the velocity of financial transactions, and therefore of credit-creating business, can be attributed to the manner in which IT has come to be the basis for the provision of these services. Financial services have been the largest user of IT since the early years of mainframe computing. This seems likely to be so until well into the 1990s with current IT cost at around 20% of all non-interest costs in U.K. finance, and rising (*Financial Times*, 10 November 1988). The maturity of the sector is reflected in the growing use of standardized software packages among the largest users, that is among those who also maintain large Management Services departments, or the capacity or produce their own software. The evolution of a range of configurations of service provision in retail branch outlets is also becoming clear (Channon, 1986). In the densely populated U.K. the impact of IT-based home banking has not been great. In spite of this the number of high street branches has shrunk by a third in little more than a decade, but bank employment actually increased over the same period up to 1984. Expanding markets for financial services across all social classes of consumer accounted for much of this increase.

The emergence of the infrastructure

Much the same expansion in business can be seen to have occurred over the early 1980s after the deregulation of security, commodity and currency markets. The development of IT as the primary medium for the globalization of trading combined with the movement toward 24-hour business across the world's main exchanges accounts for much of the increased activity. Of itself IT can be seen to have increased the velocity of transactions to an extent that has significantly increased global credit creation and therefore the availability of profit-taking for traders (Strange, 1986). An additional facilitating factor has been the complementarities that have long existed between the news and communications media and market traders. Thus firms such as Reuter, that owed its origins to the conveyance of news vital to market traders in the last century, have discovered a new role as suppliers of software and designers of dedicated IT systems for the analysis and monitoring of world markets.

In this regard the existence of third party agencies has, perhaps, facilitated more rapid change than in the more insulated development of inter-organizational networks between clearing banks. Although SWIFT (Society for World-wide Interbank Financial Telecommunications) has been operating with growing success between European and American banks since 1973, the movement towards cheque truncation, or the dispensation of paper exchanges for credit transfers between banks, has been much more difficult to accomplish. This is also true of the universal acceptance of customer access cards by the ATMs of competing banks and of the introduction of the automatic debiting of customer accounts in retail outlets (Electronic Funds Transfer at Points of Sale or Eftpos). The reasons are partly technical in so far as progress has been geared to the availability of the evolving elements of the IT systems. But more important constraints have derived from the inability of banking organizations within a purely instrumental coalition of interests to arrive at a price for the individual investment at which the collective gain would prove acceptable to each one of them. This has evidently been made more complex by the varying degrees to which competitors have felt it necessary to pass the benefits on to consumers through price/product competition.

The discovery of new operational logic

A second aspect of the systemization of consumer or high street banking is of course its increasingly mass marketing–mass production characteristics. Hence while a high degree of trust remains the implicit basis for spot trading in securities, currencies and commodities on the world exchanges (Loveridge, 1983), this can no longer be so of the plastic card consumer market-place. In both cases IT has provided a means to both increasing the velocity of transactions and managing them on the basis of a calculation of the aggregative risk associated with a large number of similar transactions. Hence the central executive *need* not depend on the personal judgement of the high street branch manager or the securities trader in each individual case in laying down standard procedures to be followed for creating credit or making investments on a large scale. Such decisions can, and in varying degrees are, now made with the assistance of, algorithmically programmed 'expert' systems. While there is little evidence that these are used without judgemental inputs on the part of the manager operator it is evident that for the latter to act against the advice of the computer shifts the balance of justification in decision making quite considerably.

At lower levels of the hierarchy the movement towards staff rotation across a number of functions is now normal practice (Crompton and Jones, 1984; Child and Tarbuck, 1985). The effect of IT has probably been to reinforce the gender-dictated duality of career structures available to men and women in the industry while offering women greater opportunities for short-range promotion to 'personal bankers' (salespersons for financial services) or to now partially de-skilled branch manager positions. In some ways the femininization of these jobs represents the change in role of the high street banker from that of offering a professionally neutral service to the would-be saver-investor to that of the seller of the products on which (s)he once offered advice.

By the end of the decade the State has stepped in to regulate the terms on which advice can be offered in Britain (Financial Services Act 1988). By this stage the self-regulating groups of traders operating at each sequence in the credit circuit were becoming compromised in monitoring and sanctioning their members by the presence of major 'players' on both sides of the market simultaneously. National central banks, like the Bank of England, ceased to be able to use such bodies, including more recent additions like the voluntary Take-over Panel of 1973, to adjudicate their own affairs on behalf of 'the community' or consumer. Central licensing systems had come to be enforced by the State and criminal prosecutions had been instigated by Government departments both in Europe and in the U.S.A against traders who appeared to have acted against their client's interests. This led to a paradoxical situation in which Governments pursuing a 'free-market' or 'supply-driven' policy of economic control were forced to enact and impose increased State intervention in financial transactions (Strange 1986).

RETAILING

Crisis and the origins of strategic logic

This sector is in fact the largest single employer in every developed nation. Yet its history during the last quarter century is one of rationalization and concentration of employment with real reductions of up to a third in the

U.K. and Europe generally. Productivity has increased to a point where leading British supermarket chains achieve increasing returns on capital investment from a return on sales of around only 2% per annum. The fastest growth in sales has in fact been achieved in consumer durables such as electronic household goods rather than in food and drinks. Here, as in groceries, sales expansion has been at the expense of owner-operated stores which, in the U.K., declined by 41% between 1970 and 1985.

The success of multiple chains over this period can be attributed to a number of factors. Economies of scale have been obtained from the increased use of operations research in logistics management and the use of computers in stock control and trunking (transportation) management. Advancement in food preservatives and pre-prepared foods (see Senker in this volume) had been matched by the spread of car ownership and the use of refrigerators, freezers etc. in the home. The use of cars for shopping has also boosted sales of household furniture and machines and led to the movement of hypermarkets, superstores and shopping malls out of city centres. Lastly, as already mentioned, the rules of the game that formerly protected small traders have been relaxed by most European Governments.

The relationship between retailing, or more properly, distribution, and manufacturing has always been fraught. The latter depends on the former for information on tastes and on concomitant shifts in design. By the same token distributors cannot sell goods that do not match customer tastes and have to buy forward in anticipation of these tastes. In periods of rapid changes in tastes these inter-dependencies became critical. Scarcely surprisingly therefore the emergence of branded goods in the late 19th century was accompanied by the forward integration of many manufacturers into distribution. This was especially the case in industries using continuous flow techniques (foods) or large batch production (men's clothes). The break-up and segmentation of these markets in the late 1960s brought crises to these sectors which sometimes resulted in the abandonment of directly owned production facilities in favour of the use of sub-contractors in supplying a range of niche directed outlets with so-called 'life-style' orientations. The classical example is that of Burton's move out of the mass production of men's suits in their Bradford factories into the employment of sub-contractors working to centrally produced designs from small workshops located anywhere between Birmingham and Hong Kong (Johnson, 1987).

This mode of controlling production through long-term sub-contracts has in fact been used by the clothes and household goods chain of Marks & Spencer since the early 1920s. But the Italian family firm of Benetton is an oft-cited example of sub-contracted retailing in the so-called post-industrial society (Clark and Starkey, 1988; Flowers, 1989). Formed by a brother and sister in 1965, Benetton was at first based on cottage industry production of knitwear designed by the founders. Its use of information networks to pass sales data from shops all over Europe (later globally) to its Villorba headquarters at the heart of its village based production facility has been seen as crucial to its early success. Subsequently the firm has itself established factories in several global locations and has divided its world market into 70 areas, each controlled by an agent responsible for communicating stock/sales data by personal computer along a General Electric designed information network to Italy and thence to factories in various parts of the world. Goods are made up and delivered to a 'just-in-time' system which relies heavily on an electronic sensor-controlled mode of monitoring progress.

In so far as the firm now owns many of its plants and has coupled supply and distribution so tightly it might be said to resemble a 19th as much as a 21st century value chain. However, it has also moved heavily into franchising its retail outlets. Quite clearly Benetton has shifted the balance of its risk absorption towards the development of a flexible system of integrated (internalized) manufacture conjoined with the externalization of the risks of distribution through the use of time constrained contracts (franchises). Rather than typifying either Marks & Spencer or Benetton as 'post-industrial' we might more prosaically describe both as endeavouring to obtain strategic control over their supply chain in a manner that enables them to retain the essential designer-innovator knowledge for themselves rather than being reliant on autonomous producers. Since the Benetton family *were* originally the designer-innovators one might expect them to be more concerned with direct control over the production of their designs once these had passed beyond the immediate boundaries of their community based craft-producers than were high street traders like Marks & Spencer (see Loveridge in this volume).

The emergence of the infrastructure

The creation of IT systems and more particularly, electronic-points-of-sale (Epos) offered distributors the opportunity to use their market position, and more particularly the information that it provided, in order to extend their influence over production. As with other innovations Epos derived from the USA, or more particularly from the collaboration of American retailers with IBM in transferring the lessons learnt in the electronic scanning of item coding by the National Aerospace Agency from these uses into the more mundane field of consumer selling. By capturing data in the scanning of bar codes as purchased items passed over or under a laser beam, at the point of sale, management were able to record, monitor and analyse transactions with a degree of accuracy not afforded by manually activated cash registers. By further conjoining Epos with other electronically activated sources of information along the sales chain from the despatch of goods from the producer the complete operations cycle could be monitored and automatic reordering undertaken on the basis of sales analysis performed by the master computer.

The diffusion of Epos in Europe has, in fact, been slow since its introduction in the middle (Belgium, Italy) to late (U.K., Germany) 1970s. The original reason was lack of agreement among European retailers and manufacturers on the form of bar-coding to be used. This meant that pioneers had to undertake their own printing of magnetic labels or the printing of bar-codes directly on to the item within each store or warehouse. By 1984 the number of manufacturers using bar-codes was predicted to bring Epos into the realm of commercial viability for all users. This critical proportion was in fact not reached for another two years or more, this despite the existence of a growing collaboration between major users in their support for the Article Numbering Association and consultancy agencies such as Tradacom and Tradanet.

As this process gathers pace the negotiations around the introduction of Eftpos introduces a new and complicating factor. Not only does the exaction of direct credit from the consumer on behalf of the clearing banks present technical and cost problems for distributors but also for large retail chains it provides a challenge to their use of their own in-store credit cards. Earlier we mentioned that these cards have come to be a major source of income for retailers in some countries, particularly in the U.S.A, France and Belgium. In the latter two countries personal 'smart cards' are widely used by big retail chains to record purchases and the credit availabilities for each customer. These advances by retailers into the areas of credit creation dominated by banks will not easily be relinquished in order to make way for Eftpos. However in France, as in Belgium, the State's role as third party in these negotiations has extended to that of subsidizing and otherwise easing the introduction of the new technology. In these countries the use of Eftpos in other services such as catering, hoteliering and fuel stations provides it with a possibly irresistible impetus.

The discovery of a new operational logic

Early equipment was subject to breakdown and complete loss of memory in power failures. Unsurprisingly therefore among the earliest users department stores tended to be in the majority. Their range of goods was greatest and trends in sales were most critical across this heterogeneous mix of sales. Information could be used locally and, by and large, most users abandoned attempts at on-line real-time transaction recording in favour of day-to-day batch processing of regionally based computer services.

Early resistance was displayed by chains such as Marks & Spencer on the basis that Epos recorded only what the customer *had* chosen, not what they would have *liked* to be able to choose. In the 1980s even Marks & Spencer abandoned labour intensive counter-sales for self-service layouts and this resistance crumbled. In practice many department and service chains have moved back to a Continental style of cashier-controlled Epos cash register complemented by dedicated sales staff who are supposed to feed back qualitative knowledge on customer tastes. Most chains are now introducing Epos on a store-by-store basis beginning with locations regarded as statistically representative of the market segment served. In this way changes in strategy brought about by market or other contingencies will not destroy the utility of the 'sample population' provided by the existing installations.

As in banking the price of the equipment is plateau-ing out at a much lower level than previously and known suppliers such as ICL and Nixdorf offer equipment that is still competitive with more recent Japanese entrants. Software houses abound; 250 were recorded in the U.K. in 1988 and bureaux such as the Grand Metropolitan Information Services offer to design, install and to operate complete systems. Such systems cover a range of uses with varying degrees of comprehensiveness and analytical sophistication. For the most part accounts, and more particularly invoice accountancy together with stock control has been most heavily affected. (Most large scale redundancies among sales staff appear to have taken place in the mid-1970s and preceded the introduction of IT.) Few chains (one small one in the U.K.) have adopted totally automatic reordering, although, as one French and one Japanese innovator have shown, it is now feasible to robotize the complete sales cycle. Instead most innovators spoke of 'the discipline' provided by Epos over existing buying and selling activities and goods movement. The knowledge that their activities were being monitored prevented local managers from speculating on local purchases and disguising subsequent losses in overall sales figures for example. For suppliers this 'discipline' also had implications. In the supermarket chain in which we conducted our first interviews the data obtained from the first three stores in which Epos was installed in the late 1970s enabled them to reduce their confectionery lines alone from 78 to 5.

As in banking the autonomy of branch management has been severely restricted in the substantive content of transactions handled at branch level (price, quantity, quality). Emphasis is now being placed on their ability to handle the social processes or procedures in dealing with customers on a day-to-day basis. Initial resistance came from the senior buyers and merchandisers whose judgements seemed to be exposed to challenge by the detailed feedback and analysis provided by Epos. Over time this strategic resistance has dropped and a realization of the value of detailed projecting is now fairly general both for in-group operations and across sectors. Hence the role of third party agencies such as market research firms like Neilson in aggregating and analysing sectoral data has provided both exemplars and arbitrators in the diffusion of Epos.

HEALTH CARE

Health care delivery stands as the exception to the 'reverse innovation' process (Barras, 1986). Doctors have retained ultimate control over most innovation in this area and although mainframe computers have been used for administrative purposes, since the 1960s progress has been slow compared with the diffusion of micro-electronic devices in the diagnosis and treatment of patient ailments. While early use in these latter areas was marked by resistance among more traditional medical practitioners, overall it may be said that the most enthusiastic adopters of IT have also been found among physicians and surgeons. This cannot easily be understood outside of the context of the system of health care delivery. The system is unique in providing autonomy and influence to service deliverers. While other professions enjoy great autonomy in the direct delivery of their services to clients, and are often protected by the support given to their self-regulating institutions by the State, few can command such influence over significant capital budgetary expenditures or the allocation of human resources on a large scale.

For this reason if no other, doctors have been in the position to take innovative initiatives if they so wished. Their influence within health care bureaucracies has been particularly great in the U.K., Scandinavia and other countries in which finance is channelled through Governmental procedures. In these situations administrators have tended to play a reactive role, that is to say one in which they have responded to needs defined by medical and para-medical practitioners (Klein, 1983). In other countries where more direct market pressures, or more usually the institutional power of private insurance companies, has played a role, the influence of administrators as market agents has enabled them to act as constraints on medical power. In these countries the use of IT for managerial purposes is generally more advanced than in public medicine.

Crisis and the origins of strategic logic

Over the last two decades the proportion of Gross National Product (GNP) being devoted to health care in most developed societies has risen steadily. In Britain by the late 1980s it has moved from around 3.5% spent in the early 1950s to around 7% of a much larger amount in real terms. At this level the U.K. was the lowest spender in Western Europe and compared with a level of 11% of GNP in the U.S.A. In an epoch in which capitalism was itself experiencing a crisis of confidence attempts to curtail public expenditures on health care and related social services extended across all Western democracies (Illiffe, 1988). These constraints took a number of forms including ceilings on pay and capital expenditures, the application of selected pricing policies etc. Most interest was generated in attempts to impose Scientific Management in the form of work management and performance indicators, particularly those deriving from the application of market based concepts of value on the clinical care being offered in each doctor–patient transaction (Loveridge, 1971; Williams, 1987).

The pressure for health administrators to adopt such quantitative indices as guides to operational effectiveness presupposed a concomitant authority to act on them. This authority seems often not to have been present within the fragmented operational structures of health care delivery (Ham, 1985). Essentially this structure derived from the manner in which a variety of itinerant, domestic and monastic occupations had been drawn together in the provision of medical care for urbanized populations over the last two centuries (Larson, 1977; Larkin, 1983; Turner, 1987). The result was a complex mix of semi-autonomous jobbing structures with centrally provided fixed services such as hospital hotelliering. Operational contingencies dictated a devolved form of administration, which in a nationally controlled system like the British National Health Service (NHS) indicated a complex system of coordinative governance (Stinchcombe, 1959; Thompson, 1967).

The work flow is normally organized in two cycles. The first is that of primary or community care, most often orchestrated by a general practitioner (GP) or community doctor. The secondary or hospitalization cycle handles only around 10% of recorded ailments but absorbs the major portion of health care funding. Access is normally by way of a GP or an emergency department and treatment is provided by separately specialized 'firms' or medical teams led by a senior consultant. These are clustered in disciplinary based departments which are serviced from centrally disposed services such as pathology and radiology. These latter, together with nursing, provide the rationale for the concentration of secondary treatment in hospitals in which increasingly capital intensive forms of diagnosis and treatment are housed.

In both cycles doctors retain an operational ascendancy that stems from an occupational monopoly ordained by the State over the diagnosis and treatment of life-threatening human ailments (Larkin, 1983). But the ability of physicians to provide this expertise pre-dated modern bureaucracies and appears to be present in most pre-industrial societies (Turner, 1987). Hence the moral authority enjoyed by the profession is rooted in a long-standing collective need expressed in most social structures.

The discovery of new operational logics

Innovation in modern medication and medical equipment has most often derived from the inventive collaboration of medical practitioners with corporate manufacturers and suppliers, together with teams of natural scientists working in a university or corporate environment (Reiser and Anbar, 1984; Shaw, 1986). For example one of the earliest British experiments in robotized surgery was financed by the Department of Trade and Industry, three computer manufacturers, three universities and the Leeds Infirmary where the coordinating medical consultants were employed (*Guardian*, 15 August 1989). This type of collaboration is ultimately dependent on the privileged access to the patients' body and bio-data possessed by the doctors and on the holistic nature of the knowledge that the latter can deploy within the diagnostic/evaluative situation. Hitherto there has been little attempt by suppliers to control medical outlets or by physicians as direct service providers to provide their own supplies on any scale. The purchasing power of the NHS is evidently immense even by global standards. But the sites of user-supplier collaboration in the U.K. have tended to be isolated and ill-coordinated. In part this can be attributed to the individualistic nature of the medical career system and personalized forms of cross-disciplinary networks along which innovations diffuse (Shaw, 1986). But the lack of coordination between the forms of sponsorship offered by different Government departments, by NHS boards and committees and by public trusts and charities can itself emphasize the 'pluralistic' nature of strategy formulation in the NHS at the cost of coherent direction and accountability.

The system of administrative decision making in the NHS has in fact been variously described as 'pluralistic', 'team-work' and 'consensual'. The apparent contradiction is contained in the nature of a collegial system of government in which representatives of the different medical specialisms together with those of nurses, administrators and lay members constitute executive boards at Regional (14 RHAs), District (192 DHAs) and Hospital levels, a well as on specialist disciplinary based bodies at each level. Strategic priorities and overall cash limits are set by Parliament acting through the Department of Health (formerly known as the DHSS when it also covered Social Services).

Specific financial allocations are made annually on the basis of board decisions. These are responses to bids put forward by medical and administrative departmental heads. The process is evidently a highly political one in which heads of central services such as pathology laboratories seek support from internal constituencies among user departments while the latter may attempt to mobilize public opinion behind their need for, say, monitoring equipment in post-natal care or renal dialysis equipment or other such instances where an apparently marginal capital expenditure might relieve a life-threatening situation. The public can also respond through charitable contribution. The diffusion of portable (bed-side) diagnostic equipment has derived impetus from this quarter in spite of opposition from local executives (management boards) who prefer consultant firms to use centrally provided hospital services.

The emergence of an infrastructure

National guidelines on resource allocation have been issued regularly by the DHSS and from time to time finance has been provided for initiatives. In 1967 half the expense of pilot projects in the application of mainframe computers to specific areas such as real-time records of patient administration and clinical information as well as a range of limited scientific and clinical purposes, mainly central laboratory services was provided centrally. All large mainframes had to be purchased from the major British computer manufacturer, ICL, and subsequent purchases were to be ICL compatible. By the mid-1970s most RHAs were providing bureau services for batch processing of hospital pay-rolls, accounts, supplies and child health administration usually administered by a Management Services Department or Regional Treasurer. Only 20 of around 2000 hospitals had obtained their own mainframe computers. Many of the centrally supported development programmes ran over budget and were abandoned having had little impact outside of a single hospital. By this time the use of personal and micro-computers of U.S. origin was growing often associated with clinical and laboratory systems. The favoured status of ICL therefore lost much of its significance.

The slow pace of innovation has to be set in the context of a growing urgency in the search for a consistent basis for the measurement and monitoring of operational performance in the NHS. A Government appointed committee (Körner, 1982) reported on the issue between 1982 and 1984. It also pointed to the potentially systemic characteristics contained within the proliferating elements of IT, particularly in their capacity to relate the data captured for the purpose of patient administration to that on financial control and, most sensitively, to clinical diagnosis and treatment. Systems of diagnostic related costing had already been developed in the U.S.A and their broad logic became incorporated in the NHS performance indicators published in 1983. Shortly afterwards the Thatcher Administration took decisive action to delimit consensus management through the appointment of general managers with responsibility for initiating action at each level of decision making. Acting on the advice of a retailing executive (Roy Griffiths) the DHSS urged the adoption of a cost-centre approach to budgetary allocation. In 1985 a number of exemplar hospitals delegated the management of annual budgets to senior clinicians. Some such as Guys, the internationally famous teaching hospital, became enthusiastic proselytizers of the new system of clinical budgeting.

The overall progress of administrative computing remained slow. Most doctors and paramedical staff saw themselves as agents of their patients in determining the amount and type of resource required for their treatment (Williams, 1987). The operational use of IT for record keeping was, and is, often resented even where the system has been designed and implemented by their clinical head. Often the task of designing acceptable 'neutral' data bases has defeated the best efforts of administrators (Dent *et al.*, 1987). In other cases conflicts came to light as a result of the removal of a 'mediating' clerk who had formerly resolved disputes over bed allocation between rival consultancy firms (author's observations). The reluctance of administrators to translate the local custom and practice of loosely conjoined 'organistic' systems into the 'mechanistic' vocabulary of system programs seemed widespread, especially when the informality of the former surrounded and endorsed an incipient elitism within the collegial system of medicine.

In 1989 the Government returned to its attempt to rationalize the system through an extension of the, so-called, 'internal market system' by which Regions, Districts and Hospitals had been encouraged to exchange services. In a White Paper it was proposed to allocate budgets to reorganized GP practices of sufficient size for them to buy patient access to secondary care in hospitals who would compete for their custom. Selected hospitals might even become self-financing on the basis of their ability to compete successfully. The work of management boards would increasingly focus on regulating the market thus created, and, more particularly, on 'auditing' the performance of individual firms and practices within the market.

The plan's emergence coincided with that of an 'open systems architecture' being offered by competing IT suppliers. This made the conjoining of the primary and secondary cycles of health care delivery through real-time transacting entirely feasible combined with a faculty for monitoring its workings centrally. Into this technological feasibility the Government attempted to inject a continuing stimulus for its use in the rationalization of the complex structure of the NHS. In many ways the traditional form of craft administration that underlay this structure invited a market based approach to this end. What was intended however was that the external threat of competition would draw departments together and make more real the concept of team work at the level of the hospital unit. To achieve this the Government were prepared to divide the medical profession between the providers of medical care at the primary or community level and the hitherto elite of hospital consultants and registrars in a manner that reversed the prevailing flow of influence within the system.

Pragmatic to logical incrementalism?

The pioneering adoption of process innovation can be as fraught with strategic uncertainty and technical incomprehension as new product invention. This is particularly so when it impacts upon the central work flow in a radical or systemic way. Adopting the framework used earlier in the volume (Chapter 5) the relatively fixed repertoires of routines that constitute the operational formulae of the innovating firm constitute knowledge assets that possess a unique value for the firm. Changing these in a manner that is systemic or non-incremental therefore constitutes risk that is tantamount to the re-establishment of the firm (Hannan and Freeman, 1986). (Indeed in the latter case employees in a 'green-field' situation are likely to be more highly motivated than those required to modify an existing structure.) Hence the first mover might naturally be expected to assess the expected entrepreneurial rents against the risks of novelty and uncertainty over an appropriate period. Lacking the means to do so (s)he might be expected to attempt to extend an existing operational formula in the pragmatic and piecemeal fashion of the so-called 'reverse innovation' from process to product design.

Even so the level of technical incomprehension is likely to be high especially among staff and senior executives who lack operational knowledge in both old and new process. The manner in which these stakeholders gain their understanding tends to interweave with the process of introducing and implementing the new machine or technique. Hence their frame of reference is shaped by whatever role, e.g. innovator, sponsor, sanctioner, champion, they play in the process (Rothwell, 1977). Innovations need to be interpreted into existing, familiar languages and frames of reference through the sponsors and champions. More significantly the establishment of performance bench-marks across a sectoral area of comparison enables an assessment of performance outputs to be made, even while a comprehension of knowledge inputs remains limited among strategic decision makers. This enables both the imposition of interim cost controls and a more informed evaluation of its strategic applications to be made.

The movement from a concern for operational control that marks the adoption and implementation of IT within the boundaries of the firms' hierarchy ('Whose finger on the terminal?') is translated into concern to close existing knowledge domains to competitors. Large business corporations no less than 'firms' of medical consultants have displayed a remarkable tenacity in their use of idiosyncratic software and in their reluctance to collaborate in the creation of a user-infrastructure beyond the boundaries of their existing market domains (Brady, 1989). The standardization of interfacing elements and of basic media, is obviously essential to the so-called 'maturation' of a socio-technical mode. However effective the market process may be in arriving at a model of best practice inter-firm collaboration over the design of components elements, such as the basis for bar-coding in the development of item scanning (see Epos), can lead to significant external economies which make possible the rapid diffusion of the new technology.

Fears for the specificity of the firms' knowledge assets may be misplaced in the light of the situationally embedded knowledge that is often required to *understand* and use programmed information. But fear also derives from a possible loss of product differentiation or from the belief that a shared network, such as that available through inter-bank collaboration in the operations of ATMs, may raise consumer awareness of competitors' products. This fear translates into the professional closure of initiatives in medical applications of IT and a suspicion towards central interventions by the DHSS. Technically it should be possible to exchange patient information in a standardized form across the entire national network of hospitals and practices in a manner that would bring immediate life-saving consequences. Achieving this end would require not merely an enormous investment of design and manufacturing expertise but also the commitment of multi-various groups of health care practitioners. In some countries, among which France appears as exceptional, the State has intervened on the demand side of the market to equip consumers with both knowledge and means of access to a yet-to-be created network. For example French subscribers to both private and public health care systems in France are to receive smart cards carrying their complete medical history which they can disclose to a doctor of their choice by passing him/her their entry code. In this way by promoting *effective* demand the State makes the consumer into the primary facilitating agent.

This use of smart cards touches on the issue of confidentiality which is clearly a major one that will almost certainly play a part in the future shaping of service industries. But the creation of the appropriate infrastructure for IT ultimately depends on a collective or inter-firm collaboration. As Olson (1968) pointed out it is a normal characteristic for individuals to wish to share in the outcomes of collective action by others, whilst being uncertain of the balance of advantage to be accrued from making a personal commitment. Collaboration is most likely where contributors are few and stable or where a common threat is such as to ensure that mere survival is a worthwhile private gain. Otherwise a third party can create opportunities for investment in public goods that can seem to offer a private net gain. In this manner agencies such as Reuter and Neilson add value to aggregated information from a number of individual sources (Mantel and Rosegger, 1987).

In some countries such as Germany and Japan the relationship between the State and corporate interest enables the Government to play the role of agent in crystallizing individual beliefs around particular strategic missions. What is perhaps more difficult to explain is the apparent ability of American corporations to cooperate over issues of standardization and interfacing of new products in order to create a least-cost infrastructure. Even as far back as 1924 manufacturers in the nascent automobile industry were able to agree on the standardization of vital component specifications while in the much smaller European industries this was not possible. It is a phenomenon that can be repeated in the most recent history of IT development. Such collaborative arrangements appear to have none of the permanency of inter-firm associations in Europe (Chandler and Daems, 1984; Barke, 1985). One answer might lie in the greater concentration of corporate activity and therefore the ease with which mutual agreements can be monitored by all parties. Another might lie in the ability to recognize a greater equalization of private gains from the exploitation of a home market as large as that of the U.S.A. Yet another reason might stem from the ability of American strategists to discover a strategic logic and to apply it to the incremental development of a new technology more swiftly than has been the case in the U.K. Thus the long drawn out process of individualistic politicizing and pragmatic ad-hocery through which innovation proceeds within many British organizations serves to preclude external commitments that might shape the direction of internal discussions. It is also true that the protective nature of most British sectoral associations may, paradoxically inhibit their actions in a way that the issue-centred more temporary alliances of American corporations are not.

The nature of direct services

Throughout the chapter an implicit distinction has been drawn between direct and indirect services. While this distinction is not always possible to make, especially among the self-employed, e.g. brokers, accountants, it is useful to define some transactions as deriving added value for the ultimate consumer through a face-to-face interaction with the service provider. These may be distinguished from 'back-office' contributions which, however vital, are rarely scrutinized and evaluated in the consumer market. The status of direct service work is evidently distinctive in the symbolism of wider society. Much of this derives from its origins in the maintenance of a pre-industrial elite or with acts of altruism and charity (Miles, 1987). The substantive content of direct service is most often discovered in counselling or caring for others. The former normally involves judgemental acts on the nature of the idiosyncratic needs of clients within specific situations. In modern times these are usually based on the possession of a body of generalized knowledge and professional certification, though the judgement of a trades person or shopkeeper may still be held locally in high regard. The importance of care derives from the dependency of the young and disabled upon a source of emotional stability and comfort. The two aspects of the service role coincide with the patriarchal hierarchy of the traditional division of labour and it is no coincidence that well over half of employees in all three service areas covered by our study were women. Nor is it surprising that the division of labour between professional counselling and supportive or persuasive caring followed a largely gender related pattern.

The transactions produced across the three sectors might appear to represent a spectrum of risk for the consumer which moves from the relatively routine purchase of food through to longer-term investments in consumer durables and the depositing of savings to the risk to life contained in the treatment of human ailments (Child and Loveridge, 1990). This continuum might be seen to be reproduced in the degree and type of regulation imposed by the State on these activities on behalf of the community. However as has been pointed out the manner in which a contemporary status and sponsorship has been attributed by the State can often be traced to the *historical* leverage exerted by past generations of shopkeepers, bankers and doctors within any given country. It is, however, when an occupation or corporate organization such as a bank has been able to conjoin an elaborated codified body of task-related knowledge with the tacit or situationally based knowledge required for judgements on high-risk situations, that they have acquired an influential constituency of institutional sponsors.

Hannan and Freeman (1986) suggest that accountability to the community is most required where organizations produce symbolic or information loaded outputs, where substantial risk is involved for the consumer, and/or where the transaction has long-lasting effects. The ability of providers of direct services to obtain autonomy and self regulation in these circumstances has depended on their ability to demonstrate both the immeasurability of knowledge inputs and the judgemental nature of outputs (Jamous and Pelloille, 1970). The evident intention of the British Government in the 1980s was to restrain the substantive or task-related autonomy enjoyed by closed institutions through the market evaluations of their outputs. As has been pointed out the dismantling of institutionalized boundaries to markets in finance and retailing served to bring about their integration in centralized hierarchies of procedural control. It remains to be seen how far the clearer articulation of market forces will lead to greater integration within the craft administration of the hospital and, if it does, how far the Government will once more have to intervene to restore 'Chinese Walls' to ensure the continuance of independent judgement in the application of specialized task skills in the manner that it has in the of financial services provision.

Clearly the limits to cross-trading are set ultimately by the client's opinion of the impartiality and confidentiality of the advice being offered. This constraint is one that affects a number of professions but most particularly the so-called Big Five firms of auditors and chartered accountants now offering a range of consultancy services, each offering privileged access to client information. Similarly the desire to control specialized inputs to the final product mix has led to problems deriving from the moral nature of service inputs to financial markets. The mode of appropriation adopted by the corporate management of services may have been shaped by a desire to constrain the overhead costs of contractual administration in the manner suggested by Williamson (1975, 1985) and developed by Boisot (1983, 1987). If this has been the case then one's explanation of the problems encountered in the attempted mergers and takeovers during the 1980s must account for a number of situational contingencies. These should include the difficulties encountered in translating the tacit knowledge deployed by brokers and jobbers in securities markets into the procedural forms deployed by clearing banks. The manner in which this knowledge was formerly used within roles that interfaced across a market-place was clearly different from that required in the complementary activities of the integrated hierarchy of the new parent bank into which they have been absorbed.

These developments have to be set in the context of a small but perceptible trend towards self-service in all three sectors. While not in the realm of supermarket transactions, medical diagnosis and self-medication in many circumstances of chronic ills or mass screening contain the bases of do-it-yourself health care. Liberals might describe this trend as enfranchising for those whose access to direct services has hitherto been somewhat limited and, judging by the facilities of the average NHS waiting room, possibly demeaning. The trend towards use of direct access to personal accounts and records has been reinforced by the availability of IT and the desire of service management to reduce their costs. It has been strengthened by the use of 'expert systems' in some areas of private medicine. The availability of such options is yet another constraint on the power of the professional to mediate between the consumer and sources of immediate personal uncertainty.

Against this must be set the evident expansion in the activities of self-employed financial advisers in the wake of mergers between formerly independent parts of the financial, insurance and real estate sector. Similarly the need for guidance through the increasingly specialized labyrinth of secondary health care will, whatever its market status, throw an increased burden on the counselling functions of primary care providers. Furthermore however 'expert' the judgement delivered through a computer it seems unlikely that human consideration and caring can be mediated by a machine. Hence the training in social skills now widespread in retailing and banking might eventually play a more significant part in the training of health care providers. Apart from the effects of these structural contingencies it seems unlikely that groups which have acquired strategic influence through their ability to offer direct services will allow their authority to be dissipated through the widespread substitution of self-service by manufacturers of automated equipment (Coombes and Green 1989).

CONCLUSIONS

In this chapter the process of implementing new technology has been set in the context of a reconfiguration of the organization's strategic position. It has been suggested that the process might be best seen in terms of two or more sequences of managerial learning. The first is largely concerned with the operationalization of the new tool or techniques; the second with a developing logic in its application to market processes. In general the first phase has been marked by a piecemeal pragmatism demonstrating little awareness of the strategic potential of the new technology but greater concern for disruption in the internal hierarchy. In the second phase there is a growing coherence in the systematic application of new techniques to the central workflow of the organization and to the design of its outputs. This has been especially so in banking where it seems likely that a third stage of strategic re-configuration is already under way allowing greater autonomy in IT applications to operating units with a greater emphasis on central monitoring rather than direct control.

Because of the nature of direct services management have come to place more emphasis on the monitoring and evaluation of outputs rather than inputs. Costly mistakes have been made in the attempted integration of specialized tasks within centrally controlled organizational structures. These may, over time, lead to the development of more loosely coupled operational units working within the context of market threats and promises. This being so the continued development of extra-occupational regulatory bodies to act on behalf of consumer rights appears equally inevitable in the political climate of the 1990s. The corporate concern for control over an ever more turbulent environment has inhibited the development of IT networks offering most value added to the consumer of direct services. Obstacles often appear to originate from the inability of individual organizations to arrive at an internally coherent basis upon which to enter into external agreements with others. The role of a third party in bringing about an appropriate mix of private advantages relative to costs of collective action has been stressed in this chapter.

In spite of developments in the monitoring and programming of service tasks it seems unlikely that the demand for direct service jobs will diminish, indeed the need for both judgemental and caring aspects of service work may well increase. However, across all three sectors the immediate actions of those possessing access to key bodies of knowledge has been to create new internal sub-hierarchies intended to absorb the immediate technical uncertainty accompanying innovation. Many of these may be give rise to localized jobs that will prove to have been transitory as the technology becomes more and more based on end-user control. Present evidence suggests little movement towards the establishment of closed occupations around the possession of IT knowledge of a kind comparable to that of past craft and professional groups (Loveridge, 1972; McLoughlin and Clark, 1988).

Movements in employment are, however, as much influenced by strategic considerations as by consideration of current manning requirements of IT installations. For example the withdrawal or reorganization of clearing bank subsidiaries in stock and commodity markets is reported as having created 30 000 redundancies in a little over two years up to 1990. Such significant movements around a long-term growth trend suggest the need for a more complex model of innovation than is provided by the more holistic of the explanations cited in the introduction to this chapter.

REFERENCES

Abercrombie, N., and Urry, J. (1983). *Capital, Labour and the Middle Classes*, Allen & Unwin, London.
Abrahamson, M. (1967). *The Professional in the Organization*, Rand McNally, Chicago.
Barke, R. (1985). 'Regulation and cooperation among firms in technical standard setting'. *Journal of Behavioral Economics*, **14**, Winter, 113–130.
Barras, R. (1986) 'Towards a theory of innovation in services', *Research Policy*, **15**(4), August, 161–74.
Bell, D. (1974). *The Coming of Post-Industrial Society*, Heinemann, London.
Blauner, R., (1964). *Alienation and Freedom*, University of Chicago, Chicago.
Boisot, M. (1983). 'Convergence revisited: the codification and diffusion of knowledge in a British and Japanese firm', *Journal of Management Studies*, **1**, 159–90.
Boisot, M. (1987). *Information and Organizations: the Manager as Anthropologist*, Fontana, London.
Brady, T. (1989). 'Users as Producers: Software's Silent Majority', Centre for Information and Communication Technologies, Working Paper No. 3, University of Sussex Science Policy Research Unit, February.
Braverman, H. (1974). *Labor and Monopoly Capital*, Monthly Review Press, New York.
Brown, R. (1981). 'Training for new technology', *Management Today*, April, 104–19.
Caulkin, S. (1989). 'Crippled by computers', *Management Today*, July, 84–8.
Channon, D.F. (1986). *Bank Strategic Management and Marketing*, John Wiley, Chichester.
Chandler, A.D., and Daems (eds) (1984). *Managerial Hierarchies: Comparative Perspectives on the Rise of the Modern Industrial Enterprise*, Harvard University Press, Cambridge, Mass.
Child, J. (1984). 'New technology and developments in management organization', *Omega*, **12**(3), 211–23.

Child, J., Ganter, H.-D., and Kieser, A. (1987). 'Technological innovation and organizational conservatism', in J.M. Pennings and A. Buitendam (eds), *New Technology as Organizational Innovation*, Ballinger, Cambridge, Mass.

Child, J., and Loveridge, R. (on behalf of the MESS team) (1990). *Information Technology in European Services—towards a Microelectronic Future*, Blackwell, Oxford.

Child, J., and Tarbuck, M. (1985). 'The introduction of new technologies: managerial initiative and union response in British banks', *Industrial Relations Journal*, **16**, 19–33.

Clark, C.A. (1940). *The Conditions of Economic Progress*, Macmillan, London.

Clarke, K.B. (1983). 'Competition, technical diversity and radical innovation in the U.S. auto industry', in R.S. Roosenbloom (ed.), *Research on Technological Innovation, Management and Policy*, Jai Press, Greenwich (Conn).

Clark, P., and Starkey, K. (1988). *Organizational Transitions and Innovation-design*, Pinter, London.

Coombs, R., and Green, K. (1989). 'Work organization and product Change in the Service sector: the case of the UK National Health Service', in S. Wood (ed.), *The Transformation of Work*, pp. 279–94, Unwin–Hyman, London.

Cowan, H. (1986). 'Developing and modifying existing systems in the 4th generation environment', in *Information Technology Strategies*, Conference Proceedings, International Business Communications Ltd, 3 December, London.

Crompton, R., and Jones, G. (1984). *White-Collar Proletariat: Deskilling and Gender in Clerical Work*, Macmillan, London.

Dent, M., Green, R., Smith, J., and Cox, D. (1987). 'Corporate information systems, computers and management strategies', *British Sociological Association Annual Conference, Science, Technology and Society*, University of Leeds, 6–9 April.

Flowers, S. (1989). 'Dashing fashion', *The Guardian*, Thursday, 27 June, 25.

Frankel, S.H. (1977). *Money: Two Philosophies*, Blackwell, Oxford.

Freeman, C., (1988). 'The factory of the future: the productivity paradox, Japanese Just-in-Time and Information Technology', *PICT Policy Research Paper No.3*, Economic and Social Research Council, Swindon, May.

Fuchs, V.R. (1968). *The Service Economy*, National Bureau of Economic Research, New York.

Gershuny, J.I., and Miles, I.D. (1983). *The New Service Economy*, Frances Pinter, London.

Gouldner, A.W. (1954). *Patterns of Industrial Bureaucracy*, Free Press, Glencoe, Ill.

Gouldner, A.W. (1957). 'Cosmopolitans and locals: toward an analysis of latent social roles', *Administrative Science Quarterly*, December, 281–92.

Gustavsen,B., and Hethy, L. (1986). 'New forms of work: a European overview, in P. Grootings, B. Gustavsen and L. Hethy (eds), *New Forms of Work Organization and their Social and Economic Environment*, Statistical Publishing House, Budapest.

Ham, C. (1985). *Health Policy in Britain*, Macmillan, London.

Hannan, M.T., and Freeman, J. (1986). 'The ecology of organizations: structural inertia and organizational change', in S. Lindenberg, J.S. Coleman and S. Nowak (eds), *Approaches to Social Theory*, Russell Sage Foundation, New York.

Handy, C. (1984). *The Future of Work*, Blackwell, Oxford.

Iliffe, S. (1988). *Strong Medicine: Health Politics for the Twenty First Century*, Lawrence & Wishart, London.

Jamous, H., and Pelloille, B. (1970). Changes in the French university-hospital system, in J.A. Jackson (ed.), *Professions and Professionalism*, Cambridge University Press, Cambridge.

Johnson, G. (1987). *Strategic Change and the Management Process*, Basil Blackwell, Oxford.

Johnson, T.J. (1972). *Professions and Power*, Macmillan, London.
Korner, E. (Chairperson) (1982). *Reports 1 to 6 of the Steering Group Health Service Information*, Department of Health and Social Services, London.
Kumar, K. (1978). *Prophecy and Progress*, Penguin, Harmondsworth.
Larkin, G. (1983). *Occupational Monopoly and Modern Medicine*, Tavistock, London.
Larson, M.S. (1977). *The Rise of Professionalism. A Sociological Analysis*, University of California Press, Berkeley.
Lindblom, C.E. (1959). The science of 'muddling through', *Public Administration Review*, 79–88.
Lovelock, C.H., and Young, R.F. (1979). 'Look to consumers to increase productivity', *Harvard Business Review*, May–June.
Loveridge, R. (1971). *Collective Bargaining by National Employees in the United Kingdom*, Institute of Labor and Industrial Relations, Ann Arbor.
Loveridge, R. (1972). 'Occupational change and the development of interest groups among white collar workers in the United Kingdom', *British Journal of Industrial Relations*, **10**(3), 340–65.
Loveridge, R. (1983). 'The professional negotiator: roles, resources and the run of the cards', in W.T. Singleton (ed.), *Social Skills—the Study of Real Skills*, MTP Press, Lancaster.
Loveridge, R. (1984). 'Micro electronics and the growing polarisation of service employment', *Proceedings of the Labour Process Conference*, Aston University, April.
Loveridge, R. (1989). 'Footfalls of the future: the emergence of strategic frames and formulae', in R. Loveridge and M. Pitt, *Business Strategy and Technological Innovation*, John Wiley, Chichester.
McCrae, H., and Caincross, F. (1973). *Capital City—London as a Financial Centre*, Eyre Methuen, London.
McLoughlin, I., and Clark, J. (1988). *Technological Change at Work*, Open University Press, Milton Keynes.
Manning, W.G., Leckbowitz, A., Goldberg, G.A., Rogers, W.H., and Newhouse, J.P. (1984). 'A controlled trial on the effects of Prepaid Group Practice on use of services', *New England Journal of Medicine*, **310**, 1505–10.
Mantel, S.J., and Rosegger, G. (1987). 'The role of third parties in the diffusion of innovations—a survey', Department of Economics Working Paper, Case Western Reserve University, Cleveland.
Marshall, M. (1987). *Long Waves of Regional Development*, London, Macmillan.
Maynard, A. (1985). 'Policy choices in health', in Berthoud, R. (ed.), *Challenges to Social Policy*, Policy Studies Institute/Gower Press, London.
Medcof, J.W. (1989). 'The effect of the extent of information technology of the job of the user upon task characteristics', *Human Relations*, **42**(1), 23–42.
Miles, I. (1987). 'From the service economy to the information society—and back again', *Information Services and Use*, **7**, 13–29.
Morse, J. (1981). 'How British banking has changed', Stamp Memorial Lecture, University of London.
Naisbitt, R. (1984). *Megatrends*, Futura, London.
Norman, R. (1985) *Service Management*, John Wiley, Chichester.
Olson, M. (1968). *The Logic of Collective Action—Public Goods and Theory of Groups*, Schocken Books, New York.
Perez, C. (1985). 'Microelectronics, long waves and world structural change', *World Development*, **13**, 441–63.
Pettigrew, A.M. (1973). *The Politics of Organizational Decision Making*, Tavistock, London.

Pollert, A. (1988). 'The flexible firm: fixation or fact?', *Work, Employment and Society*, **52**, 281–316.

Quinn, J.B. (1978). 'Strategic change: "logical incrementalism"', in *Sloan Management Review*, **20**, Fall, 7–21.

Quinn, S. (1986). 'An alternative strategy', in *Proceedings of Conference on Information Technology Strategies*, International Business Communications Ltd, London.

Rajan, A. (1987). *Services—the Second Industrial Revolution*, Institute of Manpower Studies, Falmer.

Reiser, S.J. (1984). *Medicine and the Reign of Technology*, Cambridge University Press, Cambridge.

Reiser, S.J., and Anbar, M. (1984). *The Machine at the Bedside*, Cambridge University Press, Cambridge.

Rogers, E.M. (1962). *Diffusion of Innovation*, Free Press, New York.

Rothwell, R., Freeman, C., Horsley, A., Jervis, V.T.P., Robertson, A.B., and Townsend, J. (1974). 'Sappho updated—project Sappho phase 2', *Research Policy*, **3**(3), 258–91.

Rothwell, R. (1977). 'The characteristics of successful innovators and technically progressive firms', *R&D Management*, **7**(3), June.

Rybezynski, T.M. (1984). 'The U.K. financial system in transition', *National Westminster Quarterly Review*, November.

Sabel, C.F. (1982). *Work and Politics—the Division of Labor in Industry*, Cambridge University Press, Cambridge.

Shaw, B. (1986). 'Appropriation and transfer of innovation benefit in the U.K. medical equipment industry', *Technovation*, **4**, 45–65.

Stinchcombe, A. (1959). 'Bureaucratic and craft administration of production: a comparative study', *Administrative Science Quarterly*, **4**, 168–87.

Stocking, B. (1985). *Initiative and Inertia: Case Studies in the NHS*, Nuffield Provincial Hospitals Trust, London.

Strange, S. (1986). *Casino Capitalism*, Blackwell, Oxford.

Teeling Smith, G. (ed.) (1987). *Health Economics: Prospects for the Future*, Croom Helm, London.

Thompson, J.D. (1967). *Organizations in Action*, McGraw-Hill, New York.

Toffler, A. (1980). *The Third Wave*, Collins, New York.

Turner, B.S. (1987). *Medical Power and Social Knowledge*, Sage, London.

Tushman, M.L., and Anderson, P. (1986). 'Technological Discontinuities and Organizational Environments', *Administrative Science Quarterly*, **31**, 439–65.

Watanabe, T. (1988). 'New office technology, labour management and the labour process in the contemporary Japanese banking sector', *Warwick Papers in Industrial Relations*, No.21, July.

Webster, J. (1986). 'Using information technology for competitive advantage', in *Proceedings of Conference on Information Technology Strategies*, International Business Communications Ltd, London.

Williams, A. (ed.) (1987). *Health and Economics*, Macmillan, London.

Williamson, O.E. (1975). *Markets and Hierarchies: Analysis and Anti Trust Implications*, Free Press, New York.

Williamson, O.E. (1985). *The Economic Institutions of Captialism: Firms, Markets, Relational Contracting*, Free Press, New York.

Woodward, J. (1965). *Industrial Organization Theory and Practice*, Oxford University Press, London.

Zelany, M. (1985). *Towards a Self-Service Society*, Columbia University Press, New York.

16

Managing the Future: Questions and Dilemmas

Martyn Pitt

University of Bath

Recent attempts to promulgate a 'strategic management paradigm' have not significantly improved our appreciation of the technology-strategy link at the level of the firm in either theoretical or practical terms. Schendel and Hofer's (1979) 'new view of business policy and planning' accommodated technological innovation as a somewhat peripheral element; Pennings *et al.* (1985) sought to integrate diverse theories into a rigorous strategic management paradigm but, excepting the contribution of Teece (1985), also accorded technological innovation little attention. Recently, academic management journals have increased their interest in the technological dimensions of strategy, albeit in volume no more than, say, the application of transaction cost economics to strategy problems. Elsewhere in the management literature, much of what has been said is either highly specialized and academic in tone or combines interesting anecdotes with simplistic prescriptions of dubious generality.

On the other side of the coin, the technology-directed literature has manifest shortcomings of its own. Conventional wisdoms over the descriptive aspects of sector-technology development (let alone prescriptive

The Strategic Management of Technological Innovation. Edited by R. Loveridge and M. Pitt
©1990 by John Wiley & Sons Ltd

aspects) are still very much open to debate and reinterpretation in the light of conflicting empirical observations gained in recent years, as the contribution of Clark and DeBresson (Ch.10) amply illustrates.

Clearly, then, much remains to be done, both in theory elaboration and in developing useful heuristics to guide firms in the management of strategically important technological positions. Of course, manifest contradictions between theory and pragmatics may be inevitable; as Weick (1969) asserted so pungently:

> 'The usefulness of a theory is not determined by its usability in the everyday business of running an organization or "making out" in one. Theoretical usefulness is not defined in terms of pragmatics. (The explication of particular cases) ... will often provide a tacit "prescription"' for getting along .... but we do not know the conditions under which that pragmatic recommendation or prescription holds.' (Weick, 1969, pp. 19/20).

This characteristically robust statement of Weick's should challenge practitioners of all kinds to enhance theory and pragmatics, whether or not their primary endeavour is to achieve a synthesis of the two.

Contributors to this book have emphasized a firm-level strategic managerial perspective on the problems of initiating technological innovation and appropriating its benefits. Their observations are well grounded in empirical research and they have tried to avoid narrow perspectives and gross generalizations. Nonetheless, we trust their comments will prove stimulating and action-generating. For expositional simplicity the chapters were sequenced under three broad headings (i) the content of effective (and ineffective) strategies in particular kinds of environmental context, taking due account of competitive and collaborative conditions over time, (ii) issues of structuring, within and across firms, involving choices affecting the boundaries of the firm, including alliance-building, means of risk containment and managerial control, and (iii) managerial processes, specifically, coping with the intrinsic uncertainty of technological innovation via adaptive, experiential-learning modes of organizational behaviour.

The need to implement an adaptive learning posture to cope with technological uncertainty is a core conclusion of the book, albeit one that poses more questions than it claims to have answered. In trying now to synthesize the foregoing chapters it seems to us that a few, quite crucial pragmatic questions must be addressed, and about which (notwithstanding Weick's observations) strategy-technology theory should have more to say:

- How can firms anticipate the advent of novel technologies and interpret their significance in a complex, fast-changing world?

- What kinds of strategies and structures can firms adopt, particularly in mature business sectors, to appropriate the benefits of new technologies without exposing themselves to unacceptable risks?
- What kinds of managerial styles and processes sustain a creative, inventive within-firm environment appropriate to a technologically innovative firm?
- How should firms (and individual actors) cope with ethical dilemmas arising from a technologically innovative posture?

ANTICIPATING NOVEL TECHNOLOGIES

Technological innovation is the commercial exploitation of new ideas and inventions (Roberts, 1987). One firm's innovation is, therefore, another's potential crisis. 'Technology' encompasses value-generating skills and capabilities and the products and services they enable. Anticipating the imminence of—and need for—technological change is a key managerial task that cuts across functional divisions of the firm, particularly in mature business sectors. It is the legitimate concern of marketers and general managers as well as technologists and engineers. Twiss (1986, p. 206) concluded that all companies must engage in technological forecasting: only the form and amount properly depend on circumstances. In general this requires firms to scan their external environments to anticipate those technological developments with the potential to create opportunities or threats relative to their existing activities and capabilities. But given the inherent ambiguity (unknowability) of the future (McCaskey, 1982), the dilemma is what to study, when and how.

The problem of anticipation poses both methodological and interpretive problems. Methodological, in terms of what *forms* of external scanning to adopt, how systematic and comprehensive these processes should be and what reliance to place on them in comparison with autonomous attempts to innovate (Bright, 1970; Dosi, 1982; Dutton and Ottensmeyer, 1987; Quinn, 1967; White and Graham, 1978). Interpretive, in that the most sensitive and comprehensive detection processes do not eliminate the need for reflective managerial sense-making (Schon, 1983).

Anticipation of innovative possibilities is an integral element of the strategy formulation process widely recognized in recent years (Brownlie, 1987; Dutton and Duncan, 1987; Ford, 1988; Frohman, 1982; Foster, 1986). However, modes of scanning and interpretation are conditioned by managerial perceptions of the importance of technological innovation, that is, as a proactive/offensive or as a reactive/defensive change mechanism. Sadly, it seems that structured, systematic approaches are by no means effective in relatively unstable, surprise-laden environments, precisely those in which good anticipation is most needed (Fredrickson, 1984; Fredrickson and Mitchell, 1984). Equally, innovating behaviour within the firm typically creates, not eliminates equivocality, since inventions 'ask' questions of decision makers to which they cannot reasonably have *a priori* answers.

Thus, for the most part, attempts to anticipate change possibilities produce equivocal data, making sense of which stimulates multiple modes of interpretation. Several have been described in this volume. The first may be viewed as 'pattern-recognition'. Implicit in this form of sense-making is a belief in evolving regularities that transcend sectoral specifics. Thus for example, Calori (Ch.1) used the now-familiar notion of an emerging industry context, characterized by rapid and sometimes discontinuous or radical technological change. In contrast, mature sectors (or at least, mature productive units) experience typically incremental change; radical innovation, when it occurs, is via new entrants or by inward diffusion of new knowledge to the sector. But, none the less, radical changes do occur and it has been useful to construe such changes as dematurizing in their effects.

Interpreting observed specifics in terms of an abstracted, generalized pattern is a significant step towards anticipating future technological possibilities. Combining sectoral patterns with assessments of the firm's competitive standing using some sort of generic conceptual framework should then yield contingent theoretical prescriptions of the kind much loved by strategy writers and consultants. Moenaert *et al.* (Ch.2) used various models of sector and technology evolution to refine concepts of sector maturity. An outcome of applying novel, exogenous (in their terms, alien) technologies to existing customer groups or applications was dematurity. It soon became clear, though, that a sector might be deemed mature by one definition, but not by another. Furthermore, whilst it is tempting to portray the appropriation of novel technologies as a conscious managerial strategy for sustaining competitive advantage, such responses are frequently a reactive response to exogenous change. Strategists, evidently, do not find sector/technology patterns easy to recognize and at best find them equivocal guides to future action. Thus, if the stage of sector evolution is hard to assess or if its evolution cannot be relied on to be unequivocally unidirectional, biological analogies—descriptive or predictive—are flawed, indicating the need for more sophisticated contingency theories of technology-in-sector development.

Still, a kind of pattern recognition was involved in the first stage of interpretation suggested by Child and Smith (Ch.14), i.e. registering changing 'objective' conditions such as declining demand, increased competition, declining profits etc. The second stage was challenging taken-for-granted assumptions about the business sector and how best to operate in it. These beliefs may be unique to the firm, though many are more widely accepted (Grinyer and Spender, 1979), sustaining the *de facto* 'cognitive arena' that is the sector. Reorienting the firm to new strategic postures and methods of work organization is painful and as Child and Smith and Pitt (Ch.11) argued, frequently occurs only when circumstances are severe enough to be perceived as crisis, emphasizing the crucial galvanizing, interpretive role of committed managers throughout the firm. The third element in the interpretive process posited by Child and Smith was the mechanism of developing and then learning actively from a network of interfirm contacts, some of which provide access to qualitatively better interpretations of future technological priorities for the firm than it could hope to achieve in isolation. In fact, the process of networking is so important as to be considered a separate mode of uncertainty reduction and interpretation. Sense-making has been characterized as a culturally-bound process of knowledge diffusion and codification (Boisot, 1986). Widespread awareness of a particular technical advance begins with vertical and lateral diffusion of knowledge among firms and their employees. But knowledge diffusion alone is not sufficient to ensure signification in new contexts. The achievement of convergent meanings, leading to systematic codification, is the outcome of multiple communications, negotiations, contracts, alliance-building, joint ventures and so on, creating *de facto* industry standards of expectation, specification, behaviour and performance.

New meanings also emerge from the convergence of more-or-less established technologies in a new pattern (Ford, 1988; Clarke and Thomas, Ch.12), as with the combination of materials handling, computing and communications technologies in flexible manufacturing systems (FMSs). As Jones (Ch.13) pointed out, the creative outlook required of system manufacturers to envisage the technical and commercial possibilities of automated work stations must necessarily be matched by their customers if they are to exploit FMSs fully. Thus, making sense of the opportunities and benefits of innovations is part analytical, part intuitive and creative.

Implicit in this argument is an assumption of managerial intentionality in and control of the processes of interpretation and anticipation. Using the illustration of the emerging bio-technology sector, Freeman and Barley (Ch.6) emphasized that building networks of inter-firm contacts was essentially a stochastic mechanism for learning. This mode of interpretive behaviour comprises numerous highly experimental, diverse trials or 'bets'. Firms establish relationships on the assumption that the more (and more varied) are the bets they place, the more likely some are to pay off by opening up unknown or otherwise inaccessible routes to growth.

Clark and DeBresson (Ch.10) proposed a similar mechanism in automobiles, a much longer established and, by inference, mature sector. Here, focal organizations have acted as 'innovation poles' around which a network of trading and non-trading relationships developed over time. The existence of a pole is inferred from the density of observed relationships in resource space and is not inevitably centred on a single firm. Firms with access to the pole enhance their innovation-design competences by sharing know-how within the network, transforming their individual and collective capabilities (anticipatory as well as executive) over time. 'Silicon Valley' and the Italian industrial districts offer other examples of this phenomenon. Whilst extensive networking appears contrary to 'conventional wisdom' on competitive strategy, it is almost certain to be highly functional in the increasingly dynamic and uncertain market conditions of the late twentieth and early twenty-first centuries.

Still, it would be naive to suppose that good anticipation through knowledge sharing is easy or open when state-of-art innovations are at stake. In industries like semiconductors patent activity has declined, precisely because firms believe it actually enhances diffusion of new concepts to competitors, reducing the innovator's advantage (Winter, 1987). Another common, if generally unwelcome diffusion process is the mobility of skilled personnel among firms in a sector (e.g. the 'Men from Mars' in the Cadbury case). Clearly, leading firms would do well to find more and better ways of retaining key staff by improved benefits, greater flexibility and above all, attention to their 'self-actualization' as well as financial needs.

Finally here, one notes the role of government agencies in funding exploratory research, diffusing knowledge and 'managing' the risks of innovating for commercial firms by highlighting prospects for commercial exploitation. Publicly-funded research in the U.K. has frequently been criticized as too unfocused and theoretical in this regard. In any event, the output of such research in the Western world is unlikely to increase greatly in the foreseeable future, so it is to be hoped that commercial firms will increasingly recognize and, on a contractual basis, exploit the skills, knowledge and facilities of universities and other private and public research institutions.

Whittington (Ch.8) suggested that many large firms increasingly subcontract their R and D to external agencies and are placing residual internal R and D efforts on a more focused, short-run, profit-oriented footing. Whilst it is misleading to suggest that the only way to exploit emerging technologies is to develop them internally, such policies could diminish firms' long-run technological capability, particularly in respect of core technologies on which their future competitive standing could depend (Ford, 1988). To avoid this outcome, firms must devote adequate resources to keeping abreast of external developments in both related and apparently unrelated fields, without unreasonable constraints on those charged with this task. Exogenous developments considered exploitable would then require the internalization of technologies via licence acquisition, joint ventures etc.

APPROPRIATING THE BENEFITS OF NEW TECHNOLOGIES

Appropriating the potential benefits of technological innovation requires the firm to shift unambiguously from an interpretive to an action mode. The strategic management paradigm integrates these modes; it rejects an essentially random or emergent behavioural view of firms' development in favour of proactive attempts to sustain a dynamic match of strategic intentions, operational structures and management processes to sector/metasectoral context. Account must also be taken of stakeholder expectations and structural possibilities for achieving strategic change. The prescriptions of this approach will vary according to the context and objectives of the firm, for example, whether in implementing change it is (or is trying to be) a sector leader (acknowledging, of course, that leadership is a multifaceted concept).

Loveridge (Ch. 5) expounded the notion of within-firm frames or logics-of-action and modes of organizational behaviour at both strategic and operational levels. The ways in which the leadership coalition imposes or negotiates its reflexive interpretations of strategic, contextual meaning on the organization is crucial, as is the content of specific proposals. Incongruence between frame and mode and between strategic and operating levels results in failure to impose and sustain these meanings, not only within the firm, but throughout the sector, leading ultimately to commercial difficulties.

Strategy implications

For emerging firms strategies based on technological innovation imply a vision of survival and growth through proactive leadership. To do so they must implement a viable configuration of competitive and/or collaborative strategy elements. For mature firms strategic change can be of two generic kinds, global or incremental (Mintzberg, 1978). Global change fundamentally reconfigures the elements of strategy, constituting a transformational crisis of corporate renewal. Further, as Abernathy et al. (1983) and Clarke (1987) pointed out, truly radical (architectural) technological changes disrupt market linkages and operational competences. They are irreversible in their consequences and therefore highly risky. Strategies grounded in radical innovation challenge taken-for-granted assumptions, aiming not to find new ways of tackling old problems, but to respecify the problems to be tackled. For fundamental change exploiting new technologies to succeed commercially—occupying new niches, sustaining new competitive advantages, increasing profitability and market share etc.—a firm must make appropriate, timely strategic choices in respect of market positioning and technological posture under conditions of considerable uncertainty and risk.

Embracing risk-conscious, adaptive strategies based on radical technological developments requires vision and courage, especially when apparently safer, incremental developments are obvious alternatives. In practice, it is extremely difficult to shrug off the effects of past choices, constituting an historically-grounded, firm-specific technological track, particularly when present successes derive from it. Incrementalist strategies (regular innovation in Clarke's 1987 terms) are akin to Kuhn's (1970) notion of 'normal science': problem-solving activity within an accepted set of paradigmatic beliefs. Strategies based on regular innovating, conserving the value of existing competences and market linkages, are not to be despised; implemented sensitively and logically they offer many synergistic advantages (Quinn, 1980) and can be a springboard for potentially more disruptive later changes. This perspective is endorsed by Loveridge (Ch.15) in his review of the application of new information technologies in various service industries as disparate as banking, retailing and health care. Specifically, he argues that innovations are typically applied to resolving well-understood operating 'problems', without there necessarily being evidence to conclude that organizations have considered or comprehended the technical, social or strategic competitive implications of these first steps. None the less, these limited, focused applications create a 'developing logic in systemic applications of I.T.' over several decades, to which the label strategy can arguably be applied in retrospect.

Strategies to exploit particular inventions or innovations are in principle separate from choices about whether to be a continuing *source* of new technological possibilities. It would be a mistake to suppose that research and development should always be done in-house (Whittington, Ch.8). Buying-in technologies does not preclude leadership postures when directed towards novel applications. For example, manufacturers of processed foods do not need to design process equipment to benefit from its use. Conversely, many manufacturers of 'high technology' products believe that they can stay ahead only if their process equipment is designed and made in house, so as to secure the highest standards of product performance. Backward extension into process plant also provides diversification options (e.g. FIAT Auto's Comau robot assembly equipment division). What is important in decisions like these is that the reason for engaging in a particular value-adding activity is clear and consistent with the overall objectives and strategic posture of the firm.

The mechanisms by which mature firms internalize exogenous technologies were reviewed by Moenaert et al. (Ch.2) and Saren (Ch.9), who noted real structural and processual difficulties in achieving viable postures and good performance returns (see also Marcus, 1988). A crucial step is the clarification of the technical applicability and commercial relevance of specific technologies, enabling the subsequent generation and evaluation of precise strategy proposals which take due account of competitive implications, the firm's prior experience, changes required in work organization, risk and so forth (Gold, 1983; Strebel, 1987; Zarecor, 1975). In the absence of such disciplines, the initiating firm often finds, too late, that it is poorly placed to appropriate its share of the potential benefits.

Our contributors have observed technology strategies in a variety of contexts ranging from technologically-novel (such as opto- and micro-electronics, biotechnology, thin films, lasers) to relatively mature (auto components, food retailing, banking and confectionery). Calori (Ch.1) is not alone in concluding that finding generalized strategic 'rules of the game' is difficult. Certainly one must challenge conventional 'strategic wisdoms': Porter's generic strategies (1980, 1985) are useful for discussion, but his assertion that they are mutually exclusive options is questionable. For example, in emerging industries survivors are very likely to be highly differentiated, yet also enjoy concomitant low-cost advantages (Calori).

Carr (Ch.3), too, noted shortcomings in conventional prescriptions of scale and market share in achieving long-run strategic success. He argued that major U.S. firms in automotive components have sought competitive advantage autonomously by exploiting bargaining power arising from scale of operation and market dominance. In contrast, Japanese component suppliers have ceded power to the auto assemblers in exchange for productive, long-run, collaborative relationships. These links have fostered a continuing stream of minor, but cumulatively significant manufacturing process innovations enhancing productivity and quality; for example reducing inventories and machine changeover times, applying concepts of 'just-in-time' delivery and 'built-in' quality. Know-how acquired over more than three decades cannot be quickly appropriated by Western firms ('uncertain imitability'—Lippman and Rumelt, 1982); in fact the gap between Japanese and Western firms may still be widening.

Jones's chapter also made cross-cultural comparisons. He concluded that the Japanese have adopted new work practices including FMS with greater willingness to experiment. It is too soon to claim they have fully appropriated the strategic benefits of flexibility, but Jones was particularly critical of moderate-sized Western firms who use FMS to emulate a quasi-Fordist mass-production strategy, rather than exploit its potential advantages in new competitive ways. His arguments are timely in that discussions about the strategic role of manufacturing (Collins, Hage and Hull, 1988; Dodgson, 1987; Hayes and Jaikumar, 1988; Jones and Webb, 1987; Kumpe and Bolwijn, 1988) are now better received in the strategy literature.

Clark and DeBresson (Ch.10) provided further evidence of the strategic importance of a firm's historically evolving competences. These are expressed in its technological path or trajectory over many decades, including its evolving inter-firm network of trading contacts and other non-trading links with research institutions etc. Effective strategy formulation and implementation depends on the firm's understanding of its competitive status as a product of this historical track and its collaborative potentialities in the network. In the long term, a firm will struggle to sustain an adequate repertoire of design capabilities if it is divorced from important connections,. just as, in the final analysis, an innovation pole ceases to be viable without reciprocal, synergistic knowledge flows amongst the firms that constitute the network. When a firm's design capabilities atrophy it will need direct technological support from elsewhere to survive, as witness Rover Cars in the 1980s, resulting in its strategic dependence on Honda of Japan.

In contrast, Senker (Ch.4) provided an optimistic illustration of how a developing technological competence on the part of some U.K. food retailers in a largely operational respect subsequently supported a novel conception of strategic advantage. Although food technology departments were mostly a defensive reaction to hygiene legislation in the late 1950s, they have led to an emerging capacity in some firms to innovate in the specification of processed foods and their methods of packaging and cold-chain distribution. Senker termed them 'interveners', somewhat analogous to Miles and Snow's (1978) 'prospectors'. Service organizations dealing with consumers frequently initiate the innovation process in the so-called back-office activities, as a means of incrementalizing and/or containing implementation risks (Loveridge, Ch.15). However, in many cases, the full potential benefits are appropriable only when the innovation is extended to the front office, involving the consumer directly, for example in the use of automatic teller machines (A.T.Ms). Such initiatives inevitably contain non-diversifiable, non-incrementalizable risks for the first-mover and for all practical purposes have to be considered irreversible changes.

In reviewing the strategic exploitation of technology, one should not overlook the outward transfer of technology from the inventing/innovating firm (Cho, 1988; Quinn, 1969; Foster, 1971; Ford and Ryan, 1981). Several different scenarios arise. First, where an organization sees its primary purpose as inventing: in other words, its main 'products' are innovations licensed to others for exploitation. Second, the role of 'brokers' (as exemplified by Centocor in Freeman and Barley, Ch.6) in facilitating the creation and diffusion of innovations throughout a community of interested parties. Science and technology parks have conventionally been considered to have this role. Third, the periodic development of an exploitable product or process which is perceived as peripheral to the inventor's main business objectives and therefore best exploited via licensing. Fourth, the recognition by a firm of a quite specific technological need which it cannot tackle alone, leading it to seek help from a competent external source, most probably on a one-off basis. The example of a company seeking help from a university college on laser gauging is such an instance (Saren, Ch.9). Fifth, the spread of less-than-state-of-art technologies from advanced to less advanced areas of the world.

Relatively few systematic lessons have so far been documented about how to manage either outward or inward licensing for strategic advantage, recognizing that the best approach is contingent on many circumstances, including the relative standing of licensor and licensee. Outbound licensing is commonly-used to secure widespread acceptance of a particular technological standard, though in business-to-business negotiations weak bargaining power on the innovator's side quickly erodes the technical advantage when competing with a well-entrenched and influential competitor. In any situation, licensing as a strategy to gain technical acceptance is fundamentally risky, albeit with high returns if successful. Philips of Holland and J.V.C. and Sony of Japan provide numerous examples of successes and failures in consumer electronics.

Structural implications

Appropriating the benefits of new technologies through strategic change also implies structural reconfiguration of the firm and/or its sector of activity. The classic work of Chandler (1962) highlighted the need for an appropriate match of firms' strategies and structures: strategies which take no account of structural implications are incomplete and probably inadequate. We see a wealth of empirical evidence illustrating this proposition in the application of I.T. to the service sector (Loveridge, Ch.15), where new technologies have fundamentally reconfigured banking and retailing organizations and may yet have a similar impact on the provision of health care in Britain.

One must also consider in a more theoretical vein the work on structural contingencies of Burns and Stalker (1961), Miles and Snow (1978), Miller (1988), Mintzberg (1979), Woodward (1965) and the configurational approach of Miller and Friesen (1978; 1984). If it is feasible to synthesize archetypes into a parsimonious taxonomy based on patterns of structuration and innovative behaviour, systematic understanding of strategic possibilities takes a big step forward. On the other hand, the sector-specific nature of technological evolution has undoubtedly hindered progress towards identifying robust, generalized configurations.

Freeman and Barley (Ch.6) argued that firm-in-sector configurations, based on resource consumption patterns and inter-organizational relations, define niches and hence, implicit strategies. Clark and DeBresson's (Ch.10) conclusions about innovation poles are similar, in short, that strategies reside in structures, rather than as Chandler proposed, that structures are a consequence of strategies. This perspective has the methodological advantage of construing strategies as realized patterns, not necessarily as intended outcomes (Mintzberg, 1978).

Radical innovations by *one* firm have the potential to restructure entire *sectors*, rendering old boundaries obsolete and introducing new sources of competition. Restructuring alters patterns of vertical aggregation and control within the value chain. It also creates new horizontal patterns as some firms diversify into proximate sectors whilst others concentrate on a single element of the value chain, as in flexible specialization (Piore and Sabel, 1984). Koenig and Thietart (Ch.7) reviewed the transaction cost perspective of Williamson (1975). In a complex, high-risk, high-technology sector like aerospace, they noted greater pluralism of inter-organizational form than Williamson documented. In these 'mutual' forms firm boundaries are fuzzy; whilst this presents managerial control problems, mutual forms may function very effectively in highly politicized environments, notably where investment risks and costs of failure are high.

However, as Koenig and Thietart indicated, managing technology alliances is no easy proposition (see also Nueno and Oosterveld, 1988). In some respects these mutual forms mirror the Japanese interfirm relationships to which Carr referred. They are relatively uncommon in North America, though according to Freeman and Barley (Ch.6), this is changing. The general lesson could be that the risk-shedding trend criticized by Hayes and Abernathy (1980) which led strategists towards vertical integration will be supplanted in the 1990s by less formal, more flexible linkages. What matters is not who owns the value-creating assets, but who influences and controls how they are used. Conventionally, ownership and control go together. But technological knowledge underpinned by patents and licence agreements is a powerful bargaining counter as the Genentech example shows. In some respects the difficulties faced by large firms are the obverse of those faced by emerging companies. Collaborations between large and small firms may be a good solution to their mutual problems, though considerable technical, cultural and resource-intensive disparities between them must be addressed if joint efforts are to succeed. Calori (Ch.1) adopted an optimistic note, suggesting a variety of workable alliance forms involving large and small firms in the 'second wave' of sector maturation, provided basic managerial skills and commitment exist.

In practice, large mature firms often acquire new technologies by buying the firms responsible for their development. Many of these are spin-offs created by frustrated entrepreneurial engineers and technologists previously employed by these same big firms. More attention should be given to keeping such people in the firm by encouraging the process of 'intrapreneurship' (Burgelman, 1983; 1985). Its feasibility in each case depends on avoiding possible mismatches of expectation over risk posture, ambitions, priorities, costs, rewards and pay-back periods arising from the small-firm culture within the big-firm bureaucracy. Sponsoring intrapreneurs does not guarantee successful appropriation of new technology, though the mechanism is increasingly being applied by Japanese and U.S. corporations.

These and other structural considerations raise more questions about the exploitation of technological innovation than they presently answer. At the sector level, the analytic constructs of strategic groups (Porter, 1980; Cool and Schendel, 1988), networks and resource niches (Freeman and Barley (Ch.6); Jarillo, 1988; Thorelli, 1986) and the innovation pole (Clark and DeBresson Ch.10) overlap conceptually. A strategic group is a set of firms adopting the same broadly-based, multi-dimensional competitive posture within or across given sectors. It presupposes internally consistent choices over product and process technologies, vertical and horizontal linkages etc. Close proximity of firms in a group will ensure shared knowledge of design competences and operational routines. But whereas strategic groups and niches conventionally imply resource allocations under conditions of scarcity via mechanisms of competition, poles and networks emphasize collaboration. Thus conceptual integration is needed to enhance our theoretical understanding of how firm-in-sector structuration affects the realization of viable strategic postures of technological innovation.

INNOVATIVE AND CREATIVE MANAGEMENT STYLES

Strategic transformation requires effective management styles and processes to cope with conditions of commercial and technological uncertainty. This was the clear message from Child and Smith (Ch.14) and others. What should interest us is how competent managers stimulate and sustain creativity and innovativeness. Ensuring an innovative internal environment parallels the 'excellence' debate of recent years, though the weakness of that literature, by and large, is a tendency to undervalue the role of effective *strategies* in securing excellent performance. With that caveat, some generalizations about managerial styles and processes now follow.

Involved and committed managers with clear objectives

In young and vibrant firms Calori (Ch.1) found relatively generalizable patterns of behaviour. He noted a strong, shared sense of purpose—an implicit or explicit vision of future direction, often the product of one individual—allied to effective team-work, personal commitment and due reward for achieving objectives at the cutting edge of technology. Each firm is unique, though they share a more-or-less pioneering ethos in which communication channels are relatively open and informal, and barriers to change are minimized. But Calori also suggested that operating routines and procedures varied widely.

Can such innovative cultures be sustained independent of gifted, often autocratic leaders over the long run? In small firms corporate innovativeness often stems from such people and in some instances, as in the case of Rover Cars (Clark and DeBresson Ch.10), it can be argued that the corporate innovation track was aligned with the vision of one man, managing director Wilkes, for several decades. Thus, while it is relevant to encourage innovation at *all* levels, it would be naive to ignore the extraordinary powers that senior executives wield to shape (and in some cases stymie) the scope for and direction of innovation.

Decentralized, close-to-market structures

What can be done as firms mature to retain an innovating environment, avoiding the natural tendency to grow less flexible and more bureaucratic? Structures affect innovativeness: an entrepreneurial style often flourishes only when bureaucratic, hierarchical controls are minimized and innovators have scope to exercise real initiative (Kanter, 1983; Peters and Waterman, 1982; Rogers and Larsen, 1984; Sherman, 1984). The preferred solution appears to be a decentralized structure allied to a strong project orientation, so as to direct limited resources where the need is most urgent and the pay-off most promising. It generally helps if the innovators are able reap the rewards of innovation by managing or otherwise being directly involved in the commercialization process.

In mass-market consumer contact service industries, such as airlines, retailing and banking, new technolgies of communication have facilitated a considerable shift toward local operating autonomy allied to increasingly centralized strategic control (Loveridge, Ch.15). From a variety of standpoints this configuration appears effective, permitting local responses to consumer needs up to but not beyond the point when central objectives might begin to be compromised, at least not without those responsible for strategic direction becoming alerted to the problems and opportunities inherent in these local initiatives. This configuration is now extending to applications in a variety of other service sectors, as diverse as, for example, health care and fast-food chain operations.

In the context of R&D management, Whittington (Ch.8) identified the structural mechanism of exposing even laboratory researchers to greater and more immediate market pressures, so that their efforts become market-directed, focused and accountable. There is much to be learned from being close to users as Peters and Waterman (1982) have argued so cogently. Encouraging innovators to develop a wide network of technical and commercial contacts in the quest for new ideas is functional, if not without associated risks.

Curiosity-driven, experimental, adaptive outlook

Receptivity to outside influences and ideas leads to variety of approach. But variety is an ecological characteristic with random, rather than systematic properties, so attention to external ideas and trends will on occasion lead down culs-de-sac. Still, even the most resource-rich organizations find it hard to sustain requisite variety of internal innovative ideas and competences, underlining the necessity of looking outside. Such open-system learning typically combines both contestual and collaborative elements (Loveridge, Ch. 15). The paradox in an application like information technology in retailing is that the appropriation of competitive benefits has required—in important ways—a collaborative outlook, for example, the negotiation of an industry-standard system of bar-coding of retail goods. The evidence is that competitors do not find it easy to *combine* both collaborative and competitive mindsets, yet failure to do so inevitably slows down or erodes the probability of successful appropriation of benefits. In this respect, Loveridge argues, U.S. retailers were much quicker to appreciate and respond to this need than were their British counterparts.

There is, however, somewhat contradictory evidence that many widely acclaimed innovations come not from the market but from the efforts of dedicated, idiosyncratic, pioneering idea-champions with visions beyond current market-place realities (Maidique, 1980; Nayak and Ketteringham, 1986; Schon, 1963; 1967). Characteristically they work long hours and use 'bootlegged' resources to develop their visions into commercial realities. Only in organizations with highly committed people and some tolerance of under-utilized or slack resources can this occur.

In this book Carr, Senker, Loveridge and Jones have all stressed the value of an experimental learning-oriented corporate outlook. Jones for example, suggested that applying FMS is not the end-point of innovating, but the beginning: FMS installations cannot be exploited to best effect without experimenting beyond obvious and routine applications. Carr distinguished between applying specific, possibly exogenous innovations and sustaining a continuing stream of innovatory behaviour over time—often by relatively junior operations staff. In Senker's account of food retailing, the unforeseen pay-off of developing resident technical skills was the emergence of new and broader options to innovate in directions that simply would not have been feasible if the firms had continued to rely on outsiders for their technological input.

From this perspective, strategic management is properly and necessarily a learning process rather than a decision process *per se*. Successful strategic management is then concerned with positioning the organization so as to be obliged to learn from challenging new experiences, and its managers to be receptive to the possibilities engendered in technological and other forms of innovation. If it is not a contradiction-in-terms, one might say it should be a judicious mix of 'proactive-reactivity', neither fully-planned, nor left to chance and the lead of others. It is surely important for initiatives to be encouraged at all levels, with the internal means for the *results* of consequent learning to be promulgated rapidly throughout the structure. Above all, it is surely advantageous for all organization members to subscribe to a cultural norm of innovativeness, rather than to expect it of only a few 'specialists'. This being so, tolerance of occasional failure is important.

Moderate threat and a sense of urgency

The challenge is to secure benefits from an adaptive, experimental style of learning by (i) sustaining close identity on the part of staff with the firm and the project, and (ii) combining personnel continuity and the regular infusion of 'new blood' to challenge established and perhaps obsolescent ways. How should general managers handle technical staff and what kind of organizational culture they should strive to create, given the fundamentally 'unplannable nature' of true inventive and innovative activities in fast-changing high-technology sectors (Maidique and Hayes, 1984; Quinn, 1979; 1985). Market-place exposure when accompanied by personal insecurity is unlikely to stimulate creative thinking; it may simply promote safe, conventional approaches. Conversely, being insulated from market imperatives is a recipe for complacency. The optimal environment seems to foster a creative, dynamic tension where people feel the urgency of market pressures and hence the need for action, without that pressure being constantly equated with personal insecurity. Moreover, as Brunsson (1982) and Peters and Waterman (1982) have said, an action-oriented environment that values and rewards action plays a crucial part in resolving problems.

Successful firms appropriate the benefits of innovations more rapidly than competitors. They are also more ruthless in cutting their losses with innovations that falter. Their expectations of commercial return are clear and realizable; whilst this does not inevitably mean a short pay-back period for new investment, an extended pay-back must be weighed carefully against likely future returns. A characteristic feature of such firms may also be their construction of periodic organizational crises (Pitt, Ch.11) as a mechanism for galvanizing radical innovation; as an Intel executive quoted by Sherman (1984) said: 'Two things drive this business, technology and paranoia'. In such environments crises and technological innovation are probably synonymous, requiring an adaptive, urgent management style.

TECHNOLOGICAL INNOVATION AND ETHICAL DILEMMAS

The products of technology pervade our lives and for the most part enrich us, personally and collectively: without technological advances we are unquestionably the poorer. But the very momentum of technologically-driven change also frightens many of us with the ever-present risk of pathological consequences, as Shrivastava et al. (1988) so vividly documented. This should be of vital concern to all who implement novel technologies, especially when the consequences arising from failure to control hazardous processes are so horrible.

Shrivastava suggested that inadequate thought has generally been given to the consequences of disasters. We concur, but also note the inadequate attention given to quite predictable implications of technological developments. Some examples (among many) would include (i) the failure of major U.K. food retailers to educate consumers to avoid careless post-purchase storage of processed foods, needlessly increasing the level of bacterial contamination leading to widespread food poisoning, (ii) the failure of many process industries to acknowledge, let alone combat the pollution of the environment, (iii) the way in which major electric power utilities have largely ignored planning for the decommissioning of nuclear power stations until it became an unavoidable reality, (iv) the systematic attempts of some U.S. and West European defence contractors to circumvent governmental restrictions on the export of sensitive technologies to embargoed corners of the world, and (v) a susceptibility of defence and other technical contractors to government agencies to take short-cuts in specifications or otherwise defraud the client when the opportunity arises. Then again, government agencies are not immune from occasional pressures to behave expediently.

Other less obvious excesses and omissions of technological innovation are ignored 'in the name of progress'. Indeed it is often possible to justify courses of action glibly on the grounds that 'it is what the market wants'. Job losses through automation in automobile manufacturing, one can argue, are compensated by cheaper and therefore more affordable cars. But the consequence is more vehicles in use—and thus, greater pollution—which is surely unacceptable from a global perspective, particularly for those human beings everywhere for whom car ownership remains only a dream.

The twin motives of technical curiosity and desire to compete spur mankind—and in the end that means individual firms and technologists—to innovate. Without these pressures, it is true, we forgo the positive benefits of innovation even more surely than we may avoid the pathologies. But to be blunt, technological innovation is big business and social conscience is all too easily submerged by a tidal wave of commercial greed. When experiencing these pressures, can a firm terminate the development of a lucrative innovation just because it *may* have potentially adverse consequences? Can individual technologists, innovators and managers refuse to implement a new process without incurring a severe personal penalty? All too often, we suspect, the answer is no.

Pirsig (1974) explored the uneasy fit of technology with higher values, or in his terms, 'quality'. He expressed the potentially dehumanizing affect of technology thus:

'Scientific materialism, which is commoner among lay followers of science than among scientists themselves, holds that what is composed of matter or energy and is measurable by the instruments of science is real. Anything else is unreal or of no importance. . . . The dictum that science and its offspring, technology, are "value-free", that is, "quality-free" has got to go.'

To invert another of his observations, 'High quality endeavour is art'. Clearly, technological innovation has as much right as any other human activity to be described as an art form involving high quality, ethical endeavour. So the honest innovator has a considerable responsibility on behalf of the rest of us to meet high quality standards. Are we naive to call in particular for a greater ethical input to the strategic management of technological innovation? We think not.

We certainly make no apology for incorporating ethics in this discussion. In an era when the undoubted benefits of technological innovation accruing to one half of the world's population are offset by so much wasteful output and when the potential benefits of genetic engineering could easily be diverted into weapons technologies, mirroring the development of nuclear and chemical weapons, there is no room for complacency.

We are not advocating Luddite rejection of technological progress, nor that oppressive regulation is the answer, though regulation clearly has a part to play. Quite simply, we propose that strategies of technological innovation should be subjected to regular, explicit scrutiny to ensure their ethical as well as technical integrity. There is now accumulating evidence that socially responsible business behaviour is perfectly compatible with good financial returns, a fact acknowledged by the emergence of 'ethically sound' investment funds in the U.K. and elsewhere.

Technologists cannot avoid ethical dilemmas. Rather, they should anticipate and recognize dilemmas and be prepared to stand for what they judge is ethically appropriate. More specifically, innovators at whatever level they operate, should not be compromised by the pressures and temptations of short-run commercial gain. We hope that detailed, open examination and debate of difficult decisions will become the norm. Above all, fundamental dilemmas must not be obscured by technical detail and rationalized away with 'weasel words'. Whistle-blowing can be an honourable intervention.

REFERENCES

Abernathy, W.J., Clark, K.B. and Kantrow, A.M. (1983). *Industrial Renaissance: Producing a Competitive Future for America*, Basic Books, New York.

Boisot, M. (1986). 'Markets and hierarchies in a cultural perspective', *Organization Studies*, **7**, 135–58.

Bright, J.R. (1970). 'Evaluating signals of technological change', *Harvard Business Review*, Jan./Feb., 62–70.

Brownlie, D.T. (1987). 'The strategic management of technology: a new wave of market-led pragmatism or a return to product-orientation?', *European Journal of Marketing*, **21**(9), 45–65.

Brunsson, N. (1982). 'The irrationality of action and action rationality: decisions, ideologies and organizational action', *Journal of Management Studies*, **19**(1), 29–44.

Burgelman, R.A. (1983). 'A process model of internal corporate venturing in the diversified major firm', *Administrative Science Quarterly*, **28**(2), 223–44.

Burgelman, R.A. (1985). 'Managing the new venture division: research findings and implications for strategic management', *Strategic Management Journal*, **6**(1), 39–54.

Burns, T., and Stalker, G.M. (1961). *The Management of Innovation*, Tavistock, London.

Chandler, A.D. (1962). *Strategy and Structure*, M.I.T. Press, Cambridge.

Cho, K.R. (1988). 'Issues of compensation in international technology licensing', *Management International Review*, **2**, 70–79.

Clarke, K. (1987). 'Investment in new technology and competitive advantage', in D. Teece (ed.), *The Competitive Challenge*, Ballinger, Cambridge, MA.

Collins, P.D., Hage, J. and Hull, F.M. (1988). 'Organizational and technological predictors of change in automaticity', *Academy of Management Journal*, **31**(3), 512–43.

Cool, K. and Schendel, D. (1988). 'Performance differences among strategic group members', *Strategic Management Journal*, **9**(3), 207–24.

Dodgson, M. (1987). 'Small firms, advanced manufacturing technology and flexibility', *Journal of General Management*, **12**(3), 58–75.

Dosi, G. (1982). 'Technological paradigms and technological trajectories: a suggested interpretation of the determinants and directions of technological change', *Research Policy*, **11**, 147–62.

Dutton, J.E., and Duncan, R.B. (1987). 'The creation of momentum for change through the process of strategic issue diagnosis', *Strategic Management Journal*, **8**(3), 279–95.

Dutton, J.E., and Ottensmeyer, E. (1987). 'Strategic issue management—systems—forms—functions and contexts', *Academy of Management Review*, **12**(2), 355–65.

Ford, D. (1988). 'Develop your technology strategy', *Long Range Planning*, **21**(5), 85–95.

Ford, D., and Ryan C. (1981). 'Taking technology to market', *Harvard Business Review*, Mar./Apr., 117–126.

Foster, R.N. (1971). 'Organize for technology transfer', *Harvard Business Review*, Nov./Dec., 110–20.

Foster, N. (1986). *Innovation: The Attackers Advantage*, Macmillan, London.

Fredrickson, J.W. (1984). 'The comprehensiveness of strategic decision processes: extension, observations, future directions', *Academy of Management Journal*, **27**(3), 445–66.

Fredrickson, J.W., and Mitchell, T.R. (1984). 'Strategic decision processes: comprehensiveness and performance in an industry with an unstable environment', *Academy of Management Journal*, **27**(2), 399–423.

Frohman, A.L. (1982). 'Technology as a competitive weapon', *Harvard Business Review*, Jan./Feb. 97–104.

Gold, B. (1983). 'Strengthening managerial approaches to improving technological capabilities', *Strategic Management Journal*, **4**, 209–20.

Grinyer, P.H., and Spender, J.-C. (1979). 'Recipes, crises and adaptation in mature businesses', *International Studies of Management and Organization*, **9**, 113–23.

Hayes, R.H. and Abernathy, W.J. (1980). 'Managing our way to economic decline', *Harvard Business Review*, July/Aug., 67–77.

Hayes, R.H., and Jaikumar, R. (1988). 'Manufacturing's crisis: new technologies, obsolete organizations', *Harvard Business Review*, Sept./Oct., 77–85.

Jarillo, J.C. (1988). 'On strategic networks', *Strategic Management Journal*, **9**(1), 31–42.

Jones, A., and Webb, T. (1987). 'Introducing computer integrated manufacturing, *Journal of General Management*, **12**(4), 60–74.

Kanter, R.M. (1983). *The Change Masters*, Unwin, London.

Kuhn, T.S. (1970). *The Structure of Scientific Revolutions*, Chicago University Press, Chicago.

Kumpe, T., and Bolwijn, P.T. (1988). 'Manufacturing: the new case for vertical integration', *Harvard Business Review*, Mar./Apr., 75–81.

Lippman, S.A., and Rumelt, R.P. (1982). 'Uncertain imitability: an analysis of interfirm differences in efficiency under competition', *Bell Journal of Economics*, **13**(2), 418–38.

Maidique, M.A. (1980). 'Entrepreneurs, champions and technological innovation', *Sloan Management Review*, **21**, 2.

Maidique, M.A., and Hayes, R.H. (1984). 'The art of high technology management', *Sloan Management Review*, **25**, 2.

Marcus, A.A. (1988). 'Implementing externally induced innovations: a comparison of rule-bound and autonomous approaches', *Academy of Management Journal*, **31**(2), 235–56.

McCaskey, M.B. (1982). *The Executive Challenge: Managing Change and Ambiguity*, Pitman, London.

Miles, R., and Snow, C.C. (1978). *Organization Strategy, Structure and Process*, McGraw-Hill, New York.

Miller, A. (1988). 'A taxonomy of technological settings, with related strategies and performance levels', *Strategic Management Journal*, **9**(3), 239–54.

Miller, D., and Friesen, P.H. (1978). 'Archetypes of strategy formulation', *Management Science*, **24**, 921–33.

Miller, D. and Friesen, P.H. (1984). *Organizations: A Quantum View*, Prentice-Hall, Englewood Cliffs, NJ.

Mintzberg, H. (1978). 'Patterns in strategy formation', *Management Science*, **24**(9), 934–48.

Mintzberg, H. (1979). *The Structuring of Organizations*, Prentice-Hall, Englewood Cliffs, NJ.

Nayak, P.R., and Ketteringham, J.M. (1986). *Breakthroughs*, Mercury, London.

Nueno, P., and Oosterveld, J. (1988). 'Managing technology alliances', *Long Range Planning*, **21**(3), 11–17.

Pennings, J.M. (ed.) (1985). *Organization Strategy and Change*, Jossey-Bass, San Francisco.

Peters, T., and Waterman, R. (1982). *In Search Of Excellence*, Harper & Row, New York.

Piore, M.J., and Sabel, C.F. (1984). *The Second Industrial Divide: Possibilities for Prosperity*, Basic Books, New York.

Pirsig, R.M. (1974). *Zen and the Art of Motor-cycle Maintenance*, Corgi Books, London.

Porter, M.E. (1980). *Competitive Strategy*, Free Press, New York.

Porter, M.,E. (1985). *Competitive Advantage*, Free Press, New York.

Quinn, J.B. (1967). 'Technological forecasting', *Harvard Business Review*, Mar./Apr., 89–106.

Quinn, J.B. (1969). 'Technology transfer by multinational companies', *Harvard Business Review*, Nov./Dec. 147–

Quinn, J.B. (1979). 'Technological innovation, entrepreneurship and strategy', *Sloan Management Review*, **20**, 3.

Quinn, J.B. (1980). 'Managing strategic change', *Sloan Management Review*, **21**(4), 3–20.

Quinn, J.B. (1985). 'Managing innovation: controlled chaos', *Harvard Business Review*, May/June, 73–84.

Roberts, R.B. (1987). *Generating Technological Innovation*, Oxford University Press, Oxford.

Rogers, E.M., and Larsen, J.K. (1984). *Silicon Valley Fever: Growth of High Technology Culture*, Basic Books, New York.

Schendel, D.E., and Hofer, C.W. (eds) (1979). *Strategic Management: A New View of Business Policy and Planning*, Little Brown, New York.

Schon, D.A. (1963). 'Champions for radical new innovations', *Harvard Business Review*, Mar./Apr.

Schon, D.A. (1967). *Technology & Change: The New Heraclitus*, Pergamon, Oxford.

Schon, D.A., (1983). *The Reflective Practitioner: How Professionals Think in Action*, Basic Books, New York.

Sherman, S.P. (1984). 'Eight masters of innovation', *Fortune Magazine*, October.

Shrivastava, P. (1988). 'Industrial crisis management: learning from organizational failures', *Journal of Management Studies*, **25**(4), 283–4.

Strebel, P. (1987). 'Organizing for innovation over and industry cycle', *Strategic Management Journal*, **8**(2), 117–24.

Teece, D.J. (1985). 'Applying concepts of economic analysis to strategic management', in J.M. Pennings (ed.) *Organization Strategy and Change*, 35–63, Jossey-Bass, San Francisco.

Thorelli, H.B. (1986). 'Networks: between markets and hierarchies', *Strategic Management Journal*, **7**(1), 37–52.

Twiss, B.C. (1986). *Managing Technological Innovation*, Pitman, London.

Weick, K. (1969) (first edition). *The Social Psychology of Organizing*, Addison-Wesley, Reading, Mass.

White, G.R., and Graham, M.B.W. (1978). 'How to spot a technological winner', *Harvard Business Review*, Mar./Apr., 146–52.

Williamson, O.E. (1975). *Markets and Hierarchies: Analysis and Antitrust Implications*, Free Press, New York.

Winter, S.G. (1987). 'Knowledge and competence as strategic assets', in D. Teece (ed.), *The Competitive Challenge*, Ballinger, Cambridge, Mass.

Woodward, J. (1965). *Industrial Organization: Theory and Practice*, Oxford University Press.

Zarecor, W.D. (1975). 'High technology product planning', *Harvard Business Review*.

Index

crisis (*cont.*)
  concepts 2, 5, 13, 167, 175, 224, 253,
    258, 315, 335, 341, 355
  confidence 112, 120, 341, 345
  control 184, 257
  constructed 13, 253, 257, 263-67,
    373, 384
  economic/market 114, 120, 256,
    297, 330, 341, 384
  health care 354
  impending 256–61
  imposed 262
  negotiated/politicized 262–64
  organizational 35, 257, 320, 343
  organized capitalism 183, 190, 192,
    200, 255
  realized/validated 256-64
  resolution 13, 257, 267
  retailing 349
  stress 255
CT scanners, *see* medical diagnostics
cultural norms 265
cultural space 119
culture, *see* organization culture
Cutter Laboratories 147

Daimler Benz 101
Dainichi Kiko 33
Dainippon 147
Dainon Systems Co. 150
DCF analysis, *see* payback
De Dion-Bouton 101
decision analysis 261, 282
decision making 3, 13, 35, 160, 172,
    174, 208, 218, 226, 229, 246, 263,
    275, 279, 281-86, 293, 295, 312,
    356, 384
decision, systemic 312, 376
defender 97
Delco (GM) 100
demand-pull (versus
    innovation-push) 4, 205, 207
demand, declining 372
dematurity (see sector)
Department of
  Agriculture (US) 140
  Commerce (US) 140
  Defense (US) 129
  Health (UK) 356
  Trade & Industry (UK) 185, 196,
    356

design
  capability 2, 105, 224, 235, 239, 247,
    329, 381
  centralized 105, 350
  changes 297, 331, 351
  dominant (robust) 22, 24, 41, 99,
    277, 280
  organizational 8, 274, 301, 313, 342
  procedure (Protocol/repertoire) 118,
    229, 240, 378, 381
  state of art 313
  vision 243
Deutsche Airbus 169, 170
Deutsche Bank 102, 117
diagnostic assays, *see* medical
    diagnostics
die casting 79
diesel engines 103
differentiation
  of personnel 322
  sources of 22
  strategies, *see* strategy
diffusion (of knowledge), *see*
    knowledge
dilemma
  managerial 279, 369
  productivity/innovation 229, 248,
    294, 302
  strategic 257, 369
disadvantage, competitive 11, 14, 78
disasters, natural 259, 261
discontinuities 4, 8, 254, 277, 287, 319
diversification 29–30, 50, 71
divestment 71–73
DNA 140
dogs 63
Domilens 25
dominant designs, *see* design
dominant logic (sector entry) 30
Dow Chemicals 184
Dresden Bank 104
Du Pont 184
Dunlop Co. 70, 113

Eastern Airlines 171
Eastman Kodak, *see* Kodak
EB Industries 27
ecological variety 383
Edison 101, 184
efficiency and effectiveness 13